FRONTIERS IN ELECTRONIC MATERIALS & PROCESSING

AMERICAN INSTITUTE OF PHYSICS
CONFERENCE PROCEEDINGS NO. 138
NEW YORK 1986

AMERICAN VACUUM SOCIETY SERIES 1

AVS SERIES EDITOR: GERALD LUCOVSKY
NORTH CAROLINA STATE UNIVERSITY

FRONTIERS IN ELECTRONIC MATERIALS & PROCESSING

HOUSTON, TEXAS NOVEMBER 1985

EDITOR: L.J. BRILLSON
XEROX CORPORATION

L.C. Catalog Card No. 86-70108
ISBN 0-88318-337-4
DOE CONF-8511138

Printed in the United States of America

Pstack SCIMON

TABLE OF CONTENTS

CHAPTER VI: VERY LARGE SCALE INTEGRATION (VLSI) ISSUES

PREFACE

The first topical conference on Frontiers in Electronic Materials and Processing was held at the Albert Thomas Convention Center in Houston, Texas on November 19–21, 1985 and in conjunction with the American Vacuum Society's 32nd National Symposium. Due to the rapid development and importance of the field of electronics materials and processing, the Program Committee of the AVS National Symposium decided that this year's meeting in Houston would be an excellent forum in which to bring together internationally recognized scientists and technologists to discuss very technological and scientific aspects of this field. Each chapter in this volume represents one such topic, and are in order of appearance—Chapter I: Overview of Critical Topics, Chapter II: Overview of Layered Materials, Chapter III: Status of U.S. and Japanese Silicon Technology, Chapter IV: Status of U.S. and Japanese III-V Compound Semiconductor Technology, Chapter V: Microelectronics Packaging, and Chapter VI: Very Large Scale Integration (VLSI) Issues.

The format of this Topical Conference was similar to that of the AVS National Symposium except for the manner in which the manuscripts are now being published. This volume is the first of an AVS series of topical conferences which will be published through the American Institute of Physics as a sub-series of AIP Conference Proceedings. The manuscripts were reviewed in accordance with standards for Journal of Vacuum Science and Technology publications and published in camera-ready format.

The presentation of a topical conference in conjunction with the AVS National Symposium represented a departure from the traditional AVS meeting. It allowed a greater number of outstanding scientists to present invited talks at the National Symposium without substantially reducing the space available for contributing authors. The Frontiers in Electronic Materials and Processing Conference brought to the AVS a new focal point for discussion of very timely and important issues which need not be addressed on a regular basis. Furthermore, by bringing to the national meeting experts from new areas of science and technology relevant to the Society's interest and capabilities, the Electronic Materials and Processing Division provided new opportunities for cross-fertilization with other divisions of the Society, many of whose members may be able to promote progress on the issues presented.

A large number of people contributed to the success of this first Electronic Materials and Processing Topical Conference. In addition to conference co-chairmen Paul Holloway and Paul Ho, particular thanks go to Robert Davis, who proposed and organized the sessions on Artificially Layered Materials, and Jerry Woodall, who proposed and organized the session on U.S./Japanese Status of III-V Compound Semiconductor Technology. These gentlemen coordinated a twelve-month process culminating in a fine collection of conference presentations. Assisting in the initial stages of this project were members of the EMPD Executive Committee: R. Z. Bachrach, L. J. Brillson, D. Collins, C. B. Duke, R. Farrow, M. Francombe, J. E.

Green, P. S. Ho, E. Krikorian, T. Mayer, R. Powell, and E. Wolfe, as well as D. Ehrlich, G. Rubloff, R. Singh, and T. Van Oostrom. Presiding over the sessions were Charlie Duke, Bob Davis, Arnie Reisman, Max Yoder, Jim Burkstrand, and S. P. Murarka, who deserve special thanks for their stimulating effect on the presentations and ensuing discussions throughout the three days of the conference.

The critical participants in the production of this Topical Conference Proceedings are of course the authors who have contributed their manuscripts. Their papers were subject to the normal requirements of papers published in the Journal of Vacuum Science and Technology—submission well in advance of the meeting and rigorous evaluation by multiple reviewers. In addition, each manuscript and all revisions were constrained to be in a camera-ready format for rapid publication. Indeed, this fully refereed volume goes to press only three weeks after the Topical Conference's conclusion. On behalf of the Electronic Materials and Processing Division, we wish to express our appreciation to the AVS Board of Directors for approving and supporting the concept of a topical conference in parallel with the National Symposium. This expertise would not have been possible without the full support of the AVS President Ed Sickafus, incoming President Don Mattox and National Program Chairman Jerry Woodall. We are grateful to Gerry Lucovsky, Editor-in-Chief of the Journal of Vacuum Science and Technology for suggesting the AIP Topical Conference Series as an alternative to JVST in publishing of this special proceedings.

Finally, we wish to thank Mrs. Cathryn Albright for her valuable assistance in assembling the manuscripts for this Proceedings volume.

Chairmen
P. Holloway (University of Florida, Gainesville)
P. S. Ho (IBM T. J. Watson Research Cntr.)

Proceedings Editor
L. J. Brillson (Xerox Webster Research Cntr.)

December 16, 1985

AUTHOR INDEX

CHAPTER I

OVERVIEW OF CRITICAL TOPICS

PATTERNING OF SUBMICRON VLSI CIRCUITS

R. Fabian W. Pease
Stanford University, Stanford, CA 94305

ABSTRACT

Patterning of circuit materials continues to be a key technology in the evolution of VLSI circuits. One figure of merit, the power-delay product of a logic gate, improves as the cube of linear dimension in a properly scaled design for dimensions down to the submicron regime. Thus, decreasing lateral dimensions continues to be crucially important. We are now at the point where lateral dimensions are of the order of the wavelength of the light used for patterning, and the choice of an economic microlithographic technology for manufacturing submicron VLSI is far from clear. A complicating factor is that lateral dimensions are no longer much greater than vertical dimensions so that pattern transfer techniques must avoid lateral spreading of the features. Anisotropic etching selective plating, and lift-off, are three approaches for solving this problem. Two other aspects of patterning which are equally important but are less publicized are achieving accurate overlay of successive patterns and avoidance of defects. With some imagination achieving accurate overlay should not be a "show stopper". However, it is not clear how we can overcome the defect problem. The whole question of microlithographic defects deserves serious attention from the research community.

PRESENT STATE OF THE ART

The patterning of functional electronic materials for VLSI circuits has become one of the most crucial technologies of modern electronics. The original pattern for a 6-inch diameter silicon wafer is usually generated using an electron beam system, employing a writing address and beam diameter each of 0.1 μm - 0.2 μm, with minimum features presently as small as 1.25 μm and soon to be less than 1 micron. The complexity of this pattern corresponds to about one million television images. To make matters even more challenging, the pattern must have a defect density of less than 0.1 per cm^2 corresponding to an error rate of about 1 in 10^{11}; in addition, a finished wafer might have up to 8 such patterns overlaid with one another to an accuracy of better than 0.2 μm.

The challenge of generating such patterns at an economical rate has led to the emergence of a number of very expensive tools. The original artwork (assembled in a computerized design station) is generated in a resist pattern about 0.5 μm thick on vacuum-deposited chromium (60 nm - thick) on a glass or quartz substrate. The electron beam pattern generator employs a writing beam of 10KeV electrons focused to 0.1 to 0.2 μm diameter and with a current of 5-30 nA. Present-day systems operate at 80MHz [1]; i.e., 8 x 10^7

0094-243X/86/1380003-10$3.00

addresses are patterned per second so a 10 x 10 cm^2 area of 0.2 µm addresses requires 1 hour of exposure time. Often there is considerable overhead that may double this time. Reducing the address to 0.1 µm makes for a higher quality pattern but quadruples the required exposure time. Such electron beam pattern generators cost over $3M and so the cost associated with the generator is about $1000/hr. Wet development follows exposure of the resist and then the chromium is etched. The mask is then inspected for feature size control, defect density and distortion. Generating a qualified set of masks for a VLSI circuit is thus a major expense - up to $50,000.

The combination of high cost and slow area coverage rate (about 1 cm^2 per minute) makes today's electron beam pattern generators impractical for directly exposing the wafer, thus wafer exposure is accomplished by projecting an ultra-violet image of the mask pattern onto the wafer. There are two main classes of projection exposure systems. One, exemplified by the Perkin-Elmer "Micralign"® series, employs doubly reflecting optics (fig. 1) which can image a pattern, essentially aberration-free, down to an f-number of 2 at 1:1 magnification over an arc-shaped field about 1 mm wide and several inches (up to 6) long. The wafer and mask are simultaneously scanned mechanically through an illuminated area restricted to this arc and so a complete image of the mask is built up onto the wafer. The process takes about 1 minute.

The other class, referred to as "steppers", also projects a restricted area but uses more conventional refracting lenses which project an essentially aberration-free image (down to f -1.4) over a "field" or "site" about 1 x 1 cm^2 (at the wafer). Each field is exposed while the wafer is stationary after which the wafer is stepped to bring the next site into the field of view and the exposure is repeated. Most steppers employ reducing optics so that the mask pattern can be, say, five times larger than the final (wafer) pattern.

Each class of systems has its supporters. Those favoring the steppers point out that the mask requirements are relaxed because only 1 field (of pattern) is needed (as opposed to a whole wafer) and the mask features are larger and hence are easier to make and repair. Furthermore, because only a compact area is being simultaneously exposed, it is easier to control distortion (to allow accurate overlay). Proponents of the scanning slit approach point out that their final patterns are much less vulnerable to mask defects because a single mask defect transfers to only one exposure site and it is possible to correct for linear isotropic distortion (between mask and wafer) on the newer scanning slit system. The refracting optics of the stepper have smaller f-numbers than do the scanning system, but this advantage is offset by the fact that the reflecting optics (of the latter) allow the use of multiple, and shorter, wavelengths (down to 250 nm). Both approaches have demonstrated linewidth in the range 0.5 to 1 µm (fig. 2) and it is expected that a manufacturing capability in this feature size range is possible.

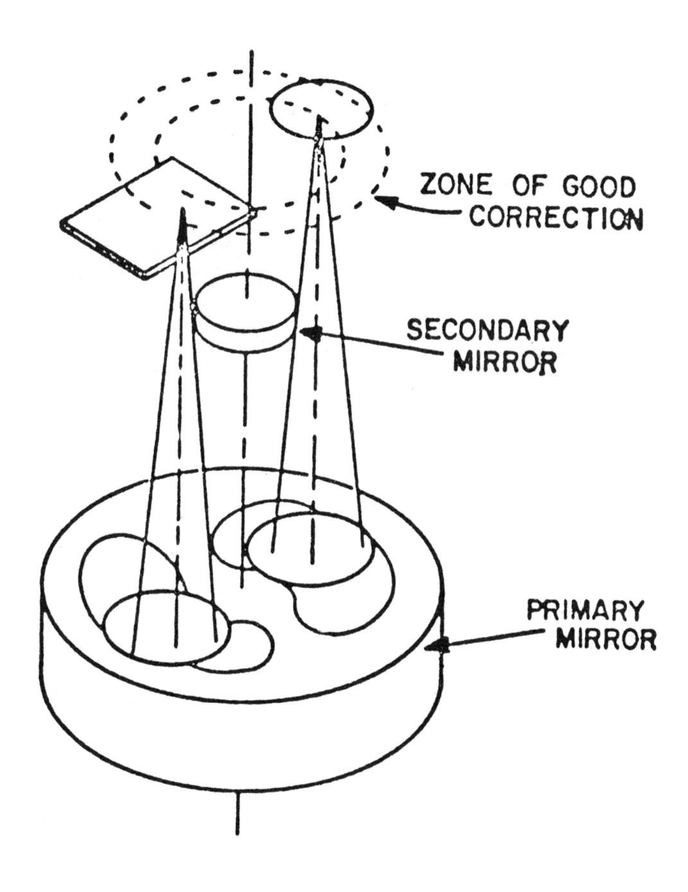

Figure 1. Schematic view of scanning slit projection used in the Perkin Elmer Micralign® series of exposure systems. The two mirrors form a high-quality 1:1 image between points in the two halves of the ring shaped zone. The mask (square) and wafer (round) are together scanned through the illuminated zone of good correction to project the complete pattern from mask to wafer.

Figure 2. Stepper-projected pattern showing 0.75 µm resist lines and 1.25 µm spaces over a substrate containing 0.5 µm high polysilicon steps (photography courtesy of the Ultratech Stepper Company).

ON-GOING DEVELOPMENTS

It is far from clear how a sub-optical (i.e. below 0.5 μm) manufacturing capability is to be implemented. Only a small proportion of the optical fraternity appear to believe that ultra violet optics can be stretched to operate in a manufacturing mode with a pattern half-pitch of 0.25 μm although special patterns have been generated in this dimensional range (e.g. interference patterns, and intimate contact printing with 193 nm radiation). However, the alternatives to optical techniques do not seem adequate either.

"High-speed" direct-write electron beam machines are now becoming available. This higher speed is accomplished often by employing a beam shaped as a complete feature (or part thereof) so that a number of addresses are exposed simultaneously. By using advanced electron optics the shape can be varied in less than 1 μs (fig. 3). However, space charge sets a limit to the total current available for a given resolution and shot noise sets a limit to the maximum practical resist sensitivities [2]. Thus such systems seem limited to less than 10 wafers per hour and are valuable only for prototyping and for "small runners".

X rays, flood ion beams and flood electron beams are all being pursued with a view to high resolution replication of mask patterns. The first such scheme, photoelectron projection, is nearly 2 decades old and employs a photoelectric (e.g. CsI) coating on a standard chromium on quartz mask which is illuminated with deep ultra violet light. The emitted electrons (from the clear areas on the mask) are accelerated and imaged into the wafer. The main problems are: (1) the whole wafer is exposed simultaneously so that accurate overlay is difficult and (2) back scattered electrons are returned, by the accelerating field, to the wafer and cause undesired exposure. Recently a scheme for avoiding both problems was described but no performance data are available [3].

X-ray lithography is the mainstream effort in sub-optical pattern replications. Here the mask technology is the main problem. The transparent substrate is a thin membrane while the absorber pattern is a heavy metal about 0.5 μm thick (at least) and hence controlling the distortion to less than 0.1 μm is particularly challenging. Although the earlier x-ray systems exposed the whole wafer simultaneously, the distortion problem can be alleviated by stepping, realigning and exposing a field of 1 x 1 cm^2 [4]. Smaller fields are believed to be impractical because VLSI chip dimensions are approaching 1 x 1 cm^2. Low throughput, because of limits on x-ray source brightness and x-ray resist sensitivity, is also a matter of concern. One, expensive, way to overcome this problem is to employ a synchrotron as a very bright radiation source.

The stencil masks usually employed with flood ion beam systems [5] and flood electron beam systems [6] appear to present even worse fabrication problems. Given the difficulties of the simplest mask technologies, it is hard to see how such exotic mask technologies will be practical in the foreseeable future.

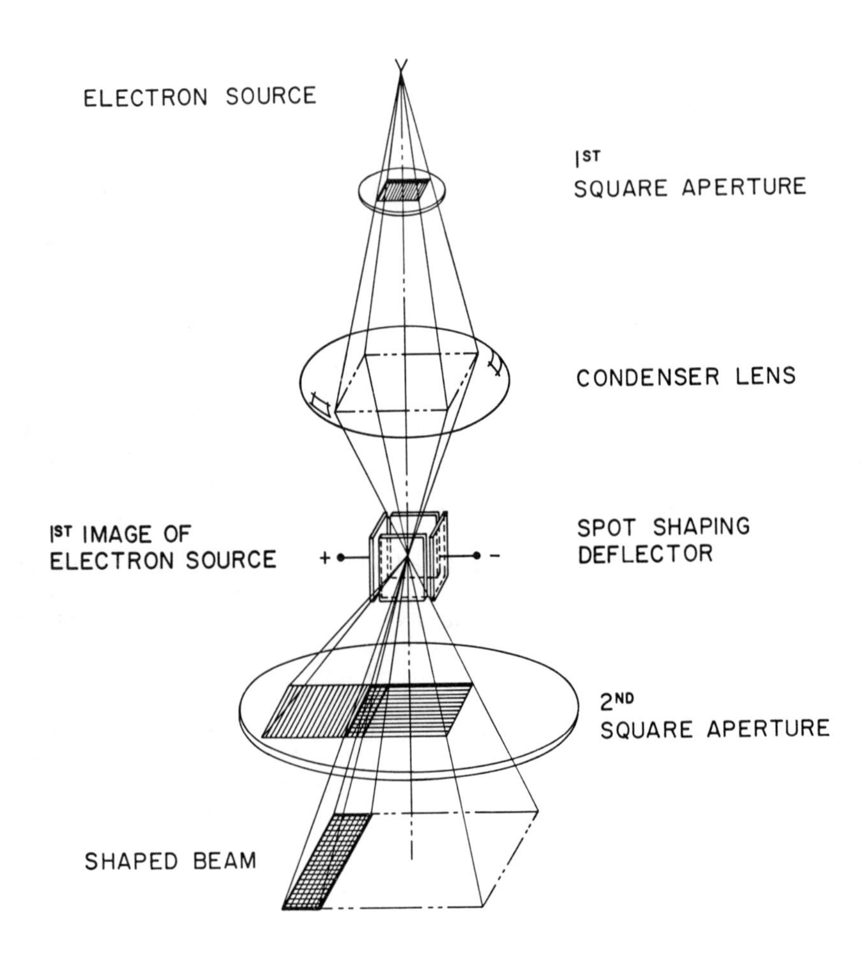

Figure 3. Principle of variable shaped beam generation used on most second-generation electron beam pattern generators. The image of the first shape can be rapidly deflected across the second shape to give a wide variety of shapes resulting from the overlap. This resulting shape can then be projected electron optically onto the wafer (courtesy H. C. Pfeiffer, I.B.M.).

A FRESH LOOK AT MASK PROBLEMS

The two biggest mask problems are distortion and defects. It appears that there are ways, borrowed from the communication disciplines, that these problems might be overcome.

Let us look first at the defect problem as some significant progress has already been reported here. If we think of the transfer of the mask pattern to the wafer as a communication problem, then the defects represent errors. It is well known that by introducing redundancy into a message, error correction can be achieved. One primitive way of doing this is by "vote taking", i.e. repeat the message three times and the correct message will be that received at least twice. In lithography such a scheme has recently been demonstrated [7]. By exposing each wafer site with, say, three, nominally identical mask patterns, a random defect in 1 pattern is a 33 and 1/3% perturbation on the integrated exposure level of the wafer. By employing a high contrast resist process on the wafer, such a perturbation can be made acceptable and both positive and negative defects can be eliminated (fig. 4). The actual exposure time is unchanged but there is a time penalty in moving the mask. However, the tremendous advantage of eliminating the effects of random mask defects should provide an incentive to minimizing this overhead. It is also necessary to achieve very accurate overlay between the three mask patterns but as we shall see below this should be achievable. It is not clear how far analagous approaches might be developed to reduce random defects generated on the wafer redundant circuits is one extreme example.

Achieving accurate overlay is presently accomplished by controlling mask and wafer distortion so that one fits the other to the required tolerance. This may be the most practical way today, but tighter design rules will make the problem increasingly difficult.

The communication engineer would point out that achieving 0.1 μm overlay accuracy does not require that mask distortion be limited to 0.1 μm over the whole area of the mask (which is what mask manufacturers are trying to achieve today). The circuit design is unconcerned whether the chip measures 10.0000 mm or 10.0001 mm; only local matching (i.e. over the area of illumination) is needed. The communication engineer would argue that we must phase-lock our incoming pattern to the existing structure whose spatial periodicity represents the carrier of the received signal. To take a specific example, consider an x-ray proximity printer in which the x-ray illumination flux is concentrated in a very small area, say 1x1 mm^2. Thus at any one instant, only 1x1 mm^2 of the mask need be accurately aligned with the wafer. The mask can be displaced with respect to the wafer by piezoelectric transducers and the error signal feeding these transducers is generated by an optical system which continuously monitors the local (to the illumination) displacement between mask and wafer. In this manner, the mask can track the wafer as the illumination is scanned. Contrary to popular myth, optical positional information can have an accuracy far better than the resolution limit of a microscope objective and tolerances as small as 0.01 μm have been reported

Vote-Taking Lithography

Pattern made from one of the fields on a 1X stepper reticle

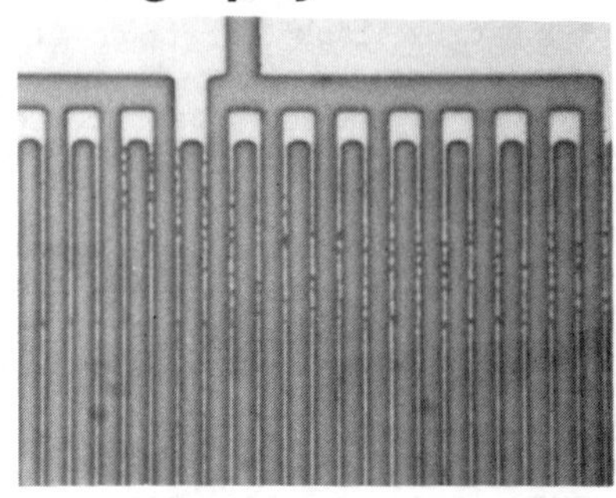

10 µm

Pattern made from three fields, including the above

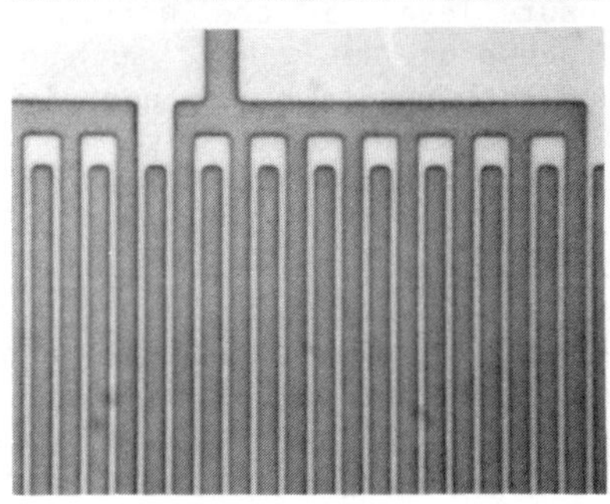

Figure 4. Vote-taking lithography; each wafer site is exposed with 3 or more nominally identical mask patterns so that a random defect in 1 mask pattern is an insignificant (i.e. 33 and 1/3% or less perturbation) part of the total integrated exposure [7].

[8]. This second scheme has not yet been implemented and, along with the vote taking lithographic scheme, is described as an example of how fresh thinking as well as heroic technology might enable us to drive down the dimensions of VLSI patterning to the sub-optical regime.

PATTERN TRANSFER

The foregoing account has concentrated exclusively on patterning of radiation-sensitive materials (resists). Although a few functional circuit materials can be patterned by direct exposure [9] in VLSI manufacturing, most require a pattern transfer step to transfer the resist pattern to the circuit material. Because lateral dimensions are now almost as small as vertical dimensions, wet, isotropic, etching is falling from favor, being replaced by various anisotropic etching techniques. Most anisotropic etching takes place in a low pressure plasma reactor such that the etchant species bombards the surface. The details of all the configurations are beyond the scope of this paper, but the basic tradeoffs are between lateral, chemical or vertical resolution. High lateral resolution means that the etching proceeds only vertically down from the resist edge. High chemical resolution means that the etchant attacks only the desired material and high vertical resolution means that the etching stops at the substrate. The available commercial instruments appear to be based more on empiricism than science and it would also appear that a variety of techniques, including selective plating and selective chemical vapor deposition, might bear serious investigation.

REFERENCES

1. See, for example, the technical specifications for the MEBES 3®, Perkin Elmer EBT, 26240 Corporate Drive, Hayward, California 94545.

2. T. E. Everhart, unpublished DARPA report, 1976. See also H. I. Smith, paper presented at 29th International Symposium in Electron, Ion and Photon Beams, Portland, Oregon, May, 1985.

3. A 1:1 Electron Projection Stepper, R. Ward, A. R. Franklin, I. H. Levin, P. Gould, M. J. Plummer, Paper C-31, presented at 29th International Symposium on Electron, Ion and Photon Beams, Portland, Oregon, May, 1985.

4. K. Suzuki, Paper H1, presented at 29th International Symposium in Electron, Ion and Photon Beams, Portland, Oregon, May, 1985.

5. J. N. Randall, et al. J. Vac. Sci. Tech. B3, 58 (1985).

6. P. Nehmiz, et al, ibid. p. 136.

7. C. C. Fu and D. H. Dameron, Elec. Dev. Lett. EDL 5, October, 1984, 97.

8. See, for example, the specifications of the Nanoline® (Nanometrics, Inc., Santa Clara, California).

9. "Laser Pantography", B. H. MacWilliams, LLNL unpublished report, 1985.

ACKNOWLEDGEMENTS

Preparation of this paper was supported by The Defense Advanced Research Projects Agency, Contract Number N00014-84-K-0077, and The National Science Foundation, Contract Number ECS 82-03296-A1,2.

OPTOELECTRONIC MATERIALS AND PROCESSES FOR COMMUNICATIONS

R. A. Laudise
AT&T Bell Laboratories
600 Mountain Avenue
Murray Hill, N.J. 07974

ABSTRACT

Optical communications has provided a driving force for much recent research in optoelectronic materials including laser, nonlinear and optical fiber materials. Present research focuses on second generation materials: very low loss non-SiO_2 based fibers, very long wavelength lasers including quaternary III-V solid solutions lattice matched to GaSb, (HgCd)Te lattice matched to (CdMn)Te and electro-optic materials including $LiNbO_3$. We review the considerations which have led to the choice of these materials and describe the current state of their preparation. In addition, we discuss process improvements in fiber preparation and improvements in understanding and manipulation of Czochralski and Bridgman crystal growth as applied to optoelectronic materials. This paper is a brief published review of an invited talk to be given at the American Vacuum Society meeting in November 1985.

INTRODUCTION

The laser has provided a driving force for nearly a generation's research and development in optoelectronic materials for optical communications. It is our thesis that the results of that research: low threshold, long lived, nearly single frequency lasers and low loss optical fibers, rather than bringing down the final curtain on optoelectronic materials research, merely set the stage for the second act: a new generation of materials and devices whose characteristics should provide a qualitatively new generation of systems.

In this paper we will briefly review past accomplishments in optoelectronic materials and processes and from these results access the vector of the future. We will focus on fibers and substrate materials for lasers and detectors, and electro-optic modulators for communications.

OPTICAL FIBERS – LOW LOSS

Figure 1 shows historical trends in fiber loss. The lowest losses correspond to impurity levels of a few ppb for Fe, and other transition metals and ≈10 ppb for H_2O.(1-5) Indeed the loss is now close to $1/\lambda_4$, the Rayleigh scattering limit. For example, at 1.5 μm the measured 0.16 db/km loss compares favorably with the calculated scattering loss. Consequently, if one wants even lower loss, he must reduce the scattering loss. This can be accomplished by moving to a longer wavelength if the lattice (multiphonon) absorption can be moved by changing the material. M. E. Lines(6-7) has modeled lattice absorption and Rayleigh scattering (Figure 2) and shown that at wavelengths >1.5 μm one might expect absorption from 10^{-2} - 10^{-3} db/km in, for instance, certain chlorides and fluorides. Van Uitert, Bruce and co-workers(8) have experimental extrapolations which suggest that the loss is in the 10^{-2} range.

The rationale for finding lower loss glasses is as follows: It is well known that Rayleigh scattering is caused by frozen-in density fluctuations which are inevitably present in the liquid state. These fluctuations are larger at high temperatures and are frozen in at the glass transition so that lower polarizability materials could be expected to have lower scattering.

The lattice absorption will move to longer wavelengths, however, only when heavy, weakly bonded atoms are present. Such atoms tend to be highly polarizable. It remained for Lines to quantitatively model these two opposing effects, show the desirability of halides, and set out the systematics for evaluating other materials.

Active research is underway on the preparation of low loss non-SiO^2 fibers in the US, Europe and Japan. Another important fiber material consideration involves maximizing band width by controlling pulse dispersion. There have been reviews of this subject elsewhere so it will not be discussed here.(9-10)

OPTICAL FIBERS – PROCESS UNDERSTANDING

The cost of fibers and hence their market penetration is to a large degree a function of how rapidly they can be prepared. Several processes have been developed to prepare the rod-like preforms of controlled composition, which are drawn into fibers, however they all depend on understanding and controlling the oxidation of the halides of Si, Ge and other materials:

$$SiCl_{4(g)} + O_{2(g)} \rightleftarrows SiO_{2(glass)} + 2Cl_{2(g)} \quad (1)$$

Thermodynamic calculations allow one to model the equilibria as a function of temperature and input gas composition.[(11)] As Figure 3[(11)] shows, the equilibrium cannot be described by a few simple reactions. However, since the Si/Ge ratio in the glasses made by the oxidation of $SiCl_4$ + $GeCl_4$ can be predicted quite accurately from thermodynamic calculations (Figure 4)[(11)], it is likely that equilibrium is obtained.

Indeed thermodynamic data for the appropriate species, together with an appropriate computer model, has allowed us to perform "experiments" in the computer with the modified chemical vapor deposition (MCVD) process used in our laboratory and factory to prepare preforms. The reader is referred to Nagel[(12)] for a brief description of the MCVD process and apparatus and for a description of fiber drawing.

Following an understanding of the equilibrium process, it is appropriate to next investigate the kinetics. Extensive investigations of the rate processes in MCVD have been conducted. Perhaps the most important finding is that the rate limiting step involves the thermophoretic movement of, for example, $(SiGe)O_2$ which is formed homogeneously in the gas phase by reactions analogous to Equation (1). Colloidal sized particles move to the cold wall where glass is formed.[(13)] Thermophoresis is the process which causes soot to collect on the cool region of kerosene lamp chimneys. Momentum exchange between colliding gas molecules and the colloidal particle is greatest on the hot side of the particle "pushing" it "down" a temperature gradient. Once thermophoresis was identified as rate limiting kinetic modeling was more straightforward. Detailed studies of rate processes have improved the deposition rate more than 10× since the discovery of MCVD with commensurate decreases in fiber costs.[(12)]

LASER MATERIALS

Optical communications lasers are needed at wavelengths where fibers are of minimum loss. In early fibers the wavelength used was 0.83 μm (to avoid absorption due to OH), in later fibers the wavelengths are 1.3 and 1.5 μm and we believe eventually wavelengths >1.5 μm will be used for systems where lattice absorption permits operating at longer wavelengths so that Rayleigh scattering is reduced. Figures 5 and 6 show lattice parameters and band widths of a number of III-V[(5)] and II-VI[(14)] compounds. These figures enable one to choose single crystal substrate materials and solid solution epitaxial materials for heterostructure lasers. It should, of course, be emphasized that additional considerations including homogeneous phase stability, band structure (direct band gap needed) and ease of preparation must also be considered in choosing a material system for a particular wavelength. Table 1 lists some systems appropriate for lasers.

Most present optoelectronic attention focuses on wavelengths >1.3 μm so we will not discuss GaAs based systems.

Because of the high pressure (29.9 atoms) at the melting point (10.72°C) of InP, liquid encapsulated (B_2O_3) Czochralski growth (LEC) is the most common InP growth method and has been practiced for over a decade since the original work of Mullin et al.(15) and Seki et al.(16) Typical carrier concentrations are approximately $10^{15}cm^{-3}$ for intrinsic or undoped n-type material. Generally, in InP, there are more than enough impurities to account for conductivity, but direct correlations with particular impurities are difficult. By analogy with GaAs(18), we might expect defects such as dislocations to be sensitive to P pressure. Semi-insulating material can be made by, for example, Fe doping. Resistivities as high as $10^9 \Omega cm$ can be obtained. Seki et al.(16) and Cockayne et al.(17) have shown that dislocation density is reduced by the "hardening" of the lattice by doping. Dopants for hardening include Zn(p), Te(n), Ge(n) and S(n) which affect the conductivity when present at levels sufficient to strengthen the lattice ($10^{18-10^{19}}$ atoms cm^{-3}) so as to get very low dislocation ($<10^3 cm^{-3}$) material. Too high a concentration of dopant can produce precipitates which are themselves sources of dislocations. Some typical growth conditions for InP are given in Table 2.(5)

Very large ($\approx$1 kg) Czochralski crystals can now be grown with dislocation density $\approx 10^4$ when undoped and lower when doped. Present problems and opportunities in InP include the need for reproducible techniques for producing very low dislocation density, large, undoped crystals.

Jordan, Caruso, Von Neida and Nielsen(19) have analyzed the genesis of thermal stress induced dislocations in Czochralski pulled GaAs, InP and Si using a quasi-steady state heat transfer/thermal stress model for glide dislocation generation.

GaSb growth does not require liquid encapsulation because Sb pressure is $<10^{-3}$ Torr at the melting point, $\approx$706°C. Growth conditions are given in Table 3.(20) A typical crystal is shown in Figure 7. We report elsewhere on the characterization including the distribution of impurities(20) incorporation. Table 3 gives typical GaSb growth conditions.(20)

Epitaxial growth is by either liquid or vapor phase processes or molecular beam epitaxy. These procedures have been reviewed elsewhere(5) so they will not be discussed here.

For wavelengths beyond 1.5μ CdTe based systems seem appropriate. (HgCd)Te is well known as a detector in this region, but it might also find application as a laser. It can easily be lattice matched to (CdMn)Te over the wavelength region for 1-20 μm.(22) The distribution coefficient for Mn in CdMnTe is $\approx$0.96 so that most of a crystal boule is homogeneous enough in lattice parameter that it may be used as an epitaxial substrate.[22] Growth conditions are given in Table 4.

ELECTRO-OPTIC MATERIALS

Electro-optic materials in matrix arrays are of interest for switches and perhaps ultimately for stand alone optical processors. Table 5 lists some important electro-optic materials. Although $(SrBa)Nb_2O_3$ and $K(TaNb)O_3$ have half-wave voltages much less than $Li(NbO_3)$ strain-free optically homogeneous crystals are very hard to prepare. For much higher frequencies, e.g., for second harmonic generation (SHG) materials like KTP, $KTiOPO_4$, are attractive. In general, high ionic polarizability is important for lower frequencies and high electronic polarizability for higher frequencies. The high speeds attainable with the photorefractive effect suggest its use for parallel optical processing. Large effects in III-Vs are especially interesting since they might allow complete integration of the laser and other functions on a single chip of one material. Resistance to optical damage is important for SHG, a special advantage for KTP. Temperature stability is poor in ferroelectrics close to the Curie temperature, T_C.

$LiNbO_3$ emerges as perhaps the most attractive present day electro-optic material not only because of its relatively large e.o. coefficient but also because we now know how to grow large high quality crystals. Very important to gaining control of the material was understanding of the effect of stoichiometry on properties.[23] $LiNbO_3$ melts incongruently and crystals are usually non-stoichiometric.[23-24] Clearly, further exploration of e.o. materials will be an important future direction.

CONCLUSIONS

Improvements in SiO_2 based fiber technology have allowed us to prepare cheap fibers at the loss limit at 1.3-1.5 μm. Further improvements in fibers will require non-SiO_2 based materials. Modeling studies suggest that halides glasses at wavelengths greater than 1.5μ could have losses at or under 10^{-2} db/km.

Advanced research on lasers is moving from InP based substrate systems to GaSb and (CdMn)Te based systems to accommodate the longer wavelengths needed.

As information in optical form becomes more pervasive, the driving force for optical switching and direct optical processing will grow. $LiNbO_3$ is already a useful e.o. material but new materials with significantly larger e.o. coefficients, stability and ease of preparation would be extremely useful.

REFERENCES

(1) See for instance, W. G. French, J. B. MacChesney, P. B. O'Connor and G. W. Tasker, *Bell System Tech. J. 53* (1974) 951, for an early description of one process, MCVD, for making fiber.

(2) Tingyi Li, *IEEE J.* Selected Areas Commun. 1 (1983).

(3) S. R. Nagel, J. B. MacChesney and K. L. Walker, *IEEE J. Quantum Electron* QE-18 (1982) 459.

(4) W. G. French, J. B. MacChesney, P. B. O'Connor and G. W. Tasker, *Bell System Tech. J.* 53 (1974) 955.

(5) R. A. Laudise, *J. Cryst. Gr.* 65 (1983) 3.

(6) M. E. Lines, *J. Appl. Phys.* 55 (1984) 4052.

(7) M. E. Lines, *J. Appl. Phys.* 55 (1984) 4058.

(8) L. G. Van Uitert, A. J. Bruce, W. H. Grodkiewicz and D. L. Wood, Presented at 3*rd* International Symposium on Halide Glasses, Rennes, France, June 1985.

(9) A. Carnevale and U. C. Paek, *Bell System Tech. J.* 62 (1983) 1937.

(10) A. Carnevale and U. C. Paek, *Bell System Tech. J.* 62 (1983) 1415.

(11) K. B. McAfee, K. L. Walker, R. A. Laudise and R. S. Hozack, *J. Am.*Ceram.*Soc.*" 67 (1984) 420.

(12) S. R. Nagel, J. B. MacChesney and K. L. Walker, *Optical Fiber Comm.* 1 (1985) 1.

(13) K. L. Walker, F. T. Beyling and S. R. Nagel, *Am. Ceram. Soc. Bull.* 63 (1980) 552.

(14) P. M. Bridenbaugh, private communication.

(15) J. B. Mullin, A. Royle and B. W. Staughan in: Proc. *3rd* Intern. Symp. on GaAs and Related Compounds, Aachen, 1970, Inst. Phys. Conf. Ser. 9 (Inst. Phys. London, 1972) p. 72.

(16) Y. Seki, J. Matsui and H. Wantanabe, *J. Appl. Phys.* 47.

(17) B. Cockayne, G. T. Brown and W. R. MacEwan, *J. Cryst. Gr.* 54 (1981) 9, and references mentioned.

(18) J. M. Parsey, J. Lagowski and H. C. Gatos, *J. Electrochem. Soc.*, in press.

(19) A. S. Jordan, R. Caruso, A. R. Von Neida and J. W. Nielsen, *J. Appl. Phys.* 52 (1981) 3331.

(20) W. A. Sunder, R. L. Barns, T. Y. Kometani, J. M. Parsey and R. A. Laudise, submitted to *J. Cryst. Gr.*

(21) E. Buehler, *J. Cryst. Gr.* 43 (1978) 584.

(22) P. M. Bridenbaugh, *Materials Letters*, in press.

(23) J. R. Carruthers, G. E. Peterson, M. Grasso and P. M. Bridenbaugh, *J. Appl. Phys.* 42 (1971) 1846.

(24) A. Rauber, *Current Topics in Matls. Sci.*, Ed. by E. Kaldis 1 (1978) 481.

TABLE 1

SOME LASER MATERIALS FOR COMMUNICATIONS

SUBSTRATE	ACTIVE LAYER	WAVELENGTH
GaAs	(GaAl)As	0.7 - 1.0 μm
InP	(GaIn)(AsP)	1.1 - 1.6 μm
GaSb	(GaIn)AsSb	1.5 - 3.5 μm
CdTe	(HgCd)Te	0.9 - 20 μm*
(CdMn)Te	(HgCd)Te	0.9 - 20 μm**

* Some lattice mismatch strain over range shown.

** Possible to get perfect lattice match.

TABLE 2[5]

SOME GROWTH CONDITIONS FOR InP CZOCHRALSKI GROWTH

CRUCIBLE	LIQUID ENCAPSULANT	SEED	PULL RATE	PRESSURE
SiO_2 or N	B_2O_3	(100), (111)B	1.5 cm/hr.	N_2, 30 atms.
BN	B_2O_3	(111)B	1.5, 2.5	Ar + He, 30-47 atms.
SiO_2 or BN	B_2O_3	(111)B	1.3-1.7	He + Ar

TABLE 3[20]

GROWTH CONDITIONS FOR GaSb"

Apparatus	–	Czochralski crystal puller[21], 20 KW, 450 KHz r.f. generator
Crucible	–	Vitreous quartz or BN
Sb/Ga (moles)	–	1.001 (to compensate for volatilization of Sb)
Stirring Speed	–	50-55 rpm
Pull Rate	–	0.48 in/hr. (1.2 cm/hr.)
Seed Orientation	–	<111>B

TABLE 4

GROWTH CONDITIONS FOR CdMnTe[22]

Apparatus	–	Resistance heated Bridgman furnace
Crucible	–	Evacuated vitreous SiO_2 ampoule-interior coated with graphite
Growth Rate	–	2 mm/hr.
Temperature Gradient at L-S Interface	–	10-20°/cm.

TABLE 5

Courtesy of A. M. Glass

ELECTRO-OPTIC MATERIALS

MATERIAL	DIELECTRIC CONSTANT ϵ	HALF-WAVE VOLTAGE* V_H (kV)	CHARGE $\epsilon\epsilon V_H$ (nC)
PIEZOELECTRICS			
SiO_2	8	580	400
$Bi_{12}SiO_{20}$	47	12	50
FERROELECTRICS			
$LiNbO_3$	32	4.4	13
$LiTaO_3$	45	4.4	18
$^+Sr_{0.75}Ba_{0.25}Nb_2O_6$	3400	0.06	18
$^+Sr_{0.5}Ba_{0.5}Nb_2O_6$	450	0.39	16
$^+KaTa_{0.3}Nb_{0.7}O_3$	1000	0.19	17
SEMICONDUCTORS			
GaAs	13.2	18	21
InP	12.6	20	22
CdTe	9.4	6.5	5.4
$CdIn_2Te_4$	200	2	3.5
ORGANICS			
^{x}Metanitroaniline		2.6	

Note: The charge $\epsilon\epsilon_o V_h$ is a useful measure for the photorefractive sensitivity of materials, and the speed of modulators.

* Depends on direction of electric field and light propagation. This parameter gives a relative measure of the voltage needed for an electro-optic modulator or, alternatively, a relative measure of the inverse device length.

\+ Tungsten-Bronze solid solutions: T_c near room temperature.

x One of the best organic materials (electronic nonlinearly is dominant).

FIGURE CAPTIONS

Figure 1 Historical Trends in Multimode SiO_2 Based Optical Fiber Loss.[5]

Figure 2 Scattering Loss in Various Materials (after M. E. Lines).[6] Solid points where refractive index dispersion is a minimum, open circles minimum total attenuation estimated, i.e., at greater wavelengths lattice absorption will be appreciable. Dark lines include composition fluctuation losses estimated for mixed glasses.

Figure 3 Species Present in $SiCl_4 + GeCl_4$ Oxidation.[11] Moles of input constituents typical for Ge doped fiber reaction conditions. Si = 1.0, Ge = 0.33, Cl = 5.33, O = 28.56 moles.

Figure 4 GeO_2 Content in SiO_2 Glass Formed by $SiCl_4$ + $GeCl_4$ Oxidation.[11] Data points are experimental results. Lines are thermodynamic calculations. Reaction takes place at 1650°K.

Figure 5 Lattice Parameters and Band Gaps for Selectede III-V Compounds.[5]

Figure 6 Lattice Parameters and Band Gaps for Selected II-VI Compoundss.[14]

Figure 7 GaSb Crystal (courtesy W. A. Sunder).

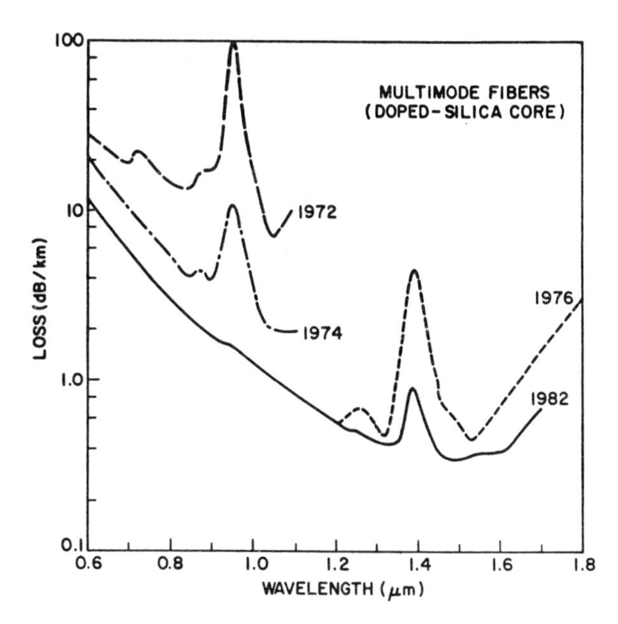

FIGURE 1

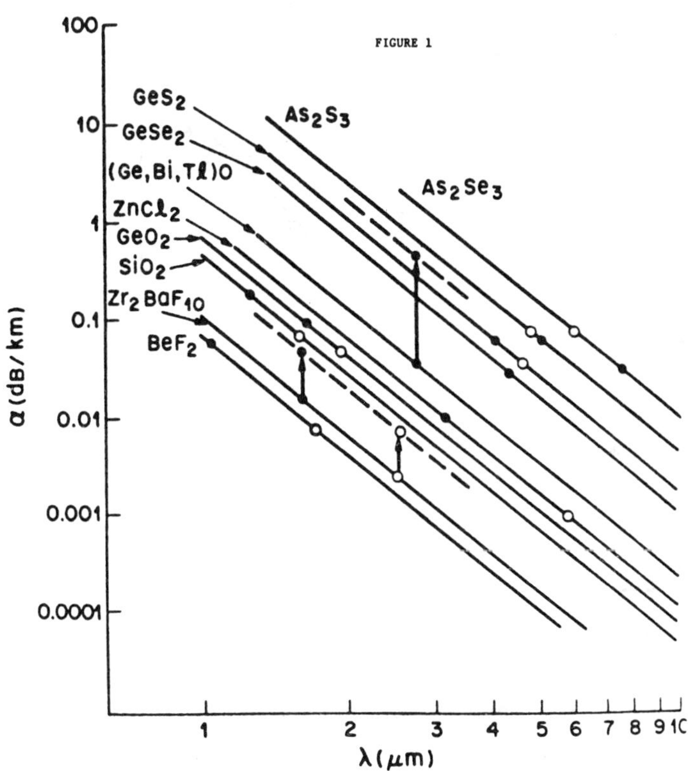

FIGURE 2

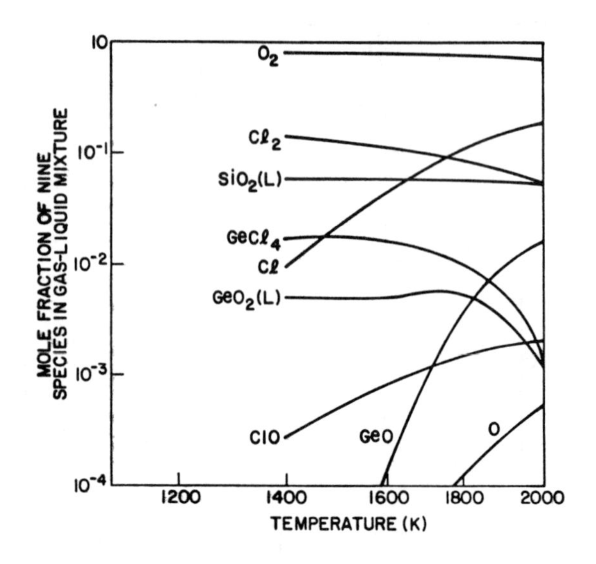

FIGURE 3

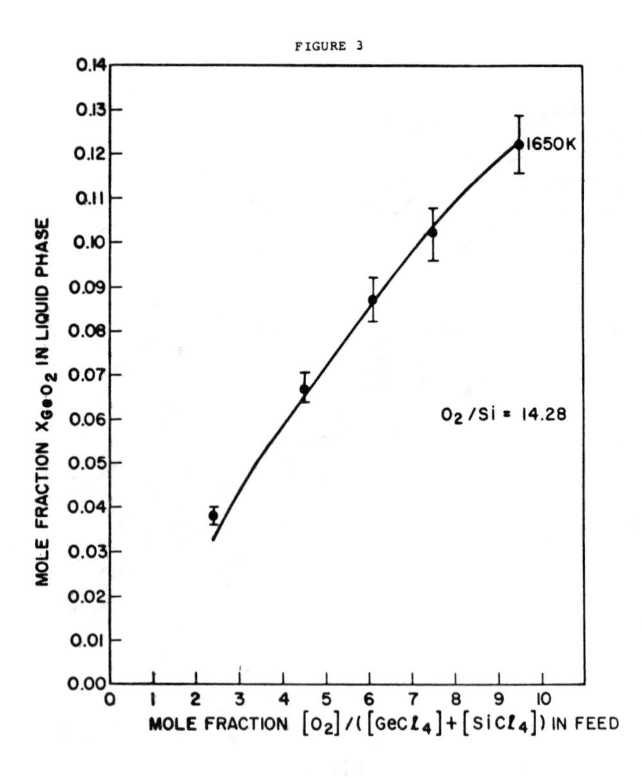

FIGURE 4

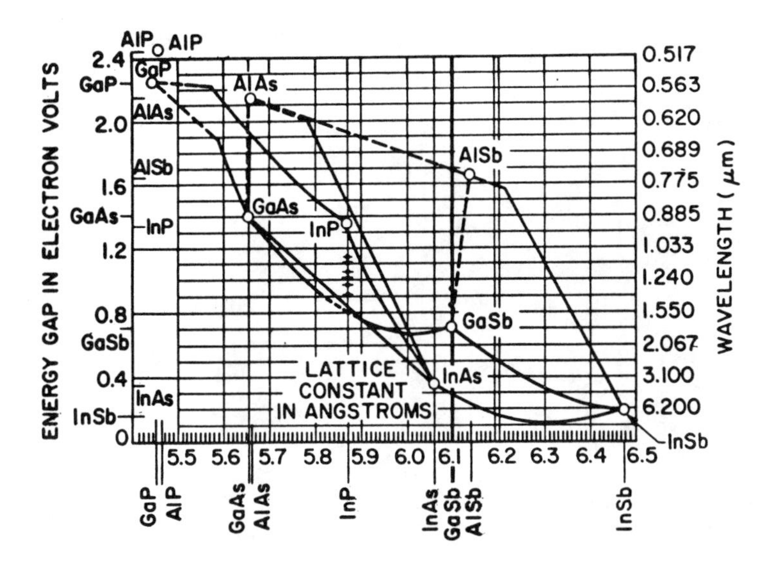

FIGURE 5

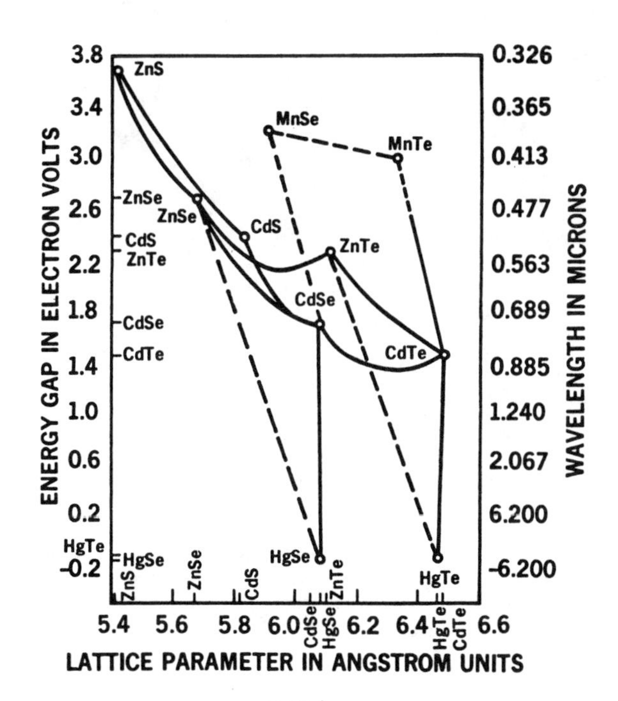

FIGURE 6

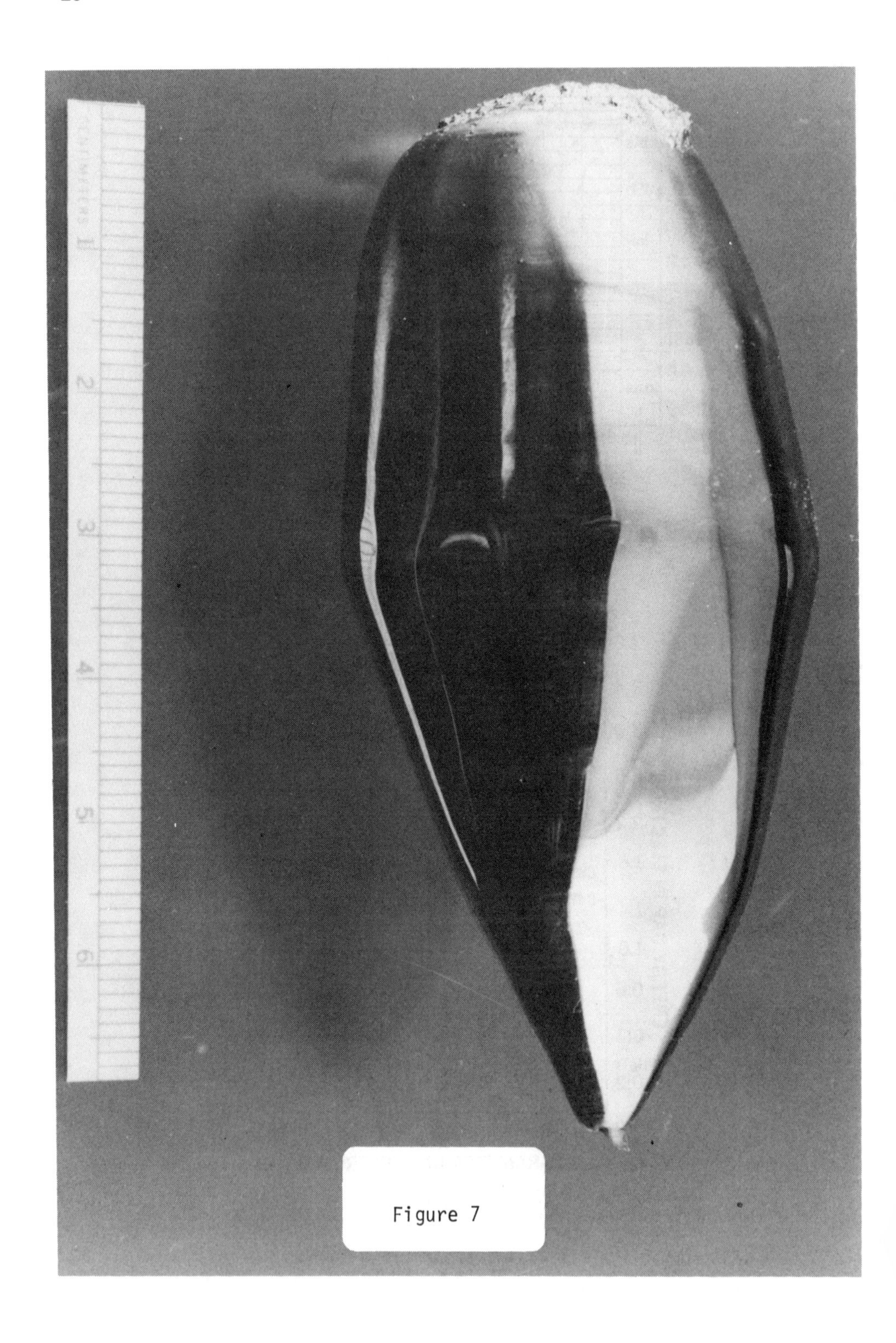

Figure 7

PROBLEMS AND OPPORTUNITIES IN ERASABLE OPTICAL RECORDING

G. A. N. Connell
Xerox Palo Alto Research Center, Palo Alto, CA 94304

ABSTRACT

Two approaches to erasable optical recording, based on magneto-optical and phase change media respectively, are discussed. The two approaches encounter different problems in achieving adequate system performance. On the one hand, magneto-optic storage media have been limited by inadequate signal-to-noise ratio in readout and insufficient corrosion resistance. On the other hand, phase change media have had limited data retention times, particularly in media with short erase times, and an inadequate write-erase cyclability. The paper reviews the progress made toward solving these problems, indicates where problems still exist, and what will likely be possible as a product in the next few years.

INTRODUCTION

Non-erasable optical recording products are now available from numerous manufacturers. They offer about 2 GBytes of storage per disk at an areal density of about 2×10^7 bits/cm^2 and data rates of about 5 Mbits/sec. The average access time is about 100 mS, slower than in current Winchester technology, but twenty adjacent tracks that contain about 1 Mbyte of information can be accessed in about 1 mSec. The disks themselves are removable. While there are many opportunities for further innovations in this area, there is keen competition to build systems that add erasability to the above, already impressive, list of features. Two approaches, based on magneto-optic and phase change media, are the subjects of this paper.

In each case, after a brief outline of an optical head design, we address the peculiar materials engineering problems. The media must support submicron marks to allow diffraction limited recording, provide adequate signal-to-noise ratio upon readout, be capable of many write-erase cycles without change in bit error rate, and provide the means for stable off-line data storage. The two approaches encounter different problems in achieving these requirements. On the one hand, the magneto-optic media have been limited in their performance by inadequate signal-to-noise ratio in readout and insufficient corrosion resistance for long-term data storage. Methods to improve the magneto-optic conversivity[1] and corrosion resistance of the magnetic material and readout and protection efficiency of the multilayer structure in which the magnetic material is encased have therefore been the focus of attention. On the other hand, the phase change media have had difficulty in achieving a sufficient number of write-erase cycles and in preventing data loss by movement of the amorphous-crystalline boundary. Materials engineering has therefore focussed on controlling the crystalline$\leftrightarrow$amorphous transition itself.

The paper will review the current status of both approaches and what will likely be possible in the next few years.

MAGNETO-OPTICAL RECORDING

A typical magneto-optical head is shown in Fig. 1. In this implementation, collimated linearly polarized light from a GaAs laser passes through a polarizing beam splitter and is focussed on a magneto-optic medium. The focussing and tracking systems are not discussed here but are similar to those used in non-

0094-243X/86/1380029-11$3.00

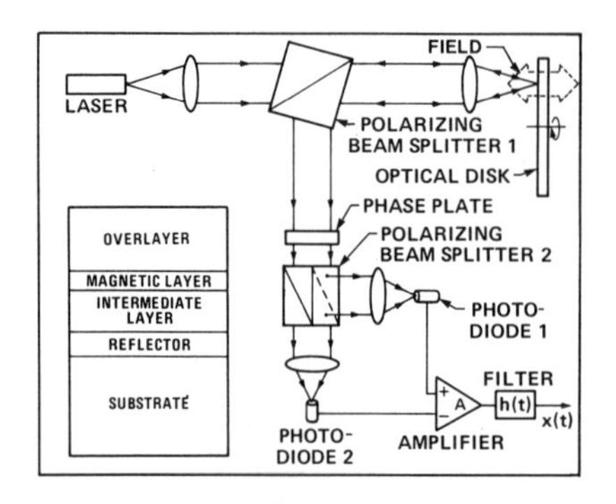

Fig. 1. Schematic representation of the layout of a magneto-optic system and, as an insaet, a blowup of a quadrilayer medium.

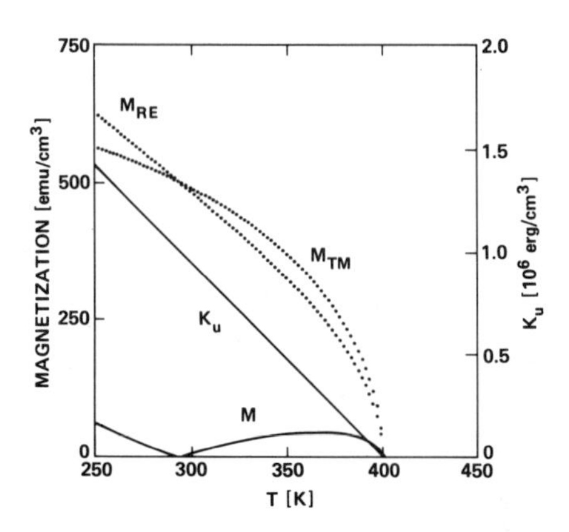

Fig. 2. The magnetization versus temperature for a mean field model of an amorphous rare earth-transition metal alloy with compensation point near room temperature. The temperature dependence of the anisotropy constant is also shown [4].

erasable optical recording systems. The reflected light is then guided to the differential detection module by the same polarizing beam splitter.

Information is recorded thermomagnetically. The laser is pulsed to high power for a short time to raise the temperature of the perpendicularly magnetized medium sufficiently for the externally applied magnetic field to reverse the direction of magnetization in the heated region. When the temperature of the medium returns to its read state, the reverse magnetized domain persists. Erasure is accomplished by the same thermal process, now aided by an oppositely directed magnetic field.

Readout of information employs the polar Kerr effect. The linearly polarized light, reflected from the medium, is rotated to the left or right, according to the direction of magnetization. Transitions in magnetization along a recorded track can therefore be sensed by the detection module. The signal power to shot noise power for such a system is[2]

$$SNR \sim P_0 r_y^2 \quad (1)$$

where P_0 is the incident laser power polarized in the x-direction and r_y is the amplitude of magneto-optically induced reflected light polarized in the y-direction. Thus r_y^2 serves as a figure of merit for readout performance. Its value has been increased both by alloy selection and by the incorporation of the magnetic layer in multilayer interference structures. It is equally important however that the medium tolerate a large P_0 during readout, subject to the writing and erasing constraints, so that the SNR is maximized.

Fig. 2 summarizes the magnetic properties of heavy rare earth-transition metal (Fe, Co) alloys[3] and demonstrates why they are ideal as magneto-optic recording media.

First, ferrimagnetic-like behavior occurs in heavy rare earth alloys because the transition metal subnetwork is strongly ferromagnetic and the coupling of the spin angular momenta of the rare-earth and transition metals is weakly antiferromagnetic.[3] Thus the heavy rare earth and transition metal moments are opposed by Hund's rule. It then follows that with a suitable choice of composition, the magnetization M will be close to zero over a wide temperature range and exactly zero at the compensation temperature. This is particularly true in Fe-based alloys.

Second, the deposition process itself results in a uniaxial anisotropy that ranges from columnar microstructure in poor material to more subtle pair ordering effects in the best material. The latter establishes a uniaxial anisotropy in the averaged local order around each atomic site.[4] Therefore, alloys containing non S-state rare earth ions, whose moments are directed by the local order through single ion anisotropy, have a mechanism for developing a large macroscopic uniaxial magnetic anisotropy ($K_u>0$) in the growth direction that is not present in S-state alloys.[6] The exact orientation of each non S-state rare earth moment is set by a competition between single ion anisotropy and exchange[5] and this leads to a scatter of the moments from site to site. The ensuing fluctuations possibly provide the mechanism for increasing coercivity and stabilizing domains at sub-micron dimensions. Both of these effects are absent in the S-state rare earth alloys, such as GdCo.

This combination of properties makes the heavy non S-state rare earth alloys suitable for magneto-optic recording for two reasons: the demagnetizing field will be small and the coercivity $H_c \sim K_u/M$ will be very high, particularly near the compensation temperature. Upon heating above the compensation temperature, the coercivity decreases because of the increase of M and decrease of K_u, and thermomagnetic recording becomes possible.

Finally, because the alloys are amorphous many problems originally encountered with crystalline alloys can be circumvented. For example, the Curie and compensation temperatures can be controlled independently and continuously by alloying. The former is used to optimize the recording sensitivity and tolerance to readout laser power, while the latter provides the necessary control over magnetic domain size and stability. As a second example, magnetic media noise in readout is negligible in the non S-state alloys, because the domain walls appear to be pinned intrinsically rather than on grain boundaries, etc. [6] As a result, fluctuations in the location of written transitions are minimal and a much closer approach to shot noise limited performance can be achieved in well-designed systems. Finally, the magneto-optic effect itself depends almost entirely on the transition metal magnetization[7] and therefore has the important property of being finite and smoothly varying at the compensation temperature.

The specific alloys currently of most interest are based on the $(Tb, Gd)_x(Fe, Co)_{1-x}$ system with $0.2 \leq x \leq 0.3$. Of the many deposition methods available, multi-hearth electron beam evaporation, rf and dc diode sputtering, and magnetron sputtering are most often used. While many of the properties of the alloys so produced are similar, there can be important differences that result from differences in microstructure. For example, Hong[8] has reported that the oxidation of some magnetron sputtered Tb-Fe films is self limiting, in contrast to the pervasive oxidation that occurs in some otherwise-similar rf sputtered alloys. Therefore, while the remainder of this section focusses on materials engineering through alloying, the equal importance of microstructure control through the choice of the deposition process and parameters must not be overlooked.

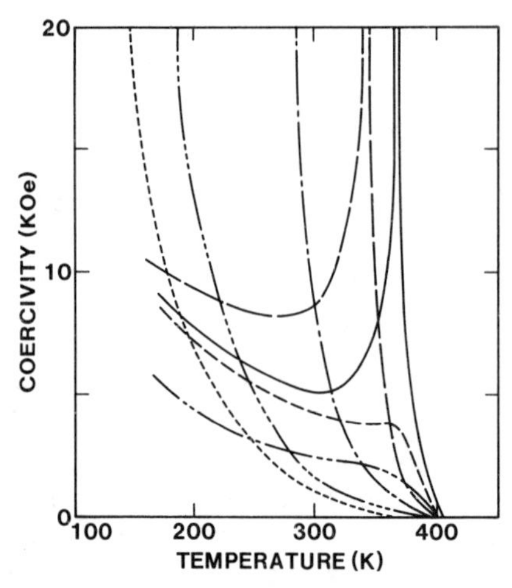

Fig. 3. Coercivity versus temperature for $Tb_{0.18}Fe_{0.72}$ (···) $Tb_{0.20}Fe_{0.80}$ (—···), $Tb_{0.23}Fe_{0.77}$ (——), $Tb_{0.25}Fe_{0.75}$ (— —), $Tb_{0.27}Fe_{0.73}$ (——), $Tb_{0.29}Fe_{0.71}$ (- - -), and $Tb_{0.32}Fe_{0.68}$ (- ·· -).

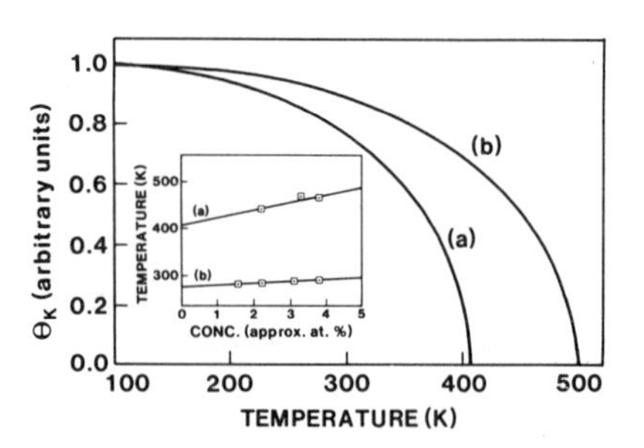

Fig. 4. Polar Kerr effect versus temperature for (a) $Tb_{0.23}Fe_{0.77}$ (b) and $Tb_{0.23}Fe_{0.72}Co_{0.05}$. The insert shows the change in (a) Curie and (b) compensation temperatures with Co substitution in $Tb_{0.23}$ $(Fe_{1-y}Co_y)_{0.77}$ alloys.

Fig. 3 shows the variation of compensation temperature with composition in Tb_xFe_{1-x} alloys.[9] When $T_{comp} < T_{cur}$, the coercivity is highly peaked near T_{comp}, but when $T_{comp} \sim T_{cur}$, the coercivity is greater than about 5kOe over the complete range of operation. It is found that material with reasonable characteristics for writing and erasing and for stability of written information is obtained with compositions for which $200K \leq T_{comp} \approx T_{cur}$. Which composition is actually selected is therefore driven by the particular needs of the rest of the system. Whatever the decision however, careful control of the composition is essential to obtain uniform writing characteristics.

One immediate problem with these binary alloys is demonstrated by Fig. 4. Namely, the magneto-optic signal at ambient temperature (i.e., the temperature that is present during readout by the laser beam) is much less than that at low temperatures. Since the magneto-optic effect itself is already small, a loss such as this is of concern and methods of increasing the low temperature magneto-optic signal and removing its temperature dependence for temperatures at least up to the readout temperature are needed. Fig. 4 shows that in $Tb_x(Fe_{1-y}Co_y)_{1-x}$ alloys the Curie temperature increases with Co-concentration without any significant change in the low temperature value of the magneto-optic signal.[10] A similar result is obtained for $(Tb_{1-y}Gd_y)_xFe_{1-x}$ alloys, but in this latter case, a deleterious effect on domain stability might be expected because of the reduction in the average single ion anisotropy. Nevertheless, both of these ternary alloys provide much improved readout characteristics relative to the binary alloy through their tolerance of laser power and provide the means for system-medium optimization.

Intrinsic increases in the low temperature value of the magneto-optic effect have been harder to achieve. Recently, however, it has been shown[7] that transitions within the Fe minority spin band dominate the optical properties because the density

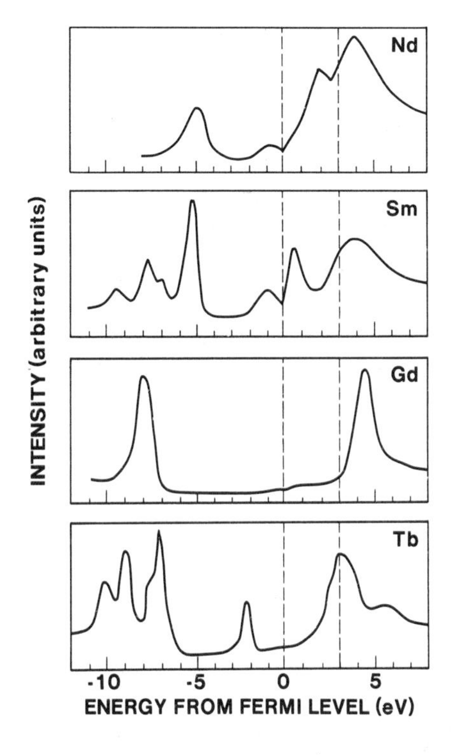

Fig. 5. XPS and XPS^{-1} data for crystalline Tb, Gd, Sm, and Nd metals [from ref. 11].

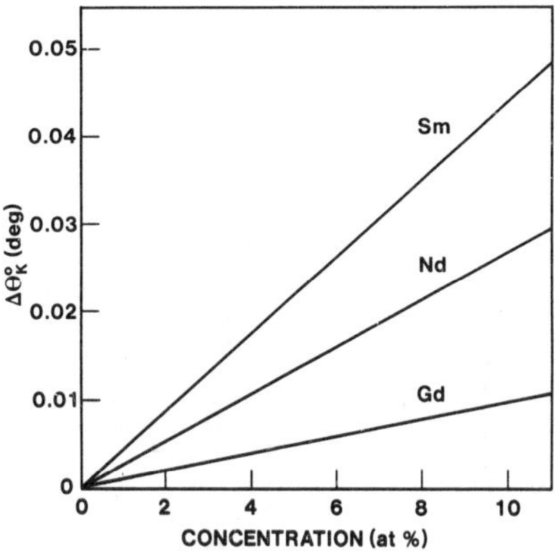

Fig. 6. Increase in low temperature polar Kerr rotation with Gd, Nd, and Sm substitution in $(Tb_{1-y}R_y)_{0.23}Fe_{0.77}$ alloys.

of Tb d-states is low near the Fermi level and Tb and Gd f-states are not accessible by optical transitions at photon energies of interest. The results obtained suggest that enhancements should be possible in $(Tb_{1-y}R_y)_xFe_{1-x}$ alloys when R is a suitable light rare earth. In this case, the contributions of transitions between the light rare earth d and f states can add to those between Fe states. Fig. 5 shows XPS and XPS^{-1} data for Tb, Gd, Sm, and Nd metals.[11] While neither Tb nor Gd have f-states that are accessible with photon energies of 1.5 eV, as mentioned earlier, it appears that Nd and Sm substitution should produce useful enhancements. Fig. 6 therefore shows the change in the low temperature value of the polar Kerr rotation, measured at 2 eV, versus Nd and Sm substitution. As expected $\Delta\theta_k^\circ$ is positive, with Sm substitution being especially effective and accounting for a maximum intrinsic enhancement in the SNR of about 40 percent. The almost zero value of $\Delta\theta_k^\circ$ for Gd substitution provides further support for the argument.

Even in the best quaternary alloys, the intrinsic magneto-optic effect is still far below that for crystalline MnBi or PtMnSb, and another enhancement technique is needed to have useful system operation. The approach taken has been to incorporate the magnetic layer into multilayer interference structures[2] such as the quadrilayer shown in Fig. 1. These structures couple more incident light into the

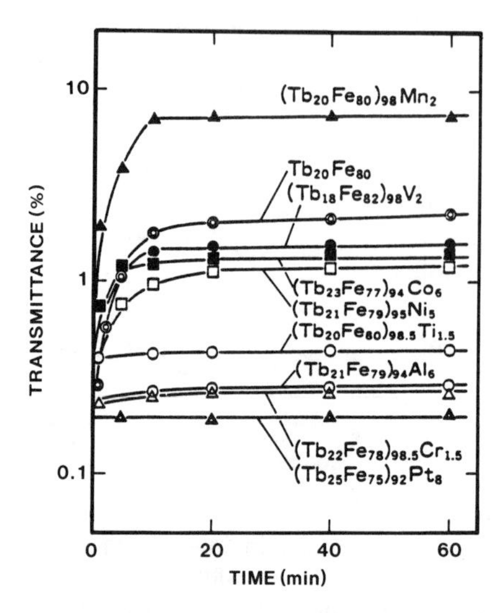

Fig. 7. Corrosion resistance of numerous TbFe-based alloys, as measured by transmittance versus time for films immersed in 2N-NaCl [from ref. 13].

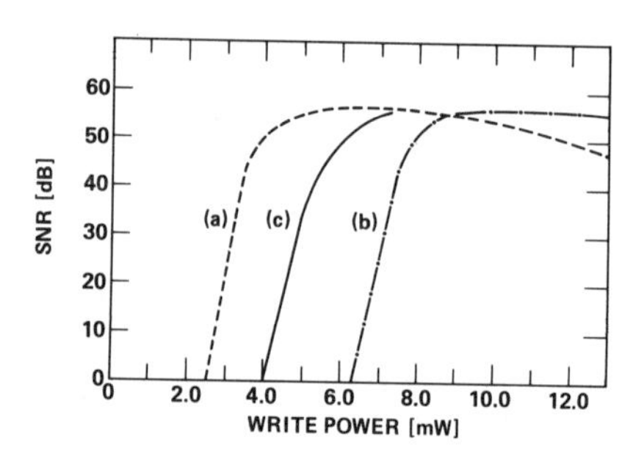

Fig. 8. SNR (in 30 kHz) versus write power at disk for three different magneto-optic media designs.

magnetic layer and couple more of the magneto-optically induced light back out. For the quadrilayer shown, the gain so created, relative to a thick uncoated film, is

$$G \sim 8\alpha t/(1\text{-}R)^2 \qquad t \lesssim 10 \text{ nm} \qquad (2)$$

where R and α are the reflectance and absorption coefficient of the semi-infinite sample and t is the thickness of magnetic layer.[12] Thus, for reasonable parameters ($\alpha \sim 5\times10^5$ cm^{-1}, t~10nm, R~0.5) G ~ 10. Less complicated structures, such as bilayers, of course produce smaller gains but, with the proper choice of the refractive index of the overlayer, are capable of providing sufficient enhancements (G~5) for the best alloys.

The problem of oxidation was briefly mentioned earlier, particularly as regards its dependence on microstructure. Fig. 7 shows that improvements in the corrosion resistance can also occur when TbFe is alloyed with Ti, Al, Cr and Pt.[13] At the concentrations added, there is little alteration in the size of the magneto-optic effect, yet there is a tremendous gain in the longevity of the material. In spite of this, there is still a need to encapsulate the magnetic layer between dielectric films, such as SiO, AlN or TiO_2 to provide additional protection. The interference structures therefore play two crucial roles.

Finally, Fig. 8 shows the SNR versus write power for three magneto-optic media designed to match different system requirements.[14] The value of over 55 dB in a 30 kHz bandwidth, achieved in each case, is sufficient to sustain a 5 Mbit/sec data rate. Work in progress must therefore now demonstrate that the longevity, as indicated by copious bit error rate tests, is sufficient for product applications.

PHASE CHANGE RECORDING

Fig. 9 demonstrates the principle of phase change recording. The medium is assumed to be initially in its crystalline state. Upon pulsing a tightly-focussed laser

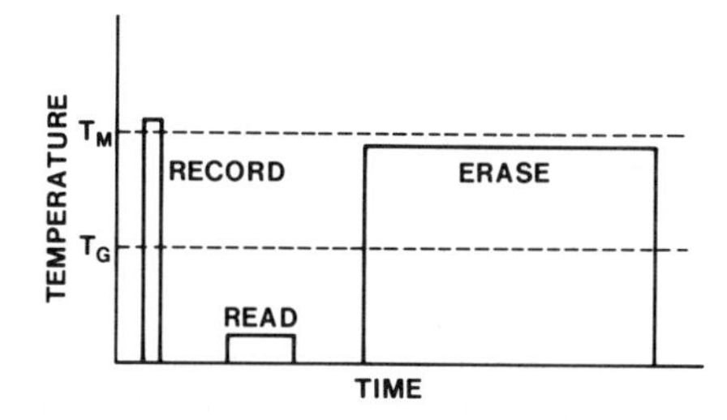

Fig. 9. The temperatures occuring in phase change media during recording, erasing and readout. T_M and T_G are the melting and glass transition temperatures respectively.

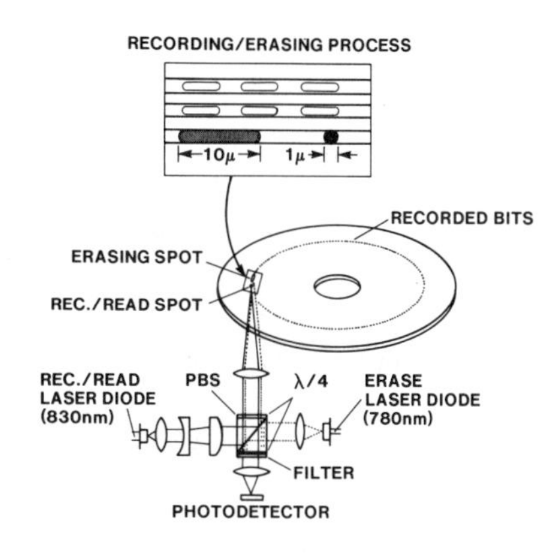

Fig. 10. A phase-change recording system. The insert shows the shapes of the record/readout and erase spots on the disk surface [from ref. 15].

to high power, the temperature of a small region is taken above the melting point T_M of the medium. The laser power is then rapidly decreased and the ensuing thermal quenching results in a recorded amorphous domain. Readout is accomplished by observing the change in reflectance that occurs at the crystalline-amorphous transition, and to this extent, phase change recording systems are directly related to non-erasable ones. Erasure, however, is accomplished by annealing the amorphous region above the glass transition temperature T_G, but below T_M, until crystalline regrowth has fully occurred and this requires the addition of new features to a non-erasable optical recording head.

Fig. 10 shows one implementation of a phase change recording system.[15] A polarized recording and readout laser is directed by a polarizing beam splitter toward the recording medium. After passage through a quarter-wave plate, the circularly-polarized light is tightly focussed on the medium. Reflected light returns via the quarter wave plate, and the resulting linearly polarized light is deflected by the beam splitter to the photodetector. Erasure is accomplished by a second laser diode that operates at a different wavelength. In this case, no correction is made to the elliptical shape of the output beam, so that after reflection at the back surface of the polarising beam splitter, the circularly polarized light that is focussed on the medium has a highly elliptical beam shape. A recorded spot that passes under this beam will therefore be heated for about 1 μs when recording is made with 100 ns pulses. Also note that the light that is reflected during erasure does not return to the photodetector.

There are many families of materials that are currently under investigation. In most cases, the work grew out of earlier studies of non-erasable recording materials, in which it was realized that amorphous films had superior recording characteristics for ablative recording. During these studies, it was recognized that phase-change recording was possible at powers short of those for ablation and under these conditions erasability could be obtained. Since then the materials engineering has focussed on maximizing the data retention time and increasing the number of record-

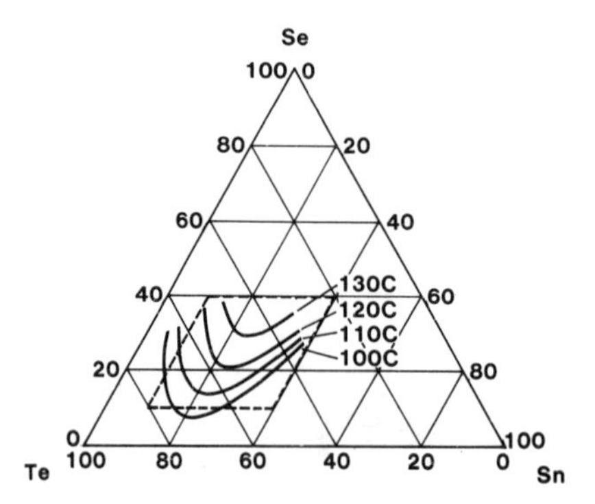

Fig. 11. Contours of equal crystallization temperature in the Se-Te-Sn system [from ref. 16].

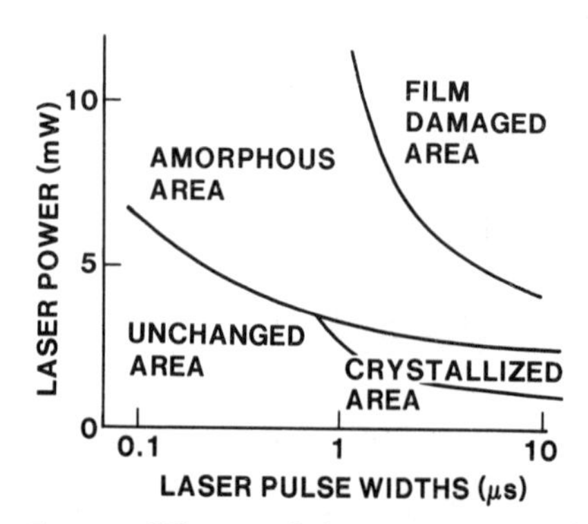

Fig. 12. Effects of laser power and dwell time on a SiO_2-coated Se-Te-Sn phase change medium [from ref. 16].

erase cycles. At present, the most interesting materials appear to be based on the Se-Te-Sn,[16] Se-In-Sb,[17] and TeO_x:Ge, Sn[18] families and results on the former will be used to bring out the issues for this technology.

Deposition of these alloys is conveniently accomplished by multihearth thermal evaporation, and scale-up methods for production have already been developed through the earlier efforts in non-erasable media. The alloys within the boxed region of Fig. 11 have been investigated in detail,[16] particularly with respect to the crystallization and oxidation of the amorphous state. It is found that the former is the preeminent factor in establishing the data retention time. For alloys with $T_C \approx 110°C$, the activation energy for crystallization of the amorphous phase is 2.3 eV and data will be retained for 10 years at 40°C. At 50°C, however, the retention time drops to less than a year, pointing up one possible area in which improvement is needed.

Fig. 12 demonstrates other trade-offs that must be made.[16] For laser dwell times of less than 1 μs, the structural state of the alloy is unchanged by laser powers of up to 3 mW, thus providing a convenient readout window. Furthermore, conversion from the crystalline to amorphous state is readily achieved with a laser power of 10 mW that is sustained for 100 nS. Reversal to the amorphous condition however requires a laser power of about 2-3 mW over a period of 2-10 μs. If the alloy is adjusted to speed crystallization, then data retention is lost. Moreover, if better data retention is required, then erasure is even slower. For the particular system implementation described earlier, multiple passes would be required for erasure before writing data, and a potential advantage of phase change recording over magneto-optic recording would be lost. This area is clearly one of significant challenge.

Figure 13 shows how the reflectivity of the alloy varies with the number of write-erase cycles for two different protection layers.[16] Not surprisingly, interfacial effects can occur at the elevated temperatures of the write and erase processes and it appears that the particular organic polymers chosen are quite unsuitable for this reason. In contrast, magnetron sputtered SiO_2 appears to offer excellent protection.

Finally, Fig. 14 shows the SNR versus the number of write-erase cycles for a TeO_x:Ge,Sn alloy.[15] For up to 10^6 cycles, the SNR in a 30 kHz bandwidth changes

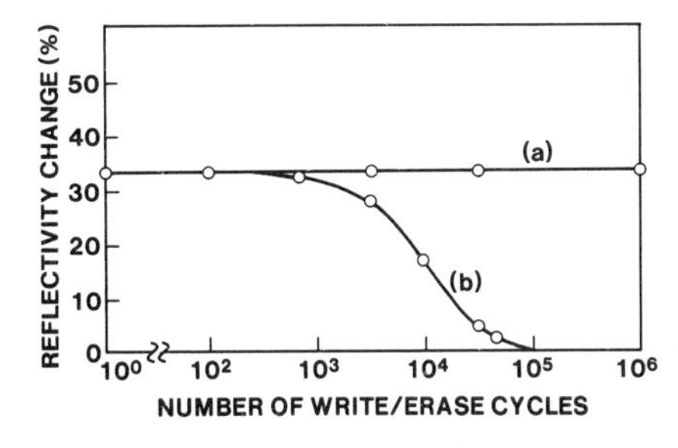

Fig. 13. Cyclability of (a) SiO_2 and (b) polymer-coated Se-Te-Sn phase change media [from ref. 16].

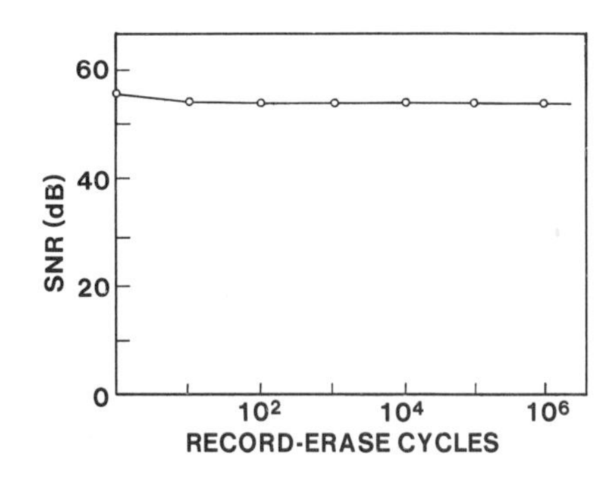

Fig. 14. SNR (in 30 kHz) versus the number of write-erase cycles for TeO_x:Ge,Sn [from ref. 15].

little and remains above 50 dB. Work on these materials must both not only confirm these promising results by full bit error rate tests over many write/erase cycles and also ensure that the competing needs of fast erasure and data retention can be met in a commercially satisfactory way.

PROSPECT

Magneto-optic and phase change recording each have their advantages and disadvantages. The magneto-optic effect is small, even with the most sophisticated media structures, and great care must be taken to suppress electronic noise. In contrast, transitions are usually observed as large changes in reflectivity in phase change media, making the suppression of electronic noise less necessary. Under certain circumstances, however, such as in Se-Te-Sn alloys with high Sn content, the change in reflectivity is small for simple structures and more complicated multilayer structures are needed.[16] This is an example of how the medium design path that leads to alloy composition which permit a more rapid erasure time also impairs the readout signal of a simple structure.

At present, most of the work on magneto-optic recording media is devoted to establishing adequate lifetime characteristics. In particular, studies are centered on the measurement of bit error rates for media on preformatted substrates. SNR, cyclability, and manufacturability all appear to be now well established. The status for phase change media is qualitatively different. Many families of materials are still under investigation and the struggle is to balance the competing needs for data retention and fast erasure, while maintaining high SNR and cyclability. At the time of writing, it would appear that this represents a greater challenge, and magneto-optical drives will be commercialized first.

What will be the characteristics of these first drives? There are several system parameters that constrain the performance.[14] The focus and tracking servos impose an upper limit on rotation rate of about 2500 RPM. The decrease in SNR at high bit density imposes a lower limit of about 1.2 μm on the bit length at the inner radius of the disk. Tracking servo accuracy and read spot size set the minimum track spacing to about 1.5 μm. The data rate is constrained by the inner track bit-length and radius, and by the rotation rate. And the disk size, track spacing and minimum bit-length constrain the storage capacity. The interplay of all these factors then leads to the results in Fig. 15, from which it is seen that the capacity of a 12-cm diameter disk

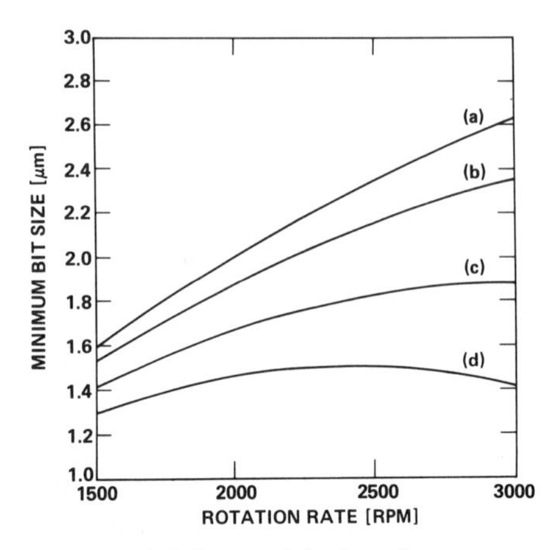

Fig. 15. Minimum bit length versus rotation rate for a 6 cm diameter disk, with 1.5 μm track spacking and 5 Mbit/sec data rate. The unformatted capacity is (a) 125 MB, (b) 187.5 MB, (c) 250 MB, and (d) 312.5 MB.

will approach 300 Mbyte/side when operated at a data rate of 5 Mbit/sec. Erasable optical recording will therefore be useful in a large variety of applications, from memories for small portable computers to mainframes with large on-line data bases. It will be far superior to magnetic tape and flexible drives for backup of large and small Winchesters, respectively. It will provide a viable alternative to the higher capacity, small Winchesters, and it will also be required to back itself up. In the end, however, its greatest opportunity will grow from the availability of inexpensive Mbyte RAM complemented by inexpensive Gbyte disk storage, a combination that will spur designers to as yet unenvisaged uses.

ACKNOWLEDGMENT

I gratefully thank Dan Bloomberg for many illuminating conversations and insights into much of the work discussed.

REFERENCES

1. D. O. Smith, Opta Acta 12, 13 (1965).
2. M. Mansuripur and G. A. N. Connell, SPIE 420, 222 (1983).
3. J. J. Rhyne, in Handbook on the Physics and Chemistry of Rare Earths (ed. K. A., Gschneider and L. Eyring, North Holland 1977), Ch. 16.
4. T. Mizoguchi and G. S. Cargill, J. Appl. Phys. 50, 3870 (1979).
5. J. M. D. Coey, J. Appl. Phys. 49, 1646 (1978).
6. R. Alben, J. J. Becker and M. C. Chi, J. Appl. Phys. 49, 1653 (1978).
7. G. A. N. Connell, Proc. Inter. Conf. Magn.1985 (San Francisco), to be published in J. Magn. & Magn. Mat.
8. M. Hong, D. D. Bacon, R. B. van Dover, E. M. Gyorgy, J. F. Dillon and S. D. Abiston, J. Appl. Phys. 57, 3900 (1985).

9. G. A. N. Connell, R. Allen and M. Mansuripur, J. Appl. Phys. 53, 7759 (1982).

10. G. A. N. Connell and R. Allen, J. Magn. & Magn. Mat. 31-34, 1516 (1983).

11. J. K. Lang, Y. Baer and P. A. Cox, J. Phys. F: Metal Phys. 11, 121 (1981).

12. G. A. N. Connell, Appl. Phys. Letts. 40, 212 (1982).

13. N. Imamura, S. Tanaka, F. Tanaka and Y. Nagao, IEEE Trans. Magn. (to be published, 1985).

14. D. S. Bloomberg and G. A. N. Connell, Proc. IEEE Compcon Spring 1985 (1EEE Comp. Soc. Press 1985), pp 32.

15. M. Takenaga, N. Yomada, S. Ohara, K. Nishiuchi, M. Nagashima, T. Kashihara, S. Nakamura and T. Yamashita, SPIE 420, 173 (1983).

16. M. Terao, T. Nishida, Y. Miyauchi, T. Nakao, T. Kaku, S. Horigome, M. Ojima, Y. Tsunoda and Y. Sugita, SPIE 529, 46 (1985).

17. N. Koshino, M. Maeda, Y. Goto, K. Itoh, and S. Ogawa, SPIE 529, 40 (1985).

MAGNETIC THIN FILMS:
THE CHALLENGE AND THE OPPORTUNITY

Kent N. Maffitt
3M Company, St. Paul, MN 55144

ABSTRACT

Current knowledge of film properties and deposition conditions required for high recording density, magnetic thin-film media will be reviewed. The opportunity to record at linear densities beyond 40,000 flux changes/mm. was demonstrated as early as 1978. The challenge has been, and still is, to identify and control parameters in the deposition process that produce films suitable for high-density recording at deposition rates practical for commercial production. In general, magnetic thin films suitable for high density recording must be magnetically anisotropic and polycrystalline with coercivities of 300-1000 Oe. Films with their easy axis of magnetization either parallel or perpendicular to the plane of the film have been shown to be suitable for high density recording. As examples, the recording performance of evaporated CoNi films with their easy axis in the plane of the film will be compared with sputter-deposited CoCr films, in which the easy axis of magnetization is perpendicular to the film.

INTRODUCTION

Magnetically anisotropic, vacuum-deposited thin films suitable for magnetic recording have been the subject of laboratory investigations for some time. Longitudinal recording films with their easy axis of magnetization in the plane of the film were reported in 1970.[1]

Iwasaki reported the advantages of single-layer perpendicular recording using films of CoCr with the easy axis of magnetization normal to the film plane in 1977,[2] followed by dual-layer perpendicular recording with CoCr over NiFe in 1979.[3] The preferred film construction depends upon the application. In all three cases, linear recording densities well beyond 40,000 flux reversals per cm. have been obtained. In the limit of high density recordings, the theoretical difference between longitudinal and perpendicular recording media is the length of the transition region between oppositely magnetized zones.

This transition region is controlled by the microstructure of the film and demagnetization effects. The microstructure is strongly dependent upon the deposition process and will be discussed later. Demagnetizing fields of the recorded bit are a fundamental of magnetostatics, and they ultimately limit the maximum recording density possible in longitudinal recording media. However, as will be shown later, the output difference between perpendicular and longitudinal media is relatively small, even at record densities greater than 40,000 flux changes/cm.

In the case of perpendicular recording media, demagnetizing fields are reduced as the recording density is increased. At very high recording densities, beyond 80,000 flux changes per cm., the length of the transition region is dominated by the microstructure of the film. It is thought to be approximately equal to the width of the boundary layer between crystallites.

For dual-layer perpendicular recording media, the properties of the recording layer are essentially the same as those of single-layer recording media. The underlayer is a low coercivity layer that acts as a magnetic keeper on the side of the recording layer opposite the recording head. The advantage of dual layer is improved record efficiency and increased output at high recording densities when used in conjunction with monopole heads.

This paper will focus on magnetic properties of single layer films and will not treat head and media interactions or the effects of head design on recording performance. Particular emphasis will be placed on the magnetic microstructure required for good record performance.

FILM MICROSTRUCTURE

Microstructure is the product of film composition, deposition process, and the complex dependence of film formation on both. Since it is not possible to observe directly the dynamics of film formation, measurement and control of relevant parameters during deposition are extremely important. This is especially true in the case of magnetic films, since small variations in composition or deposition conditions can cause considerable changes in magnetic properties.

In both the longitudinal and perpendicular cases, there is growing evidence that the basic unit best suited for high density recording is a small crystallite surrounded by a non-magnetic sheath,[4-11] which results in independent magnetic reversal of crystallites rather than reversal by inter-crystalline domain wall motion. Further, the crystallites should be well oriented and preferably acicular in shape. In the ideal case, the magnetocrystalline anisotropy and shape anisotropy cooperate to maximize the magnetic anisotropy with the easy axis in the film plane for longitudinal media, and normal to the film plane for perpendicular media. The ideal crystallite diameter is magnetically stable and somewhat larger than the volume and shape defined by the superparamagnetic limit. For most materials, this diameter is a few hundred angstroms.

To permit independent magnetic reversal of crystallite magnetization, exchange forces between crystallites must be decoupled. If adjacent crystallites are not decoupled, the boundary between recorded bits forms a zigzag pattern.[1,12-15] At high recording densities, the amplitude zigzag pattern can be comparable to the length of the recorded bit, thus limiting the maximum recording density. In media of this type, noise also increases as the record density increases.[16-20]

Polycrystalline microstructures in thin films have been known for some time. For example, the microstructure of sputtered films has been well characterized as a function of substrate temperature and gas pressure during deposition,[21,22] and the microstructure of evaporated films has been studied as a function of the angle at which the deposited atoms strike the film surface.[23] However, obtaining a non-magnetic sheath around individual crystallites introduces the need to induce and control local variations in composition at the boundaries between crystallites.

MATERIALS & COATING METHODS

From a materials point of view, Fe, Ni, Co and many of their alloys are candidates as magnetic recording media. For example, Fe, Co, Ni, CoNi CoCr, CoCrNb, CoCrTa, CoCrFe, CoRe, CoV, Co-CoO, CoGd, CoHo, CoFe, CoSm, FeGd, FeSm, GdTbFe, $BaFe_{12}O_{19}$ and Fe_3O_4 have all been used to produce magnetically anisotropic thin films.[2,14,24-39] Of these materials, CoNi and CoCr have been the most intensively studied as longitudinal and perpendicular recording media respectively.

Coating options include evaporation, RF sputtering, magnetron sputtering, target facing target system (TFTS) sputtering and ion sputtering. All have been used to produce magnetically anisotropic films.[1,2,40-43]

Since the actual microstructure obtained is strongly dependent upon film composition and deposition technique, two examples will be used as illustrations. Sputtered films of CoCr and evaporated films of CoNi will be used as examples of perpendicular and longitudinal recording media respectively.

CoCr MICROSTRUCTURE

Sputtered films of CoCr have been intensively studied since 1977.[2] The need for films with well oriented hcp crystallites has been established as a necessary condition for high density recording. The <0002> axis must be normal to the film plane. X-ray rocking curves are used to measure film orientation. The width of the <0002> peak at half maximum, - 50, is two to eight degrees for well oriented films.

Columnar structures such as those shown in the SEM micrograph of Figure 1 are typical of well oriented films. The columns are 1500-3000A in diameter. However, the individual crystallites shown in the bright field TEM image of Figure 2 are approximately 250-500A in diameter, indicating that the columns are considerably larger than the crystallites. A closer examination of Figure 2 reveals that crystallites within a cluster have similar image density; i.e., these clusters scattered the image-forming electrons in a similar manner. These observations, plus the work of others, suggest that the columns observed by SEM are composed of clusters of crystallites.[5,21,34]

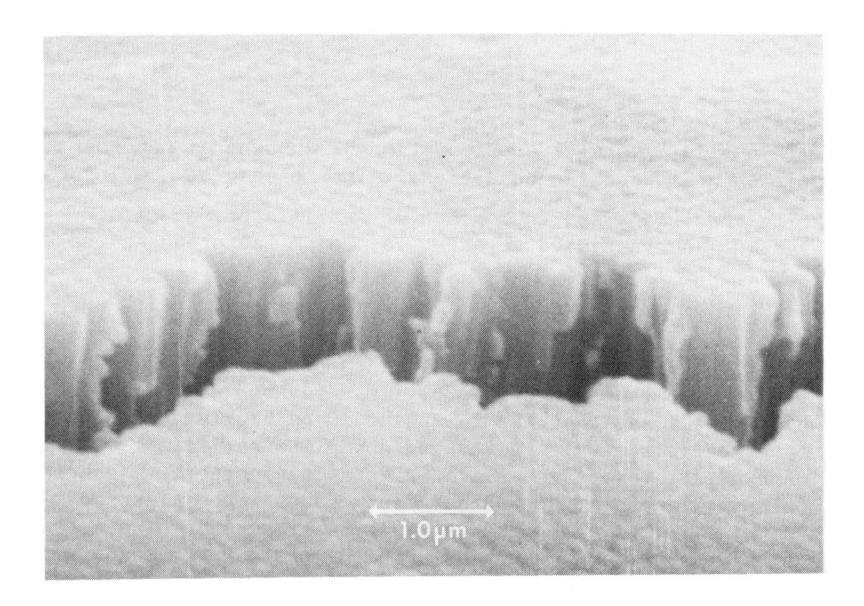

Fig. 1. Columnar microstructure in sputtered CoCr film.

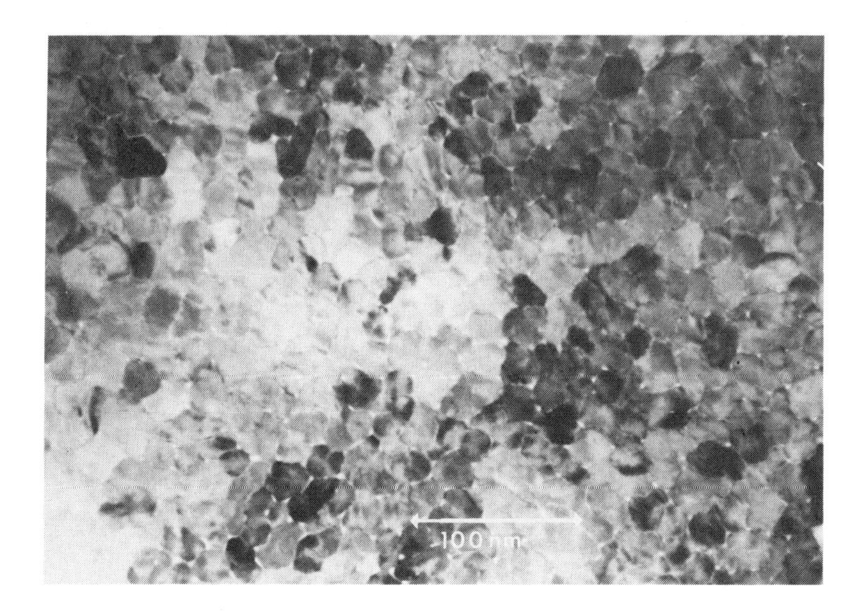

Fig. 2. TEM image of CoCr film.

Since the idealized crystallite boundary is less than 100A wide in the case of CoCr, direct measurement of the boundary composition is not possible with present day techniques. However, there is evidence to suggest that the boundary region is Cr rich relative to the crystallites and consequently of lower magnetic moment if not non-magnetic.[4,7,9,40,41,44,45,46] The driving force resulting in Cr segregation is not well understood. Possible causes include strain-induced diffusion and phase segregation.[45]

Thin-film crystallite orientation is in general dependent upon deposition and substrate conditions. In the case of CoCr, substrate "seeding" has been used to improve crystallite orientation.[47-50] Of the seeding layers tested to date, germanium has produced the best orientation.[50]

CoCr MAGNETICS

The magnetic properties required to obtain good read/write performance from homogeneous CoCr films are not well understood. It is well established that the films should be highly anisotropic and that the magnetic easy axis must be normal to the film plane. It is well known that good crystalline orientation increases magnetic anisotropy through its dependence upon crystalline anisotropy. However, the relative importance of crystalline and shape anisotropy in determining perpendicular and in-plane coercivity is not clear. Further, the dependence of read/write performance upon coercivity has not been clearly demonstrated. Typically, films with 700-1000 Oe coercivity and an Hkeff of 1500-5000 Oe have been shown to produce high output at high recording densities, but a systematic study relating output to magnetic properties has not been published to date.

Some CoCr films are not magnetically homogeneous. Hysteresis[51] and ferromagnetic resonance measurements[52] have been used to detect the presence of "transition" layers. The coercivity of this transition layer varies from 10 Oe to 200 Oe depending upon film composition and deposition conditions.[32,51] That is, the film is magnetically a dual-layer film.

The case of the film with a transition layer of 10 Oe coercivity is especially signficant since the D_{50} density (the density at which the output drops by 50%) was 105,000 flux charges/cm., the highest reported to date.[32]

In at least some cases, the presence or absence of a transition layer is dependent upon sputtering conditions. The details of this dependence have not been published, however.

CoCr COATING DYNAMICS

It is instructive to examine the dynamics of sputtered film formation from an atomic point of view. Consider the simple geometry of a diode sputtering system with a CoCr target opposed by a heated substrate. In addition to sputtered Co and Cr atoms, the substrate surface is exposed to sputtering gas ions, impurity gases in both neutral and ionized form, Co ions, Cr ions, secondary electrons and intense ultraviolet irradiation. All impart energy to the film surface, affecting the surface chemistry and mobility of arriving Co and Cr atoms. Although the relative importance of these factors is a strong function of the specific sputtering system being used, some generalizations are possible.

Biasing the target with RF results in significant heating of the substrate by secondary electron bombardment, and crystallite orientation is improved. Hc and Hkeff are, in general, higher in RF-sputtered films.

DC biasing the substrate to induce positive ion bombardment is reported to have only a slight effect on Hc and Hkeff.[53]

Impurity gases such as oxygen and nitrogen reduce Hc and Hkeff[8,54,55] probably by reducing the ad atom surface mobility thereby inhibiting both crystallite growth and Cr segregation at the crystallite grain boundaries.

In the case of polymer substrates such as polyethylene terathalate (PET) and polyimide, the substrate itself can be a major source of impurity gases. This is especially true in the case of continuous coating, where the large surface area of the roll to be coated represents a major source of gas. The most practical way to compensate for this increase in background pressure is to increase the deposition rate thereby maintaining the required concentration of Co and Cr atoms relative to impurities during film formation.

Techniques for increasing the deposition rate of CoCr include TFTS sputtering,[56] magnetron sputtering,[46] and thermal evaporation[57] All three techniques have been used to produce films with good magnetic properties, and all three increase the deposition rate over that of conventional sputtering. In principle, the highest rate would be thermal evaporation. In practice, the best technique will depend not only upon rate, but also upon factors such as rate control, deposition efficiency and film uniformity.

MICROSTRUCTURE OF CoNi

Vacuum-evaporated, longitudinal magnetic recording media are most often produced by arranging shields between the source and substrate so that the deposited atoms strike the surface at an angle of 70-90° from the normal.[12,58] In the case of continuous coating, the nucleation layer is formed by atoms striking the surface at grazing incidence. As the substrate moves, the atomic angle of incidence decreases to a minimum of approximately 70°. Figure 3 shows a TEM image of a microtome section of CoNi evaporated at oblique incidence in a partial oxygen atmosphere in a continuous coater. The oxygen is deliberately introduced to inhibit adatom surface mobility during film formation and thereby limit crystallite size. Selected area electron diffraction revealed that the dark nodules within the columns of Figure 3 are randomly oriented crystallites. The lighter region surrounding the nodules is oxygen rich. The curved columnar structure is typical of continuously coated films evaporated at oblique incidence and is the result of self-shadowing after the initial nuclei are formed.[23]

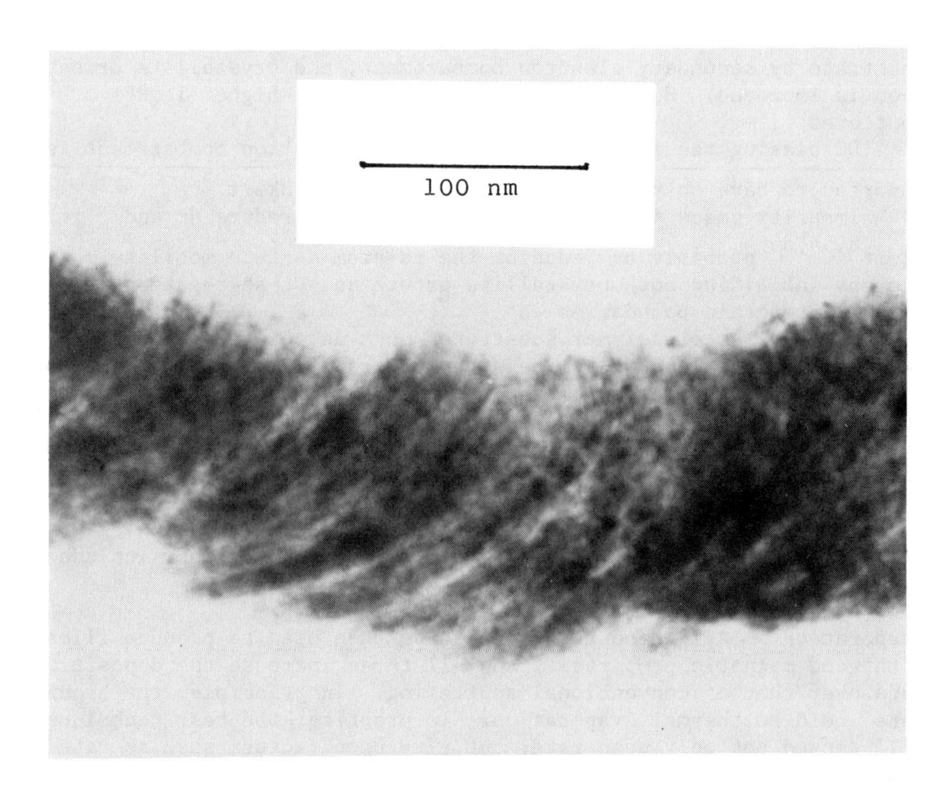

Fig. 3. TEM micrograph of microtomed section of CoNi Film. Top of micrograph was original film surface.

MAGNETICS OF CoNi

The coercivity of films evaporated at oblique incidence is anisotropic with the easy axis in the plane of the film. The dependence of coercivity on angle of incidence is very similar for Co, Ni and Fe[24] which represent respectively the hcp, bcc, and fcc crystal structures thus indicating that the anisotropy is principally due to shape anisotropy.

RECORD PERFORMANCE

Predicting high density record and playback performance on the basis of macroscopic magnetic properties is uncertain at best. This is especially true in the case of perpendicular recording.

The final test of any recording media is the read/write performance. One test of performance is output vs. record density. As the data in Figure 4 shows, the differences between thin film longitudinal and thin film perpendicular media are relatively small when tested under the same conditions.

Figure 4 shows examples of output versus record density for both CoNi and CoCr. These recordings were made with the head in contact in a stretched surface configuration. The same head was used for read and write. Track width was 37 micrometers, and gap length was 0.38 micrometers. The read signal was fed into a spectrum analyzer, and the fundamental frequency component was measured with a 30 kHz bandwidth. Other details of the record set up are described elsewhere.[59]

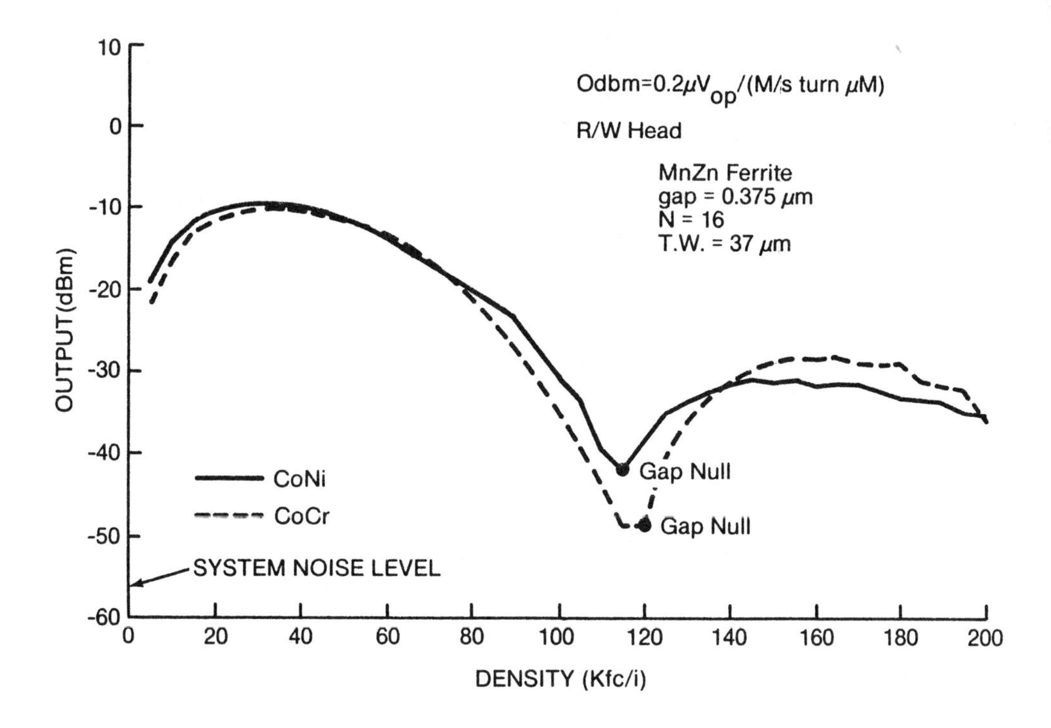

Fig. 4. Output versus record density.

Table I shows the magnetic properties of the two films.

TABLE I. MAGNETIC PROPERTIES OF Co AND CoCr FILMS

	CoNi	CoCr
Ms (emu/cc)	630 a	425 b
Hc (Oe)	800 a	820 b
Hk (Oe)	-	4300 b
Squareness	0.67 a	0.14 b, c
Thickness	0.12	0.50

a = longitudinal b = perpendicular c = not corrected for demagnetization

The minimum in output at 115-120 kfci is the density at which the recorded wavelength equals the length of the record gap and is head dependent rather than media dependent. More detailed analysis has shown these media to be nearly equivalent in terms of spacing loss, transition length and digital bit shift.[59,60]

Differences in the read/write performance of longitudinal and perpendicular media is still a subject of study and discussion by recording experts. Consequently, relating read/write performance to magnetic and material characteristics is also a subject for further study. However, one major factor contributing to the performance in both the CoCr and CoNi cases is thought to be the magnetic isolation of the crystallites and the crystallite size which is small compared to the recorded wavelength even at 80,000 flux reversals/cm.

DURABILITY AND STABILITY

Although magnetic thin films must have the magnetic characteristics just described, physical durability and corrosion resistance are of equal or greater importance. Typically, this means the thin film must tolerate many million passes of a record head at temperatures as high as 60°C and at relative humidity as high as 90%.

As far as corrosion resistance is concerned, CoCr is more stable than CoNi. However, neither CoCr nor CoNi alone is durable enough for most applications. Most attempts to solve this problem have employed some sort of protective and/or lubricating layer. To prevent spacing losses, this layer must be no more than a few hundred angstroms thick. Sputtered carbon, solution-coated fluorocarbon, and duPont 804 lubricant are typical of overcoats tested.[61-63]

The mechanical properties of films can also be changed by varying deposition conditions. For example, SEM studies of CoCr film fracture have shown that fracture changes from intergranular to intragranular as sputtering pressure is reduced and/or sputtering power is increased.[7] Another approach has been to modify the durability of CoCr by small additions of Ta, Fe and Zr.[36]

To date, no completely satisfactory solution has been found for the case of magnetic thin films deposited on flexible substrates. It should also be pointed out that thin-film corrosion and durability have not been studied nearly as extensivley as magnetic properties.

SUMMARY

Both longitudinal and perpendicular thin-film recording media have high output at high recording density. Magnetically, the films must be anisotropic, and magnetic reversal of crystallites should be independent of its neighbors. The anisotropy is primarily due to shape in the case of CoNi, while crystalline anisotropy is dominant in CoCr. Reversal by rotation is best facilitated by crystallites that are isolated from one another by a non-magnetic boundary layer. To obtain such a microstructure requires understanding and control of such factors as film nucleation, adatom mobility during film formation and impurity gases. The actual conditions required for a given material are very specific to the material and deposition technique used. The interested reader is referred to the references cited for further details.

ACKNOWLEDGMENTS

The author wishes to acknowledge D. C. Koskenmaki, D. P. Stubbs, J. Skorjanec, C. D. Calhoun, C. L. Bruzzone, and J. Storer for their help in collecting materials for this paper, as well as for many stimulating discussions. Without their help, this paper would not have been written.

REFERENCES

1. N. Curland and D. Speliotis, J. Appl. Phys. **41**, 1099-1101 (1970).
2. S. Iwasaki and Y. Nakamura, IEEE Trans. on Magn., vol. **MAG-13** No. 5, 1272-1277 (1977).
3. S. Iwasaki, Y. Nakamura, K. Ouchi, IEEE Trans. on Magn., vol. **MAG-15** No. 6, 1456-1458 (1979).
4. E. Wuori, J. Judy, J. Appl. Phys. **57**, 4010-4012 (1985).
5. M. Ohkoshi, H. Toba, S. Honda, T. Kusuda, J. Magn. & Magn. Mater. **35**, 266-268 (1983)
6. R. Sugita and K. Honda, **National Technical Report 28**, 996-1006 (1982).
7. T. Suzuki, IEEE Trans. on Magn., vol. **MAG-20** No. 5, 675-680 (1984).
8. J. Thompson and D. Stevenson, "Comparison of the Effects of Oxygen on Cobalt and Cobalt-Chromium Thin Films," Intermag 1985.
9. M. Ali and P. Grundy, IEEE Trans. on Magn., vol. **MAG-19** No. 5, 1641-1643 (1983).
10. T. Chen, G. Charlan, T. Yamashita, J. Appl. Phys. **54**, 5103-5111 (1983).
11. K. Ouchi and S. Iwasaki, IEEE Trans. on Magn., vol. **MAG-18** No. 6, 1110-1112 (1982).
12. K. Okamoto, K. Hara, H. Fujiwara, T. Hashimoto, J., Phys. Soc. Jap., **40**, 293-294 (1976).
13. T. Chen, IEEE Trans. on Magn., vol. **MAG-17** No. 2,1181-1191 (1981).
14. H. Tong, R. Ferrier, P. Chang, J. Tzeng, K. Parker, IEEE Trans. on Magn., vol. **MAG-20** No. 5, 1831-1833 (1984).
15. K. Yoshida, T. Okuwaki, N. Osakabe, H. Tanabe, Y. Horiuchi, T. Matsuda, K. Shinagawa, A. Tonomura, H. Fujiwara, IEEE Trans. on Magn., vol. **MAG-19** No. 5, 1600-1604 (1983).
16. Noise in High Performance Magnetic Recording Media, N. R. Belk, P. K. George, G. S. Mowry, Intermag 1985.
17. N. R. Belk, P. K. George, G.S. Mowry, J. Appl., Phys. **57**, P. 3946 April (1985).
18. H. Tanaka, G. Goto, N. Shiota, M. Yanagisawa, J. Appl. Phys. **53**, pp. 2576-2578.
19. R. A. Baugh, E. S. Murdock, B. R. Natarajan, IEEE Trans. on Magn. **MAG-16**, pp 1722-1724 (1983).
20. N. R. Belk, P. K. George, G. S. Mowry, Intermag 1985.
21. J. Thornton, J. Vac. Sci. Technol., **11**, 666-670 (1974).
22. R. Messier, A. Giri, R. Roy, J. Vac. Sci. Technol., **A2**, 500-503 (1984).
23. J. van de Waterbeemd, G. van Oosterhout, Philips Res. Repts., **22**, 375-387 (1967).
24. D. Speliotis, G. Bate, J. Alstad, J. Morrison, J. Appl., Phys., **36**, 972-974 (1965).
25. D. Smith, M. Cohen, G. Weiss, J. Appl. Phys., **31**, 1755-1762 (1960).
26. M. Ohkoshi, K. Tamari, S. Honda, T. Kusada, IEEE Trans. on Magn., vol. **MAG-20** No. 5, 788-790 (1984).
27. M. Khan and J. Lee, J. Appl. Phys., **57**, 4028-4030 (1985).
28. T. Chen and G. B. Charlan, J. Appl. Phys., **50**, 4285-4291 (1979).

29. J. Desserre, IEEE Trans on Magn., vol. **MAG-20** No. 5, 663-668 (1984).
30. A. Dirks & H. Leamy, J. Appl. Phys., **49**, 1735-1737 (1978).
31. K. Fukuda, Y. Kitahara, F. Maruta, J. Ezaki, IEEE Trans. on Magn., vol. **MAG-18** No. 6, 1116-1118 (1982).
32. N. Watanabe, Y. Ishizaka, K. Kimura, E. Imaoka, "High Density Recording with Perpendicular Media Co-Cr-Nb and Ring Head," Intermag 1985.
33. T. Arnoldussen, E. Rossi, A. Ting, A. Brunsch, J. Schneider, G. Trippel, IEEE Trans. on Magn., vol. **MAG-20** No. 5, 821-823 (1984).
34. M. Matsuoka, M. Naoe, "Sputter Preparation and Read/Write Characteristics of Ba-Ferrite Thin Film Disk," Intermag 1985.
35. Y. Hoshi, M. Matsuoka, M. Naoe, "Preparation of Iron Oxide Films with Large Perpendicular Magnetic Anisotropy," Intermag 1985.
36. M. Naoe and M. Matsuoka, J. Appl. Phys., **57**, 4019-4021 (1985).
37. S. Ohnuma, Y. Nakanouchi, T. Masumoto, "Magnetic Properites of Fe-Base Amorphous Ultrafine Particles," Intermag 1985.
38. T. Tanaka, T. Miyazaki, E. Kita, A. Tasaki, IEEE Trans. on Magn., vol. **MAG-19** No. 5, 1650-1652 (1983).
39. M. Matsuoka and M. Naoe, J. Appl. Phys., **57**, 4040-4042 (1985).
40. M. Naoe, S. Yamanaka, Y. Hoshi, IEEE Trans. on Magn., vol. **MAG-16** No. 5, 646-648 (1980).
41. J. Smits, S. Luitjens, F. den Broeder, J. Appl. Phys., **56**, 2260-2262 (1984).
42. K. Ouchi and S. Iwasaki, J. Appl. Phys., **57**, 4013-4015 (1985).
43. S. Iwasaki, IEEE Trans. on Magn., vol. **MAG-16** No. 1. 71-76 (1980).
44. W. G. Haines, J. Appl. Phys., **55**, 2263-2265.
45. S. Iwasaki, K. Ouchi, T. Hizawa, IEEE Translation Journal On Magnetics In Japan TJMJ-1, No. 1, April (1985).
46. Y. Fujii, K. Tsutsumi, T. Numata, Y. Sakurai, J. Appl. Phys., **55**, 2266-2268 (1984).
47. H. Gill and M. Rosenblum, IEEE Trans. on Magn., vol. **MAG-19** No. 5, 1644-1646 (1983).
48. H. Gill and T. Yamashita, IEEE Trans. on Magn., vol. **MAG-20** No. 5, 776-778 (1984).
49. C. Byun, E. Simpson, J. Sivertsen, J. Judy, "A Study of the Effects of Thin Oxide Seed Layers on RF-Sputtered CoCr Films", Intermag 1985.
50. M. Futamoto, Y. Honda, H. Kakibayashi, K. Yoshida, "Microstructure and Magnetic Properties of CoCr Thin Film Formed on Ge Layer," Intermag 1985.
51. C. Byun, J. Sivertsen, J. Judy, J. Appl. Phys., **57**, 3997-3999 (1985).
52. P. V. Mitchell, A. Layadi, N. S. Vander Ven, J. O. Artman J. Appl. Phys., **57** (1), 3976-3978.
53. S. Kadokura, and M. Naoe, IEEE Trans. on Magn., vol. **MAG-18** No. 6, 1113-1115 (1982).
54. J. Daval and D. Randet, IEEE Trans. on Magn., vol. **MAG-6** No. 4, 768-773 (1970).
55. S. Honda, N. Yamashita, M. Ohkoshi, T. Kusuda, IEEE Trans. on Magn., vol. **MAG-20** No. 5, 791-793 (1984).

56. Y. Hosh, M. Matsuoka, M. Naoe, S. Yamanaka, IEEE Trans of Magn **MAG-20**, pp. 797-799 (1984).
57. R. Sugita and F. Kobayashi, IEEE Trans. on Magn., vol. **MAG-18** No. 6, 1818-1821 (1982).
58. K. Shinohara, H. Yoshida, M. Odagiri, A. Tomago, IEEE Trans. on Magn., vol. **MAG-20** No. 5, 824-826 (1984).
59. D. P. Stubbs, J. W. Whisler, C. D. Moe, J. Skorjanec J. Appl. Phys., **57**, 3970-3972, 1985.
60. J. W. Whisler, J. R. Hoinville, D. P. Stubbs, C. D. Moe, J. Skorjanec, J. Appl. Phys., **57**, 3973-3975, 1985.
61. P. Buttafava, V. Bretti, G. Ciardiello, M. Piano, G. Caporiccio, A. M. Scarati, "Lubrication & Wear Problems of Perpendicular Recording Thin Film Flexible Media," Intermag 1985.
62. S. Agarwal "Structure & Morphology of RF Sputtered Carbon Overlayer Films," Intermag 1985.
63. E. Rossi, G. McDonough, A. Tietze, T. Arnoldussen, A. Brunsch, S. Doss, M. Henneberg, F. Lin, R. Lyn, A. Ting, G. Trippel, J. Appl. Phys., **55**, 2254-2256 (1984).

CHAPTER II

OVERVIEW OF LAYERED MATERIALS

SUMMARY ABSTRACT: ARTIFICIALLY LAYERED SUPERCONDUCTORS

Charles M. Falco
University of Arizona, Tucson AZ 85721

The coherence length in superconductors is sufficiently long (typically 50 Å ~ 1 μm) that it is possible to significantly alter, in a highly controlled manner, a number of physical properties (T_c, $H_c(\Theta,T)$, j_c, etc.) by artificially layering the superconductor with another material. The second material in the layered structure can be chosen to be another superconductor, a normal metal, or a magnetic material, depending upon the phenomena of interest. We have prepared by sputtering techniques[1,2] artificially layered superconductors consisting of Nb/Cu and Ta/Mo, and studied a number of their superconductive properties.[3,4] This abstract summarizes some of our work on these superconducting metallic superlattices.[5]

High-rate magnetron[1] and magnetically-assisted triode[2,6] sputtering techniques are capable of depositing many interesting superconductive metals with purity equivalent to that of the starting target.[7] Our present system[6] uses a liquid nitrogen-baffled diffusion pump to reach a base pressure of ~$7x10^{-8}$ Torr. By use of a micrometer-adjustable variable orifice valve in the main pumping line in combination with mass flow controllers to regulate the flow rate of high purity Ar gas, stable, dynamic equilibrium is reached in the range 3-5 mTorr. Microprocessors are used to control the motion of the substrates above the sputtering guns as well as the power to each of the guns (typical rates ~20-50 Å/sec) in order to achieve control of ±0.3% over the amount of material deposited in each layer of the multilayered material.[2] That this level of control is possible has been verified by a combination of optical interferometry, x-ray diffraction and Rutherford Backscattering measurements.[6] A variety of substrates have been used for this work, including several orientations of single-crystal sapphire, float glass, mica and Si. Typical samples have total thickness approximately 0.5-1 μm, with individual layers in the range 4 Å to 5000 Å.

Inductive and resistive T_c measurements have been made on Nb/Cu samples of individual layer thicknesses in the range 4-5000 Å, with agreement to within a few mK of the two techniques.[3,8] These data can be analyzed using the DeGennes-Werthamer[9,10] proximity effect theory, enabling the T_c as a function of layer thickness of very thin films of pure Nb to be extracted. The dependence of the energy gap upon layer thickness has also been determined from quasiparticle tunneling measurements,[11] and is in agreement with the T_c data. Phonon energies observed in Nb/Cu superlattices appear bulk-like down to layer thicknesses ~30 Å. Dimensional crossover has been observed in both the temperature dependence and the angular dependence of the critical fields, $H_c(T,\vartheta)$, of these materials.[1,2]

Detailed studies of pair tunneling to a series of these artificial metallic superlattices have been conducted.[4] A unique crossover effect is observed as a function of temperature in the magnetic field diffraction patterns of Josephson tunnel junctions made with Nb/Cu superlattices, plus an overlay of ~200 Å pure Nb, as one of the electrodes. The diffraction patterns change from the characteristic Fraunhofer pattern of a Josephson junction at low temperatures to bell shaped curves above the T_c of the superlattice where only the Nb overlayer is superconducting.

Recently, Ta/Mo superlattices of very high structural quality have been prepared and characterized using a variety of x-ray diffraction techniques (Bragg, low-angle Bragg, transmission and reflection Laue, and Debye-Scherrer) as well as Rutherford Backscattering Spectroscopy.[6] The superconducting properties of these samples, including Josephson tunneling characteristics, will be reported elsewhere.

It is a pleasure to acknowledge the collaboration of Dr. I. K. Schuller for work on Nb/Cu, of Profs. R. Vaglio and A. M. Cucolo on Josephson tunneling studies of Nb/Cu and Ta/Mo, and of W. Bennett, A. Boufelfel and Prof. J. Leavitt on structural studies of Ta/Mo. This work was supported by the Director, Office of Energy Research, Office of Basic Energy Sciences, Materials Sciences Division of the U. S. Department of Energy under Contract No. DE-AC02-83ER45025.

1. I. K. Schuller and C. M. Falco, in **Inhomogeneous Superconductors-1979**, edited by D. U. Gubser, T. L. Francavilla, J. R. Leibowitz and S. A. Wolf (American Institute of Physics, New York), p.197.

2. C. M. Falco, J. de Phys. 45, C5-499 (1984).

3. C. M. Falco and I. K Schuller, in **Superconductivity in d- and f-Band Metals**, edited by W. Buckel and W. Weber (Kernsforschungsanslage Karlsruhe, Germany, 1982), p. 283.

4. R. Vaglio, A. M. Cucolo and C. M. Falco, submitted.

5. For a recent comprehensive review of other synthetically layered superconductors see, S. T. Ruggiero and M. R. Beasley, in Synthetic Modulated Structures, edited by L. L. Chang and B. C. Giessen (Academic, New York, 1985), p. 365.

6. C. M. Falco, W. R. Bennett and A. Boufelfel, in Dynamical Phenomena at Surfaces, Interfaces and Superlattices, edited by F. Nizzoli, K.-H. Rieder, and R. F. Willis (Springer Verlag, Berlin, 1985), p.35.

7. C. M. Falco, J. Appl. Phys. 56, 1218 (1984).

8. I. Banerjee, Q. S. Yang, C. M. Falco and I. K. Schuller, Solid State Commun. 41, 805 (1982).

9. P. G. DeGennes and E. Guyon, Phys. Letters, 3, 168 (1963).

10. N. R. Werthamer, Phys. Rev. 132, 2440 (1963).

11. Q. S. Yang, C. M. Falco and I. K. Schuller, Phys. Rev. B 27, 3867 (1983).

12. I. Banerjee, Q. S. Yang, C. M. Falco and I. K. Schuller, Phys. Rev. B 28, 5037 (1983).

BASIC PROPERTIES OF MAGNETIC SUPERLATTICE STRUCTURES; COLLECTIVE EXCITATIONS AND SPIN REORIENTATION TRANSITIONS

D. L. Mills
Department of Physics
University of California
Irvine, California 92717

Abstract

We discuss some basic properties of superlattice structures formed from films, with one or both constituents a magnetically ordered material. After a brief review of the nature of magnetic interactions in crystals, and a discussion of spin waves in a thin film, we discuss the collective modes of a superlattice formed from ferromagnetic films separated by a nonmagnetic material. Then we turn to a summary of recent theoretical studies of spin reorientation transitions and spin waves in a superlattice structure formed from alternating layers of ferromagnetic and antiferromagnetic films.

1. Introductory Remarks

One reason why solid state physics remains active and vigorous several decades after the fundamental concepts of the field were set forth is that new classes of materials are synthesized or studied, and these are sufficiently distinct from those encountered previously that theories with new content are required to explain their properties. These theories must then be tested against the appropriate sequence of experiments. Materials which are quasi-one-dimensional or two-dimensional in nature provide an example, since statistical mechanics applied to less than three dimensions show that matter behaves very differently when the space dimensionality is reduced.

During the past decade, composite materials have proved of great interest, because they may exhibit macroscopic properties which differ from those of either constituent, and furthermore these properties may be modified or tailored in advance of synthesis. A special case of a composite material is the superlattice structure which in its simplest form consists of alternating layers of material A, and material B, as illustrated in Fig. (1a). A very large effort has been devoted to the fabrication of semiconducting superlattices, and at this point for suitable combinations of materials, one may construct samples perfect on the atomic scale, in that the transition from one material to the next occurs on the scale of one atomic layer.

Superlattice structures may also be constructed from a variety of other materials. In the view of this writer, superlattices in which one or both constituent is magnetically ordered offer rich possibilities. One has a new variable which may be utilized to tune or modify their properties after they are fabricated. This is the strength of an externally applied, static magnetic field. Quite generally, magnetically ordered materials possess strong absorption lines in the microwave and infrared frequency range, and these may be shifted by application of an external field. The physical origin of the strong absorption lies in the spin wave modes of the material, which in general couple to electromagnetic radiation by the magnetic dipole mechanism. In a superlattice, one may design a structure with resonances in desired frequency ranges. We shall also see here that application of a modest magnetic field to a suitable superlattice may lead to dramatic rearrangements in the ground state spin configuration and, of course, the absorption spectrum in the microwave and infrared as well. The magnetic field induced spin reorientation have the character of phase transitions, so in fact we may realize very interesting phase diagrams which also may be tailored in advance.

We have been engaged in a set of theoretical studies of spin waves in superlattice structures, and the influence of an externally applied field on the ground state spin configuration. In this paper, we present a summary of the results of this work, along with a discussion of the relevant experiments. We first consider spin waves in a superlattice formed from films of ferromagnetic material, with each ferromagnetic film separated by a nonmagnetic metal that serves as a spacer. Very beautiful light scattering experiments have elucidated the nature of the spin waves in these structures, and confirmed theoretical predictions which outlined the sensitivity of the mode structure of the superlattice to its microscopic structure. Recently, we have been engaged in studies of model superlattices formed from alternating layers of ferromagnetic and antiferromagnetic material. It is in such systems that application of modest magnetic fields may lead to spin reorientation transitions. We begin our discussion with a brief review of the nature of the interactions between spins in magnetic crystals, then we turn to the topics outlined above.

II. Magnetic Interactions in Crystals; A Brief Review

When the nature of the spin wave spectrum of magnetic crystals is discussed, or the character of the spin order, it is necessary to understand the strength and relative importance of the basic coupling mechanisms. Here, we provide a brief summary of those interactions which depend on the spin orientation of the constituents of the crystal. For simplicity in exposition, we confine our attention to ferromagnetic crystals, though of course the concepts are very general. The basic interactions are:

(a) The Zeeman Energy of Interaction with an External Magnetic Field:

Typically, we shall suppose that a DC external magnetic field of strength H_o is applied parallel to the $\hat{z}$ direction. This introduces into the Hamiltonian the term

$$H_z = -\mu_o H_o \sum_i S_z^{(j)} \quad , \tag{2.1}$$

where $\mu_o \vec{S}^{(i)}$ is the magnetic moment of the spin on lattice site i.

(b) The Dipole-Dipole Interaction:

If the spins are excited, as a spin wave propagates through the system, a spin at site i feels a dipole field $\vec{h}_d^{(i)}$ whose strength is

$$\vec{h}_d^{(i)} = \mu_o \sum_{j \neq i} \frac{1}{r_{ij}^3} [\delta\vec{S}^{(j)} - 3(\hat{n}_{ij} \cdot \delta\vec{S}^{(j)})\hat{n}_{ij}] \tag{2.2}$$

with $\delta\vec{S}^{(i)}$ the deviation of the spin from its equilibrium orientation, $\hat{n}_{ij}$ a unit vector directed from site i to site j, and r_{ij} is the distance between the two sites.

Rather than perform the sum over sites in Eq. (2.2), the dipole field may be calculated from macroscopic considerations, in the limit where we consider spin waves with wavelength very much longer than a lattice constant. If n is the density of spins, we replace $\vec{S}^{(i)}$ by a continuous function of $\vec{x}$, $\vec{S}(\vec{x})$, and we define $\vec{m}(\vec{x}) = n\mu_o \delta\vec{S}(\vec{x})$ as the precessing component of magnetization density associated with the spin wave. Then we know that $\vec{h}_d(\vec{x})$ satisfies

$$\nabla \times \vec{h}_d = 0 \qquad (2.3)$$

for disturbances which vary slowly with time, so we may write $\vec{h}_d = -\nabla\phi(\vec{x})$. Also, one has

$$\nabla\cdot\vec{h}_d + 4\pi\,\vec{\nabla}\cdot\vec{m} = -\nabla^2\phi(\vec{x}) + 4\pi\nabla\cdot\vec{m} = 0 \quad , \qquad (2.4)$$

and we may relate the magnetic potential ϕ to the magnetization density $\vec{m}$.

Simple arguments show that the typical strength of the dipole field experienced by a spin in a ferromagnetic crystal is $4\pi M_s$, where M_s is the saturation magnetization.

(c) The Exchange Couplings

This interaction is both very strong and very short ranged, so in contrast to the dipolar interaction it is limited typically to nearest or possibly next nearest neighbors. It is quantum mechanical in origin, arising from the overlap of wave functions of neighboring ions. In general, it has the form

$$H_x = -\tfrac{1}{2}\sum_{ij}{}' J_{ij}\vec{S}^{(i)}\cdot\vec{S}^{(j)} \ . \qquad (2.5)$$

A measure of the strength of the exchange coupling is H_E, the effective exchange field felt by a spin. With z nearest neighbors, and only nearest neighbor exchange, one has $\mu_o H_E = zJS$. In most magnetic materials, the exchange field is larger than H_o or the dipole field ($4\pi M_s$) by a few orders of magnitude.

However, it influences long wavelength spin waves only modestly. If $\Delta\theta_{ij}$ is the angle between $\vec{S}^{(i)}$ and $\vec{S}^{(j)}$, then $\vec{S}^{(i)}\cdot\vec{S}^{(j)} = S^2\cos(\Delta\theta_{ij}) \cong S^2\{1 - \tfrac{1}{2}(\Delta\theta_{ij})^2 + \cdots\}$. If a long wavelength spin wave of wave vector $\vec{k}$ is excited, and a_o is the lattice constant, $(\Delta\theta_{ij})^2 \cong (ka_o)^2 \ll 1$. The contribution to the excitation energy of the wave is then not H_E, but the very much smaller quantity $H_E(ka_o)^2$. Indeed, at long wavelengths, such as those encountered when spin waves are excited by optical probes, or infrared and microwave sources, we are into the limit where $H_E(ka_o)^2 \ll H_o$, and $H_E(ka_o)^2 \ll 4\pi M_s$. Thus, the properties of such waves may be analyzed within a framework where only the influence of the Zeeman and dipolar fields need be considered, and the role of the exchange is modest.

III. Long Wavelength Surface Spin Waves in Thin Films; The Damon-Eshbach Wave

We move one step closer to the discussion of superlattices by discussing the nature of certain striking spin wave modes which propagate down the surface of a semi-infinite ferromagnetic material, or along the surfaces of a thin film. When we consider a superlattice formed by stacking a set of ferromagnetic films separated by a nonmagnetic material, these waves shall center the discussion in a central matter.

The geometry of interest is illustrated in Fig. (1b). We consider a ferromagnetic film of thickness d, and in the usual geometry the magnetization M_s lies parallel to the film surface, as shown. We suppose an external Zeeman field H_o is applied parallel to the magnetization. The surface waves of interest propagate perpendicular to the magnetization, as shown. In the long wavelength limit, we have seen in Section II that only the Zeeman and dipolar contributions to the energy of the spin motion need be considered, and furthermore we may introduce a magnetic potential ϕ to describe the magnetic fields generated by the spin motion.

Consider first a semi-infinite medium, $d = \infty$, with the surface in the xz plane of the coordinate system of Fig. (1b). Then when the surface wave is excited, the magnetic potential has the form

$$\phi(x,y) = \begin{cases} \phi^{>} e^{ik_\parallel x} e^{-\alpha_>(k_\parallel)y} & , \; y > 0 \\ \phi^{<} e^{ik_\parallel x} e^{+\alpha_<(k_\parallel)y} & , \; y < 0 \quad . \end{cases} \tag{3.1}$$

Elementary considerations require $\phi^{>} = \phi^{<}$, and for this special direction of propagation, one has $\alpha_>(k_\parallel) = \alpha_<(k_\parallel) = k_\parallel$. Constraints imposed by the boundary condition at $y = 0$ lead to an expression for the frequency Ω_s of the wave[1]

$$\Omega_s = \gamma[H_o + 2\pi M_s] \quad , \tag{3.2}$$

with the origin of the contribution $2\pi M_s$ in the demagnetizing fields generated by the spin motion. In Eq. (3.2), γ is the gyromagnetic ratio.

The above surface spin waves are often referred to as Damon-Eshbach surface spin waves; these authors elucidated their properties many years ago.[2] One sees from Eq. (3.1) that the spin motion generates oscillatory magnetic fields above the crystal. These allow one to couple to and excite Damon-Eshbach waves by means of a wire or meander line placed on or near the crystal surface. Such waves, excited in a thin film (see below) rather than the thick crystal considered here, are used by device physicists for a variety of applications. In the device literature, they are referred to by the acronym MSSW (magnetostatic surface wave). These waves are true elementary excitations of the medium, and as a consequence are present as thermally excited entities. Very beautiful studies of thermally excited Damon-Eshbach waves have been carried out by the method of inelastic light scattering.[3,4]

This mode has properties that are most striking. Its principal feature is that it is a "one way wave". By this we mean that it can propagate from right to left across the magnetization, but not from left to right! A unidirectional wave such as this, or one whose propagation characteristics differ for these two senses of propagation, is referred to as a nonreciprocal wave. One might think that this odd behavior has its origin in the breakdown of time reversal invariance by the presence of the spontaneous magnetization M_s. The matter is subtle because bulk spin waves in extended magnetic media do not exhibit these nonreciprocal characteristics. It has been argued[5] that the nonreciprocal characteristics have their origin in the breakdown of reflection symmetry by the presence of the surface.

In thin films, one has surface waves localized about the upper and the lower surface of the films; in general, the wave fields associated with each overlap, and excitations localized on the two surfaces interact. For the Damon-Eshbach wave in a film with upper surface located in the plane y = 0, and the lower surface at y = - d, the magnetic potential has the form

$$\Phi(x,y) = \begin{cases} \phi^{>} e^{ik_\parallel x} e^{-\alpha_{>}(k_\parallel) y} & , \; y > 0 \\ \phi^{(+)} e^{ik_\parallel x} e^{+\alpha(k_\parallel) y} + \phi^{(-)} e^{ik_\parallel x} e^{-\alpha(k_\parallel) y} & -d < y < 0 \\ \phi^{<} e^{ik_\parallel x} e^{+\alpha_{<}(k_\parallel) y} & , \; y < -d \;. \end{cases} \tag{3.3}$$

For the special case of propagation normal to the magnetization, we have $\alpha_{>}(k_\parallel) = \alpha(k_\parallel) = \alpha_{<}(k_\parallel) = k_\parallel$, when one examines details. Now the frequency of the wave is

$$\Omega_s(k_\parallel) = \gamma[(H_o + 2\pi M_s)^2 - 4\pi^2 M_s^2 \exp(-2|k_\parallel|d)]^{1/2} \;. \tag{3.4}$$

When one examines the nature of the spin motion associated with the Damon-Eshbach wave in the thin film, when $k_\parallel > 0$ and one lets $d \to \infty$, the wave becomes localized to the <u>upper</u> surface. But if $k_\parallel < 0$, the wave is localized on the <u>lower</u> surface in this limit. Thus, in the thick film limit, we have waves localized on the upper surface which propagate from left to right, and waves on the lower surface that propagate from right to left.

Now that we have in hand an understanding of the nature of long wavelength surface spin waves in an isolated ferromagnetic film, we are ready to discuss collective modes of a superlattice structure formed by arranging such films in a periodic stack such as that illustrated in Fig. (1a).

IV. The Long Wavelength Collective Modes of Magnetic Superlattice Structures

The first superlattice structures synthesized which incorporated magnetically ordered materials as one constituent consisted of alternating films of ferromagnetic nickel, and the metal molybdenum, which for our purposes is regarded as a magnetically inert material. These samples were farbricated at Argonne National Laboratory, by a sputtering technique developed by Falco, Schuller, and their colleagues.[6] Long wavelength spin wave modes in these materials, which are collective excitations of the superlattice structure as a whole, were studied by light scattering spectroscopy.[7,8] The mode structures explored in these experiments were demonstrated to be strongly influenced by the relative film thickness of the two constituents, and the results agree remarkably well with theoretical expectations.[8,9] Thus, we have in hand examples of superlattice structures which support collective modes whose propagation characteristics are subject to modification by altering the dimensions of the films from which the final sample is synthesized.

Consider a superlattice such as that illustrated in Fig. (1a), where constituent A is a film of Ni metal, and constituent B is Mo. For the moment, the magnetizations of all the Ni films are parallel to the interfaces, and all are parallel to the z axis. From the discussion of Section III, it is evident that a Damon-Eshbach wave may propagate along each of the interfaces between the Ni film, and the neighboring Mo film. In a schematic manner, in Fig. (1a), we show the spatial variation of the magnetic field associated with two such modes, each on an adjacent interface. The magnetic fields overlap, and this overlap results in an interaction between the two modes, very much as the overlap between wave functions of electrons localized on adjacent lattice sites leads to cross coupling between the two states.

There is in fact a spin wave localized on each of the Ni-Mo interfaces, and the interaction between them produces an elementary excitation which is an excitation of the superlattice as a whole. Again there is a direct analogy between the problem of interest, and the motion of electrons in a one dimensional periodic lattice, in the tight binding limit, where overlap between adjacent wave functions leads to the formation of energy bands. In our case, the surface waves on the various interfaces link together to form a spin wave band capable of transporting energy <u>normal</u> to the interfaces.

The mathematical description of such waves has been developed,[9] and we proceed as follows. The magnetic potential of the wave localized on the interface at $y_n = n(d_1 + d_2)$ is written

$$\Phi_n(x,y) = e^{ik_\parallel} \phi_{k_\parallel}(y-y_n) \qquad (4.1)$$

where $\phi_{k_\parallel}(y-y_n)$ may be deduced from Eq. (3.3) by replacing y by $(y-y_n)$ everywhere, and omitting the factor of $e^{ik_\parallel x}$ in Eq. (3.3).

Bloch's theorem may be invoked to construct the magnetic potential associated with the collective mode of the superlattice structure. We denote this by $\psi^{(B)}_{k_\parallel k_\perp}(x,y)$, and then

$$\psi^{(B)}_{k_\parallel k_\perp}(x,y) = e^{ik_\parallel x} \sum_n e^{ik_\perp y_n} \phi_{k_\parallel}(y-y_n) \qquad (4.2)$$

where $k_\perp$ lies within the first Brillouin zone of the superlattice structure, $-\pi/(d_1 + d_2) \leqslant k_\perp \leqslant +\pi/(d_1 + d_2)$.

Camley, Rahman and Mills[9] have derived the dispersion relation of these waves. For the case where $\vec{k}_\parallel$ is normal to the magnetization (this is the geometry assumed here), the dispersion relation is given by

$$\Omega_B^2(k_\parallel,k_\perp) = \frac{H_o(H_o + 4\pi M_s)}{1 + \Delta(k_\parallel,k_\perp)} + \tfrac{1}{2}\left[H_o^2 + (H_o + 4\pi M_s)^2\right] \frac{\Delta(k_\parallel,k_\perp)}{1 + \Delta(k_\parallel,k_\perp)} \qquad (4.3)$$

where

$$\Delta(k_\parallel,k_\perp) = \sinh(k_\parallel d_1)\sinh(k_\parallel d_2)/[\cosh(k_\parallel d_1)\cosh(k_\parallel d_2) - $$

$$- \cos(k_\perp[d_1 + d_2])] \qquad (4.4)$$

The subscript B in Eq. (4.2) and Eq. (4.3) indicates that the modes under consideration are bulk modes, which propagate through the superlattice structure very much as Bloch electrons propagate through a crystal lattice. Once again, keep in mind as one sees from Eq. (4.2), that these "bulk" waves are in fact linear superpositions of modes, each localized on one interface. It is striking that the dispersion relation in Eq. (4.3) is invariant under interchange of d_1 and d_2. Of course, the dependence on d_1 and d_2 shows that the propagation characteristics of these waves may be "tuned" by altering d_1 or d_2. Also the magnetic field is another experimental variable at the disposal of the experimentalist, so the collective modes of the magnetic superlattice offers one remarkable flexibility.

The discussion so far has focused its attention on an idealized superlattice of infinite spatial extent. In fact, any real structure is terminated, as indicated in Fig. (1c). In this case, we find a new mode, again a collective mode of the superlattice structure, which is localized to the surface. The magnetostatic potential associated with this mode is again a linear supposition of potentials localized on the various interfaces, but now with an envelope which decays exponentially as one moves into the superlattice:

$$\psi^{(s)}_{k_\parallel k_\perp}(x,y) = e^{ik_\parallel x} \sum_n{}' e^{-\alpha(k_\parallel)y_n} \phi_{k_\parallel}(y-y_n) \quad . \tag{4.5}$$

We have a surface mode, which is in fact a linear combination of surface modes!

The theory shows[9] that the condition which must be satisfied for the surface mode to exist is that $d_1 > d_2$, i.e. the ferromagnetic films in the superlattice structure must be thicker than the nonmagnetic "spacers". What is remarkable is that the frequency of the superlattice surface wave is <u>exactly</u> the same as the Damon-Eshbach surface spin wave, on a semi-infinite sample. Thus, for the superlattice the surface wave frequency is given by Eq. (3.2), with M_s the magnetization within one of the Ni films, not the average magnetization $\bar{M}_s = M_s d_1/(d_1 + d_2)$ of the structure. For propagation perpendicular to the magnetization, we find[9]

$$\alpha(k_\parallel) = k_\parallel \frac{(d_1 - d_2)}{(d_1 + d_1)} \quad , \tag{4.6}$$

so as $d_1 \to d_2$, the envelope function penetrates deeply into the stack of films.

The light scattering experiments cited earlier verified explicitly that when $d_2 > d_1$, the superlattice surface mode is present, while it is absent when $d_2 < d_1$. The bulk modes thus give us an example of modes whose propagation characteristics may be "tuned" by variations in the microstructure of the superlattice, while the surface mode may literally be created or destroyed by such modifications.

All of the above discussion is confined to the case where the moments in the Ni films are all parallel. In a recent publication, Mika and Grunberg[10] have explored the collective modes of a magnetic superlattice within which the magnetization alternates in direction, between the +z and the -z direction as one moves down the stack. They find a rich spectrum; if the superlattice consists of a finite number of magnetic films, there is then an "even-odd" effect, i.e. a superlattice fabricated from an even number of magnetic films has a mode spectrum rather different from one with an odd number of layers.

V. Magnetic Field Induced Spin Reorientation Transitions in Magnetic Superlattices

In the previous section, we saw explicit examples of how variations in the microstructure of a magnetic superlattice, along with applied field, could be used to control the spin wave spectrum of the sample, and hence its response to electromagnetic radiation in the appropriate frequency domain. In these examples, as the external magnetic field H_0 is varied, the ground state spin configuration remained unchanged, and only the spin wave spectrum was altered.

It is possible to devise superlattice structures within which application of a modest external field leads to major changes in the ground state spin arrangement. Of course, it follows that there are also qualitative changes in the spin wave spectrum which follow. We have completed a rather extensive study of a particular structure in which this behavior is obtained,[11] and this section summarizes our conclusions. At the time of this writing, a structure such as that we envision has yet to be fabricated in the laboratory. We have focused on the properties of a superlattice formed from alternating layers of ferromagnetic and antiferromagnetic material; while there are examples of multilayer structures of such materials which have been studied experimentally (Cr-Co superlattices, permalloy-FeMn multilayers), the samples synthesized to date do not have interfaces of high quality, on the microscopic scale, and one will need to improve sample quality to realize the behavior we discuss. We are greatly encouraged by the appearance of very high quality Gd-Y superlattices.[12]

We consider a model superlattice fabricated from a ferromagnet, and an antiferromagnet, both with a BCC structure. In the superlattice, the antiferromagnet then consists of sheets of spins parallel to the interfaces, and there is ferromagnetic alignment of spins in each sheet.

Consider the ground state in zero magnetic field, of such a superlattice when the antiferromagnetc film has an even number of atomic layers. We suppose exchange coupling J_I of ferromagnetic sign across the interface, so the first sheet in the antiferromagnet prefers to align parallel to the spins in the antiferromagnet. Let + refer to a spin sheet directed in the +z direction, and - refer to a spin sheet directed in the -z direction. Then if each film consists of four sheets of spins, the ground state of the antiferromagnet has the following character:

```
••• (+ + + +)(+ - + -)(- - - -)(- + - +) •••
       FM       AFM       FM
```

The point is that as one moves down the superlattice, the moment in the ferromagnet alternates in sign, in a fashion reminiscent of the model explored by Grunberg and Mika.[8] This will be the case in any structure within which the antiferromagnetic constituent has an even number of layers.

This ground state is unstable with respect to application of a modest Zeeman field H_o. If it is applied parallel to +z, rotating the down spin ferromagnet to the up position gains the energy $2N_F SH_o$, where N_F is the number of layers in the ferromagnet. If the spins are rotated in this manner, clearly there is a loss of exchange energy at the interface; this is independent of N_F, to first approximation. We then expect a ground state instability, with the field H_{c_1} required to induce the transition proportional to $1/N_F$. Hence, this field can be "tuned" by varying the microstructure of the superlattice, and furthermore for N_F large, H_{c_1} can be made very modest.

We have carried out extensive studies of the (classical) ground state of the spin system for this model, by minimizing the energy as a function of externally applied field. Dipole interactions are ignored, since they play a minor role here, in contrast to their influence on the long wavelength spin waves discussed earlier. In the antiferromagnet, in addition to the exchange coupling between spins, we include a crystalline anisotropy term of the form $K \sum_i' (S_z^{(i)})^2$, known to influence the bulk properties of antiferromagnets importantly.

Quite clearly, for sufficiently large H_o, all spins, including those within the antiferromagnetic films, will be rigidly aligned along the +z direction. We find in total a sequence of four distinct phases of the spin system, as one progresses from the zero field ground state, to the fully aligned high field state. These are illustrated in a schematic fashion in Fig. (2), where the double arrow indicates the orientation of the total moment within each ferromagnetic film. The four phases are the following:

(a) The low field state, $0 < H_o < H_{c1}$:

Here we realize the zero field spin configuration, and this is stable for fields which range from zero, to the first critical field H_{c1}. The calculations show that when the number of layers N_F is large (say, greater than six), H_{c1} varies inversely with N_F, as the argument given above suggests.

(b) The unsymmetric state, $H_{c1} < H_o < H_{c2}$:

When H_o is raised above H_{c1}, the spins in the down spin ferromagnet begin to twist, to lower their Zeeman energy. The net moment in the down spin films assumes an angle $\theta_1 \neq \pi$ with respect to the z direction. A torque is transmitted to neighboring ferromagnetic film (the up spin film) through the intervening antiferromagnet (within which the spins also suffer a twist), so the net moment in the up spin films is tipped away from the z axis by an angle $\theta_1 \neq \theta_2$.

(c) The supperlattice spin flop state, $H_{c2} < H_o < H_{c3}$:

With further increase in field, the "down" spins continue to twist, so the angle θ_2 of Fig. (a) decreases. The system "locks" into a state of glide plane symmetry, as illustrated in Fig. (2c), where the transverse moments in the two sets of ferromagnetic film are equal and opposite.

(d) The fully aligned state, $H_{c3} < H_o$:

With further increase in field, the angle θ in Fig. (2c) decreases, until the fully aligned state in Fig. (2d) is reached.

As one moves from one spin configuration to the next, the transition has the character of a second order phase transition. Thus, as the field is increased a value sufficient to create the fully aligned state, we have a second order phase transition at H_{c1}, at H_{c2}, and at H_{c3}. We have studied the spin wave spectrum and the absorption spectrum of the various phases, and we find a "soft spin wave mode" in the absorption spectrum at each transition.

These studies provide us with an example of the rich behavior we may expect, when high quality superlattices are synthesized from materials with diverse kinds of magnetic order.

Acknowledgments

This work was supported by the Army Research Office, through Grant No. PO 426620.

References

1. A systematic review of the properties of surface spin waves on various magnetic surfaces, with derivations in this case, are given by D. L. Mills, Chapter 3 of Surface Excitations edited by V. M. Agranovich and R. Loudon, (North Holland, Amsterdam, 1984).

2. R. Damon and J. Eshbach, J. Phys. Chem. Solids 19, 308 (1961).

3. For experimental studies, see the work of J. R. Sandercock and W. Wettling, IEEE Trans. Magn. MAG-14, 7784 (1979).

4. Theoretical discussions of the light scattering spectra, and the influence of exchange on Damon-Eshbach waves, have been given by R. E. Camley and D. L. Mills, Phys. Rev. B18, 4821 (1978), and R. E. Camley, Talat S. Rahman and D. L. Mills, Phys. Rev. B23, 1226 (1981).

5. R. Q. Scott and D. L. Mills, Phys. B15, 3545 (1977).

6. Z. Q. Zheng, C. M. Falco, J. B. Ketterson, and I. K. Schuller, Appl. Phys. Lett. 38, 424 (1981).

7. A. Kueny, M. R. Khan, I. K. Schuller, and M. Grimsditch, Phys. Rev. B29, 2879 (1984).

8. P. Grunberg and K. Mika, Phys. Rev. B27, 2955 (1983).

9. R. E. Camley, Talat S. Rahman, and D. L. Mills, Phys. Rev. B27, 261 (1983).

10. K. Mika and P. Grunberg, Phys. Rev. B31, 4465 (1985).

11. L. L. Hinchey and D. L. Mills (to be published).

12. J. Kwo, E. M. Gyorgy, D. B. McWhan, M. Hong, F. J. di Salvo, C. Vettier and J. E. Bower, Phys. Rev. Lett. 55, 1402 (1985).

Figure Captions

Figure (1): (a) A superlattice formed from films of material A, each with thickness d_1, and material B with thickness d_2. We also show, in a schematic manner, the spatial variation of the magnetic field associated with spin waves on two of the interfaces. (b) The basic geometry appropriate to the discussion of Damon-Eshbach spin waves in thin films. (c) A sketch of a semi-infinite magnetic superlattice. We also illustrate the magnetic field of the surface wave, which is a collective excitation of a superlattice structure.

Figure (2): A schematic illustration of the magnetic phases found in reference (11), for a superlattice formed from ferromagnetic and antiferromagnetic materials. The details of the structure explored in the theory are described in the text. The figure here shows, as a function of external field, only the direction of the net magnetic moment in adjacent ferromagnetic films. We have (a) the low field ground state, (b) the unsymmetric state, (c) the superlattice spin flop state, (d) the high field fully aligned state.

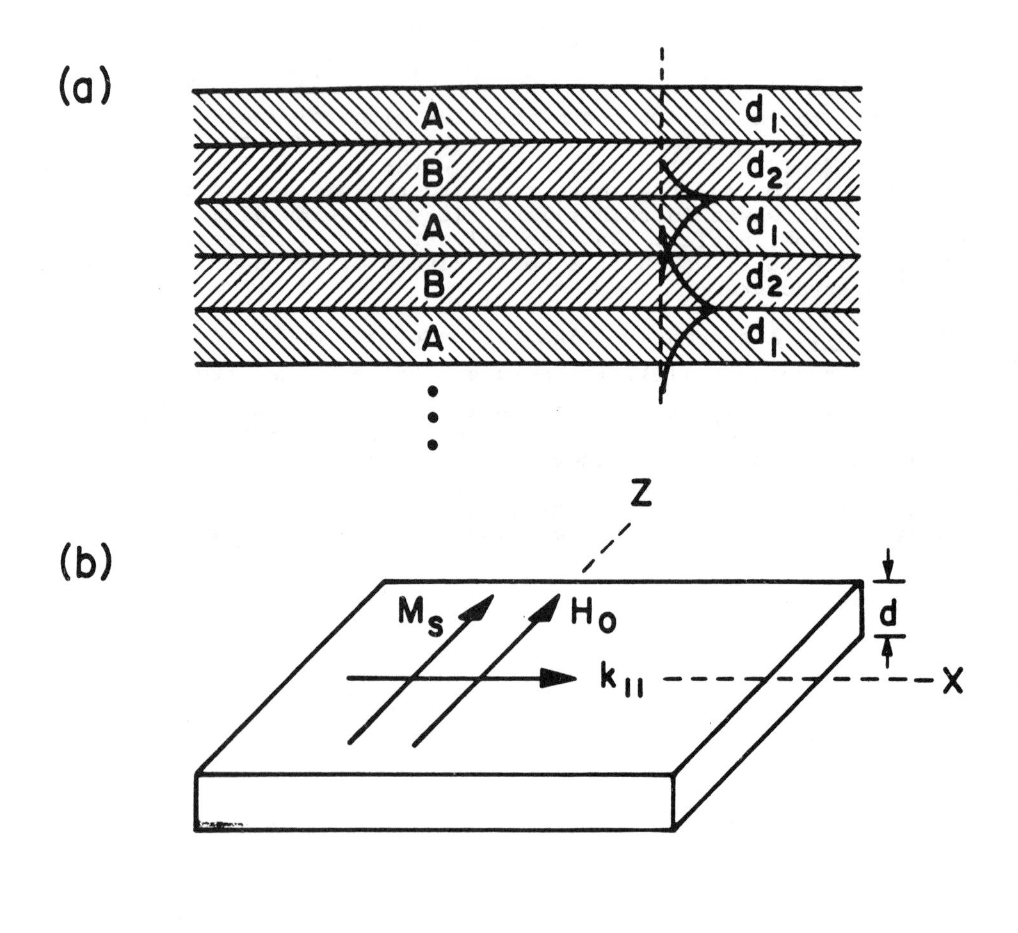
(a)
A
B
A
B
A
d_1
d_2
d_1
d_2
d_1
(b)
Z
M_S
H_0
$k_{||}$
X
d

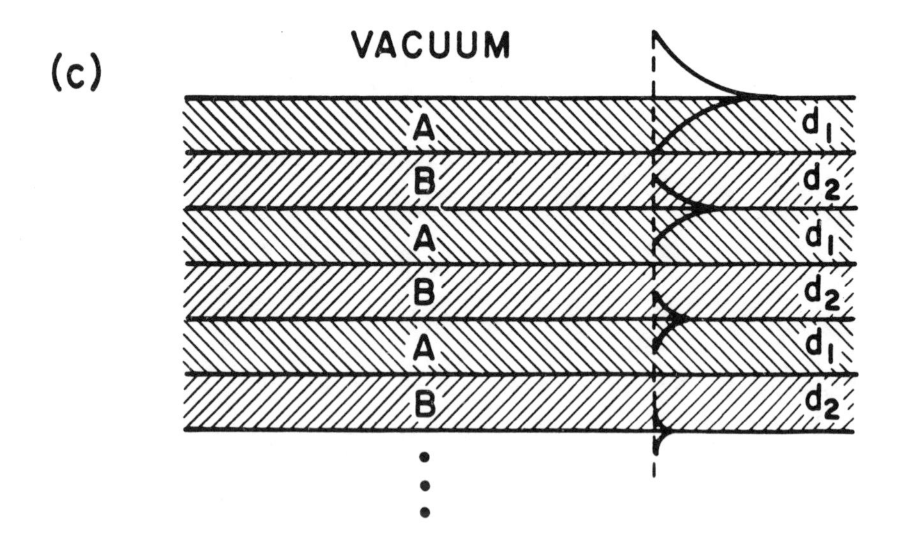
(c)
VACUUM
A
B
A
B
A
B
d_1
d_2
d_1
d_2
d_1
d_2

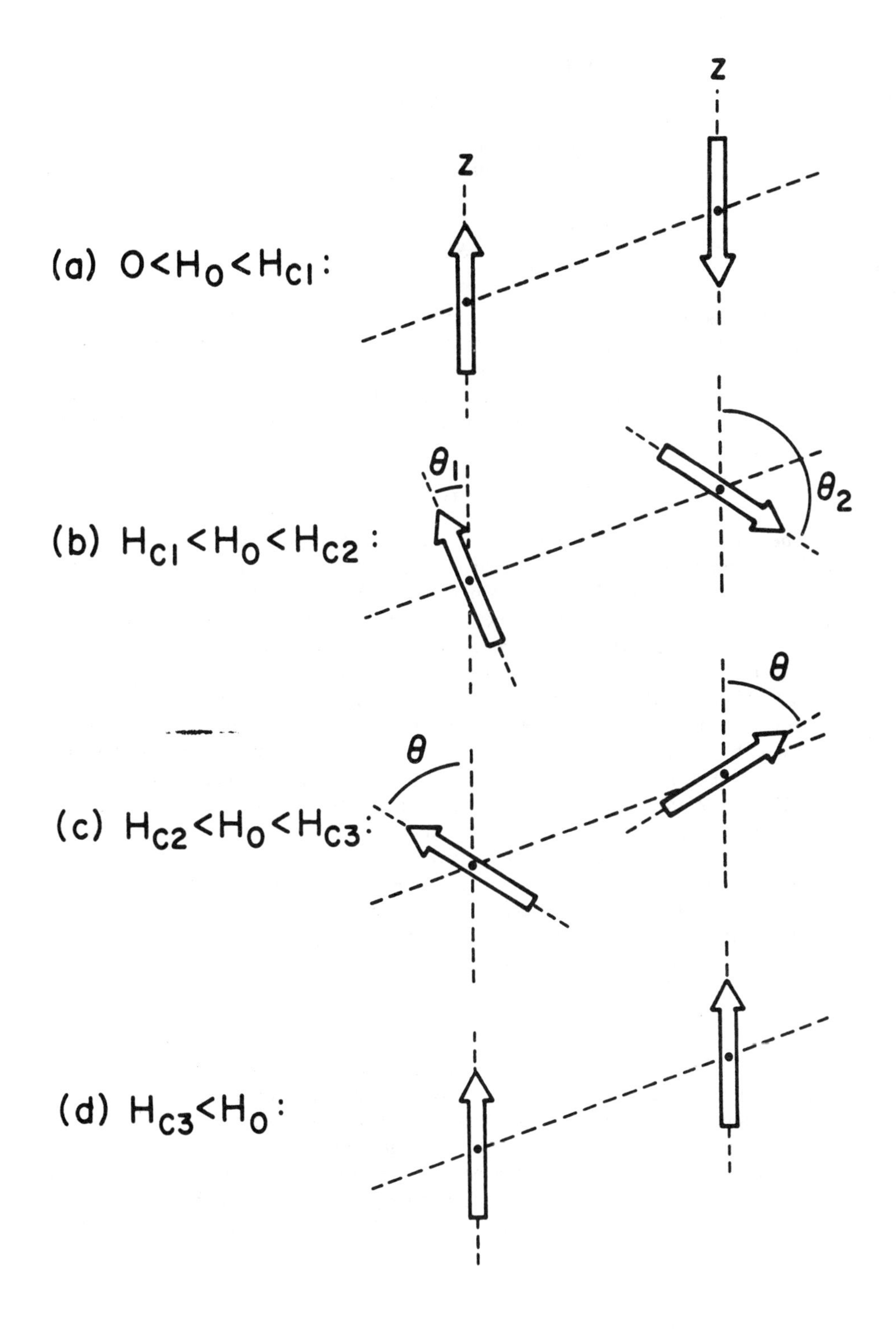

z
z
(a) $0<H_0<H_{c1}$:
θ_1
θ_2
(b) $H_{c1}<H_0<H_{c2}$:
θ
θ
(c) $H_{c2}<H_0<H_{c3}$:
(d) $H_{c3}<H_0$:

Artificially Layered Materials

J. M. Rowell

Bell Communications Research
Murray Hill, New Jersey 07974

ABSTRACT

In contrast to semiconductor superlattices, where electrical measurements (such as the carrier mobility at 78K) for characterization are common, the use of resistivity and resistance ratio measurements for other types of artifically layered materials, such as metallic multilayers, has not become routine. The advantages of such straightforward measurements in understanding the properties of both the individual layers and the interfaces in these materials will be discussed.

INTRODUCTION

The scientific and technological interest in artificially layered materials based on III-V semiconductors has spawned considerable interest and research into the properties of such multilayers in which either one or both components are not semiconductors. In particular, all metallic or metal-insulator multilayers have been created by various deposition techniques including both sputtering and thermal evaporation (which seems to be called MBE if the vacuum is $\sim 10^{-9}$ torr or better). However, the fascinating variety of physical effects, such as quantum wells and modulation doping, that have been created in semiconductor superlattices, where both the band gap and doping profiles are subject to design and control, cannot be produced in metallic systems. Instead, the properties of metallic multilayers generally reflect either thin film effects or interface phenomena. In the former case, properties are changed in identical fashion to a single thin film but the effects are easier to measure because many such film are present in the sample, whereas in the latter case the effects of the interfaces dominate the properties of the sample as it becomes "all interface" when the layers are very thin. Cooperative effects, where new physical properties result from interaction between dissimilar adjacent layers, or between next nearest neighbor identical layers separated by the dissimilar layer of the second component, have not been spectacular. The most interesting cases include new crystal phases that have been stabilized by interfaces,[1] superconducting behavior that has been modified by the proximity effect[2–5] and by coupling across an insulating layer[4] and, most recently, interesting changes in the magnetic behavior of rare earth metals that appear to be due to breaking their spin structure in the repetitive fashion allowed by multilayering.[5–7] However, radically new materials with different band structure and electronic and vibrational properties have not been produced in the way those entering the field probably hoped. In this talk I would like to suggest that the reason for this is that in many cases the electrons in these metallic multilayers are confined to each layer by interface scattering, and thus no change in electronic properties resulting from the repetitive layered structure can be expected. I will further suggest that straightforward electrical measurements should be used more routinely to characterize these materials.

ELECTRICAL CONDUCTIVITY OF METALLIC MULTILAYERS

It is routine in studies of semiconductor superlattices to quote the carrier mobility (generally at 78K) as a criterion to judge the quality of the sample. What should be used as a similar measure of excellence in the case of metallic multilayers? I suggest that it is nothing new, simply both the resistivity and resistance ratio measurements that have always been used to define the quality of bulk metals.

Consider a multilayer of two dissimilar metals with equal layer thickness $\lambda/2$ (period λ). At room temperature the resistance (hence the resistivity) of the sample can be calculated from the resistances (resistivities) of the many individual layers of each component, all added as parallel resistors. However, this is only possible as long as the resistivity (ρ) of the layers remains equal to its bulk value, which means the electronic mean free path (mfp) must be unchanged from its bulk value. Any significant increase in the resistivity of the film above that calculated from the component bulk resistivities is an indication of changes in the perfection of the multilayer. To realize this, consider how the mfp can be changed in a multilayer. There are two dominant effects, either the electrons scatter from the interfaces between the layers, so that the mfp becomes simply related to $\lambda/2$ (as in an isolated thin film of thickness $\lambda/2$), or the electrons scatter within the film. Assuming the layers are free of impurities from the deposition process, this additional scattering must be due to alloying of the two constituent metals of the multilayer or to disorder in the layers, which can be driven by strains from the interfaces or deposition of the layers at too low a substrate temperature.

These various possibilities are most easily studied if a series of samples are made as a function of λ, and a plot made of sample resistivity versus λ. In the limit of large λ, the mfp will be less than $\lambda/2$ in both films and, assuming the layers are reasonably crystalline, ρ will have the value calculated from the constituents. In the limit of small λ the possibilities are as follows:

1. The layers remain well ordered, no alloying occurs and interface scattering is small. The mfp will exceed $\lambda/2$ for small λs but the resistivity will be constant as λ is changed from large to small values.

2. The layers remain well ordered but alloying occurs at the interfaces. Eventually, with decreasing λ, the sample becomes a uniform alloy with its appropriate resistivity, about twice that of the constituents in the case of transition metals.

3. The electron scattering at the interfaces is so strong that the mfp is limited by the layer thickness, hence the resistivity rises with decreasing λ in a way

that can be calculated and reaches very high values for small λ, when the mfp may be only a few lattice spacings. This behavior is, of course, expected for metal-insulator multilayers.

4. This case is assumed from the observation that in many multilayers the resistivity reaches 100–150 $\mu\Omega$cm for λ that is small but still many lattice spacings, typically 30 – 50Å. This very high value of ρ is only achieved in bulk metals when they are amorphous, hence we must assume that these multilayers are so disordered that the mfp becomes comparable to a lattice spacing in these samples.

These four cases of electron scattering are represented in Fig. 1. (a)----(d), respectively. In practice it may be difficult to isolate each behavior, in fact it seems likely that in multilayers made of two metals the mfp limitation in case 3 may be a layer of amorphous or highly disordered material at the interface which occurs for all λ, rather than a transition of the sample as a whole to an amorphous structure, as suggested in case 4. Schematically, the four dependencies of ρ on λ are plotted in Fig. 2. Although the arguments above have perhaps been over-simplified, there are many results in the literature that resemble the trends of Fig. 2 quite closely. For example, Nb/Ta[3] resembles case 2. Unfortunately, the only other single crystal multilayers, namely Ni/Cu[8] and the rare-earth samples,[6,7] were grown on thick seed layers of Cu and Nb, respectively and resistivities could not be measured. Nb/Cu was pointed out[2] to have mfp limited by layer thickness, and reaches 150 $\mu\Omega$–cm for λ = 20Å, hence is typical of cases 3 and 4.

RESISTANCE RATIO

A second useful parameter for characterizing multilayers, as in bulk metals, is the resistance ratio R(300K)/R(4K). To my knowledge, no multilayer samples have been yet reported with a resistance ratio comparable to that which is fairly easily achieved in single films, say 30 (a conservative number) in Al, Cu and transition metals. This is simply stating that multilayers have not yet been grown with the perfection that allows the mfp to increase 30 times as the sample is cooled, say from 30Å for a transition metal at 300K to 900Å at 4K. In fact, very few multilayers have been shown to have a mfp that exceeds $\lambda/2$ at any temperature, Nb/Ta being one such case.[3]

In case 4 above the amorphous material will be identified by a resistance ratio which is close to or even slightly less than one. This is fairly common in multilayers, hence a plot of resistance ratio versus λ shows values of (say) 2 for a large λ, decreases with decreasing λ and becomes less than 1, at which λ the sample can probably be regarded as amorphous, at least as far as transport properties are concerned.

SUMMARY – TRANSPORT PROPERTIES

The main points I would like to make in this talk are:

1. Transport measurements, of both resistivity (not resistance) and resistance ratio against λ, should be made for all multilayers.

2. Many metallic multilayers have an electronic mean free path that is very short, much less than λ/2, even for small λs, hence any hoped for band structure effects resulting from the layering will not be observed.

3. It is not clear that a "new material" (in the sense of an ordered alloy such as β-brass [CuZn], but one that does not exist naturally) has been produced in a metallic multilayer.

SUMMARY – VIBRATIONAL PROPERTIES

Little work has been to study the lattice dynamics of metallic multilayers. Electron tunneling is a useful probe for superconducting systems, but has the disadvantage of measuring vibrational properties only within a coherence length of the surface of the sample. If the multilayer has a short mfp, the coherence length will be short also and the measurement may not probe beyond the top layer. Previously published studies on the relatively clean Nb/Ta multilayers[3] illustrate the limitations of the tunneling technique.

REFERENCES

1. Lowe, W. P. and Geballe, T. H. Phys. Rev. **B29**, 4961 (1984).
2. Schuller, I. K. and Falco, C. M., "Inhomogeneous Superconductors" AIP Conference Proc. 58, 197, 1979. Schuller, I. K., Phys. Rev. Lett. **44**, 1597, 1980. Banerjee, I., PhD Thesis, Northwestern Univ., 1982.
3. Hertel, G., McWhan, D. B. and Rowell, J. M. in "Superconductivity in d- and f- Band Metals", Kernforschungszentrum Karlsruhe, Germany, p. 299 1982. Durbin, S. M., Cunningham, J. E. and Flynn, C. P., J. Phys. **F12**, L75, 1982.
4. Ruggiero, S. T., Barbee, T. W., Jr. and Beasley, M. R., Phys. Rev. **B26**, 4894, 1982.
5. Greene, L. H., Feldmann, W. L., Rowell, J. M., Batlogg, B., Gyorgy, E. M. Lowe, W. P. and McWhan, D. B., Superlattices and Microstructures 1, **407**, 1985.
6. Kwo, J., McWhan, D. B., Gyorgy, E. M., Hong, M., Bower, J. E. and DiSalvo, F. J., International Conference on Magnetism, August 26-30, 1985, San Francisco, CA. p. 136.
7. Sinha, S., Cunningham, J., Du, R., Salamon, M. B. and Flynn, C. P., International Conference on Magnetism, August 26-30, 1985, San Francisco, CA. p. 136.
8. Gyorgy, E. M., McWhan, D. B., Dillon, Jr., L. R., Walker, L. R. and Waszczak, J. V., Phys. Rev. **B25**, 6739, 1982.

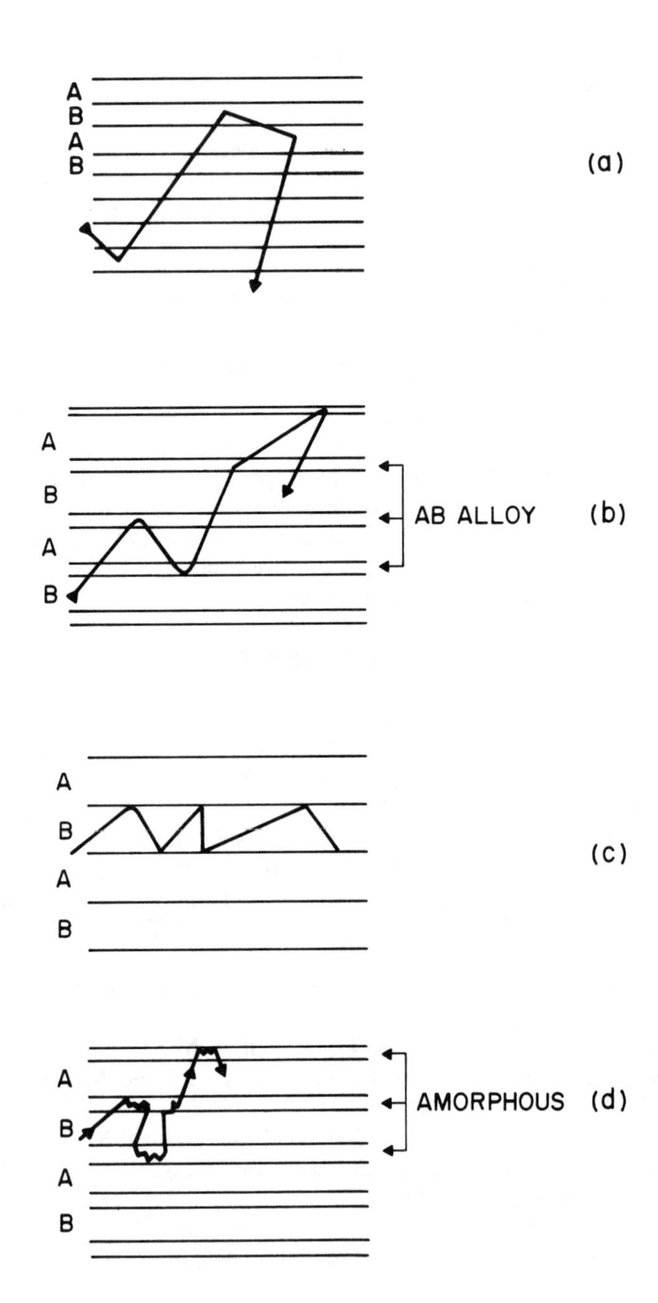

Fig. 1: Electron scattering in the limit of small λ for four possible cases of multilayers. Figs. (a)-(d) correspond to cases 1.-4., as described in the text.

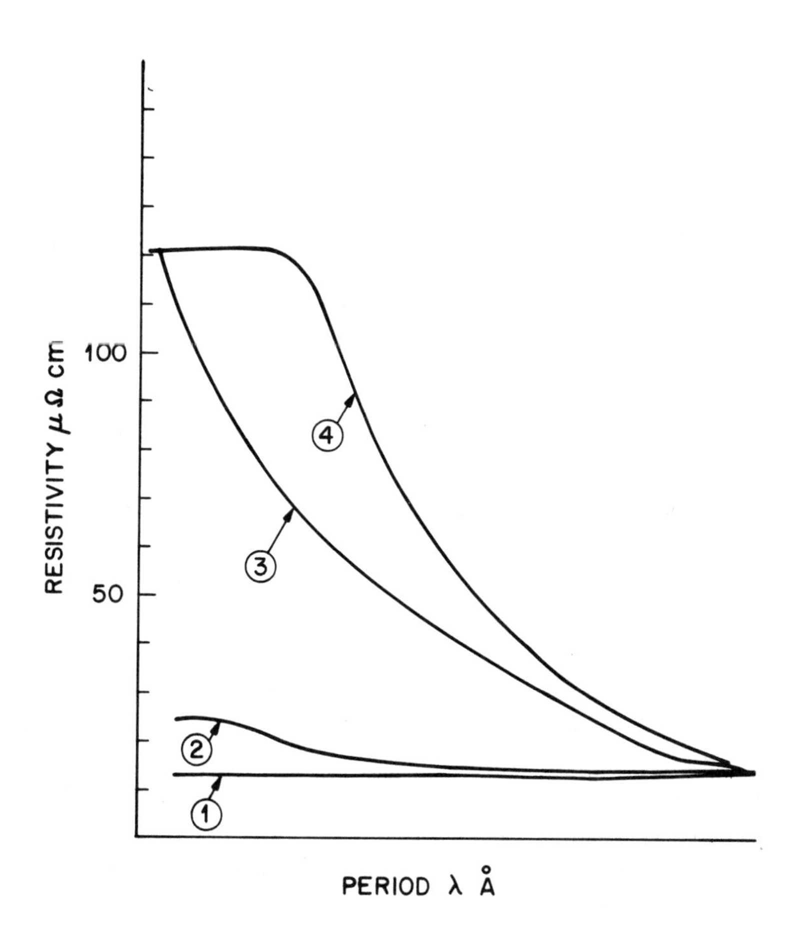

Fig. 2: Resistivity versus superlattice periodicity (λ) for the four multilayer cases described in the text. The corresponding electron scattering for curves 1-4 is shown in Figs. (a)-(d), respectively.

Interfaces in Metallic Superlattices

*D. B. McWhan and C. Vettier**

AT&T Bell Laboratories
Murray Hill, New Jersey 07974

ABSTRACT

X-Ray diffraction determinations of the modulations in the composition, interplanar spacing and magnetic moment of metallic superlattices are reviewed. The value of synchrotron radiation in the study of multilayers is emphasized.

INTRODUCTION

There are several different types of modulations in a superlattice, and they can be studied in detail using x-ray diffraction techniques at synchrotron sources. There is a modulation in both the chemical composition and the interplanar spacing, and in magnetic superlattices composed of alternating regions of magnetic and non-magnetic components there is a modulation in the magnetic moment. Of particular interest is the shape of the interfacial region between the different components which make up the superlattice. In an ideal superlattice the composition would change from 100% of one component to 100% the other component over a distance corresponding to one atomic layer. However, in practice there may be interdiffusion of the components and fluctuations in the position of the interface which lead to an average interfacial width that is several atomic layers thick. Apriori there is no reason why the elastic response of the system should lead to a modulation in the interplanar spacing which follows the composition modulation exactly. The interfaces in the interplanar spacing modulation may be substantially thicker than the corresponding interfaces in the composition modulation. Similarly in a magnetic superlattice the magnetic moments near the interface may be reduced from the value of the moment in

* Permanent address Institute Laue-Langevin, Grenoble, France.

the middle of each magnetic region i.e. the effective magnetic interface may be thicker than the composition interface. In this paper we discuss each of these modulations separately and demonstrate how they can be determined using the high flux and the tunability of the x-ray wavelength at synchrotron x-ray sources.

COMPOSITION MODULATION

The diffraction pattern from a material with a periodic composition modulation is composed of average Bragg reflections surrounded by satellites. If the modulation wavevector is parallel to a major crystallographic axis then the reciprocal lattice in this direction is given by $Q(\ell,p)=2\pi(\frac{\ell}{d}+\frac{p}{\Lambda})$ where d is the average interplanar spacing, Λ is the modulation wavelength and ℓ and p are integers. In the limit where the lattice mismatch between the two components is small the intensities of the satellite reflections with $p\neq 0$ are proportional to $C_p^2(f_a-f_b)^2$ where C_p is the p_{th} Fourier coefficient of the composition modulation in real space and f_a and f_b are the atomic scattering factors per layer of each of the components. Even if there is a small interplanar spacing modulation, its contribution to the intensities at small angles, i.e. reflections for which $\ell=0$, is negligible. Much of the early structural characterizations of superlattices involved measuring the integrated intensities of the satellites and comparing them with intensities calculated for various models for the composition modulation. For all of the x-ray diffraction studies reported in the literature, of which the authors are aware, the interfaces between the two components are never sharper than two atomic layers.[1] Even in the nearly perfect semiconductor systems such as $(GaAs)_n(AlAs)_m$ there is an apparent minimum interface thickness of two atomic layers.[1,2] In this system the structure of a series of superlattices with alternating n layers of GaAs and m layers of AlAs with (n,m)=(1,1), (2,2), (5,4), (7,4), (10,4), and (12,9) were determined. A consistent model emerged in which the superlattices can be approximated at long wavelengths by regions of 100% GaAs and 100% AlAs with a smooth transistional or interfacial region between GaAs and AlAs which is approximately two atomic layers thick. Similar interface widths have been found in the best metallic multilayers systems such as $Nb_n Ta_m$[3,4,5] and $Gd_n Y_m$.[6] The natural conclusion of this model is that at short modulation wavelengths the interfacial regions overlap and the average amplitude of the composition modulation for an alternating monolayer superlattice is ~30% instead of 100% (see fig. 5 of ref. 1).

If this apparent minimum interface thickness is correct, then it has severe consequences with regard to the study of two dimensional properties in metallic systems. Most theoretical studies of the magnetic properties of thin layers indicate that dramatic changes in the ordering temperature or in the critical exponents such as for the temperature dependence of the magnetization only occur when the magnetic material is 1 to 4 atomic layers thick.[7,8] This is exactly the region where the composition of each layer starts to drop substantially below 100% in a multilayer, and therefore it is difficult to achieve meaningful two dimensional effects because a nominal monolayer will in practice be an alloy with a 30% modulation in composition with a two layer wavelength. We will explore the validity of the x-ray diffraction results from two points of view. An estimate of the possible error in the determination of the interfacial width is made and then possible errors resulting from dynamical effects are assessed.

In a typical superlattice with a modulation wavelength of 50-100Å, somewhere between five and ten satellites are observed at low angles using a standard rotating anode x-ray source. The integrated intensities are then corrected for Lorentz, polarization and absorption factors to give a set of structure factors squared. As an example we consider a $Nb-Ta$ superlattice grown on sapphire (01$\bar{1}$2), in which seven satellites are observed.[5] From the positions of the satellites the modulation wavelength is found to be 84.7Å and from the positions of the average Bragg reflections at higher angles the average interplanar spacing is 1.65Å. This gives ~53 atomic layers per modulation wavelength, and therefore, 53 parameters have to be determined in order to completely describe the composition modulation. With only 7 measured intensities, the composition modulation is approximated by a trapezoid. This waveform describes the essential features of the modulation in terms of two parameters which are the distance over which the layers are 100% either Nb or Ta and the width of the interfacial region. The first parameter allows for the fact that the actual composition is not known exactly and that it may be assymmetric with, for example, more Nb than Ta. Finally a scale factor is needed to compare the observed and calculated absolute values of the structure factors.[9] The best fit of a trapezoidal wave to the data is obtained by minimizing the standard crystallographic R factor.[10] In order to demonstrate the sensitivity of the refinement to the parameters, contours of the R factor for various combinations of the interface thickness and of the width of the pure Nb region are shown at the top of figure 1. As can be seen the possible error in interface thickness is less than an Å ngstrom. There are two minima which reflect the fact that, from this data alone, one cannot tell if the multilayer is rich in Nb or Ta because the intensity varies as $(f_{Nb}-f_{Ta})^2$. However the possible error in the thickness of the pure Nb or Ta region is of the order of two to three Å ngstroms. The best parameters were used to calculate the phases of the Fourier coefficients, and the trapezoidal wave is compared with the Fourier transform of the measured structure factors in the bottom of fig. 1. This typical example illustrates that the average composition modulation is determined with reasonable accuracy from x-ray diffraction measurements. Furthermore there is consistency between different laboratories. X-ray studies of $Nb-Ta$ multilayers grown on the sapphire $(11\bar{2}0)$ give an interface width (2x the quoted diffusion length) of 9Å which is similar to the 10.5Å determined above for a different growth direction.[3,5] Finally the progression to shorter wavelength samples leads to the same conclusion that has been drawn in the $(GaAs)_n(AlAs)_m$ system, namely that the interfaces overlap and that the modulation amplitude at short wavelengths is greatly reduced. The intensities of the 1st order satellites of a $Nb-Ta$ sample with a modulation wavelength of 18.6Å give a composition modulation of $\approx 25\%$.[5]

In the majority of x-ray studies of multilayers done to date it has been assumed that the films are sufficiently thin that the kinematic theory is applicable. However for high quality single crystal films, especially at low angles dynamical effects may become important. In order to assess the possible importance of dynamical effects we review the original Darwin formulation

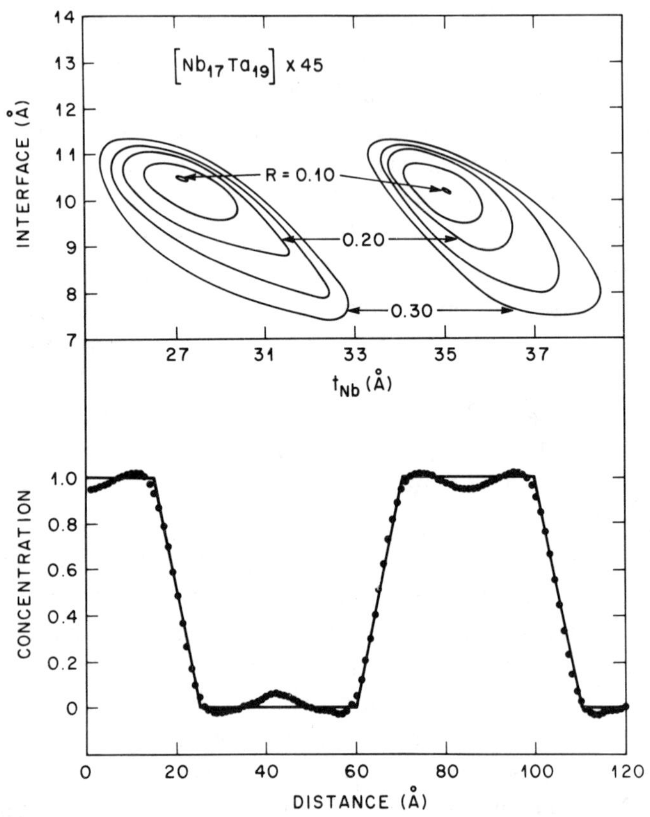

FIG. 1 (top) variation of the R factor with the interface thickness and thickness of the Nb (bottom) comparison of the trapezoidal wave with the Fourier series based on the 7 observed coefficients.

as given by James and apply it to the *NbTa* sample discussed above.[11] It is concluded that the neglect of dynamical effects in the crystal structure analysis leads to an underestimate of the interfacial width in a multilayer.

In the kinematic approximation to the dynamical theory it is assumed that the intensity of the incident x-ray beam is constant throughout the sample except for the effect of normal photoelectric absorption. In reality if the crystal is perfect and the incident beam is at the angle for Bragg diffraction from a set of planes, then part of the incidental beam is reflected at each plane. The intensity of the incident beam is decreased from plane to plane, and this effect is called primary extinction. Darwin showed that the integrated intensity for a reflection $I(\ell,p)$ is reduced from the intensity given by the kinematic theory $I'(\ell,p)$ by:

$$I(\ell,p)=I'(\ell,p)\ \frac{Tanh(mq)}{mq}$$

$$\left|q\right|=2Nd^2\ \left|F\right|\ \left(\frac{e^2}{mc^2}\right)$$

where m is the number of planes in the multilayer, N number of atoms per unit volume, d the interplanar spacing, and $\left|F\right|$ is the structure factor.[11] If mq is $<<1$ the correction is negligible, but the correction can be significant for thick films with long modulation wavelengths. Evaluating the expression for $\left|q\right|$ for the example of the $NbTa$ multilayer with Λ=84.7Å, one calculates $\left|q\right|$ = 0.3, $6x10^{-3}$ and $5x10^{-4}$ for reflections with (ℓ,p)=(0,1),(0,7),(2,0) respectively. Converting to the number of modulation wavelengths needed for $mq=1$, one gets 3, 24, and 39 respectively. In the limit where the multilayer is a perfect crystal, which is 45x84.7=3812Å thick, the extinction corrections $I(\ell,p)/I'(\ell,p)$, are calculated to be 0.07, 0.5, 0.8 respectively for the reflections above. In discussing the structure refinement above, the intensities of the first seven harmonies were measured and compared with a model for the composition modulation using the kinematic theory. As the structure factor of the pth harmonic is proportional to the pth Fourier coefficient, the sharpness of the interface is related to the relative size of the higher order Fourier coefficient to the lower order coefficients. If the observed intensity of the seventh harmonic is increased in relation to the 1st harmonic by extinction by a factor of 7 as suggested by the Darwin theory, then when corrected for extinction, the actual ratio of the Fourier coefficients would be 7x larger. The true interfacial width would be much larger than that calculated above. Therefore the kinematic theory gives a lower limit for the interfacial width.

This discussion of extinction assumes that the multilayer is a perfect crystal and most real materials will be less than perfect which means that the effective value of m over which the layers are perfect corresponds to a value much less than the total number of layers in the superlattice, and the extinction corrections become negligible. This is most likely the case for the majority of metallic superlattices grown to date where the typical mosaic spread may be many tenths of a degree, and nowhere near the Darwin width expected for a perfect crystal. However as the quality of metallic superlattices increases the difficult problem of making reliable extinction corrections will have to be faced. In fact in the study of $(GaAs)_n(AlAs)_m$ multilayer mentioned above, the reflections at $Q(\ell,m)=(2,0)$ and (4,0) were not used in the refinement because it was suspected that extinction might be a problem.[2]

INTERPLANAR SPACING MODULATION

In the example discussed in the last section the lattice parameters of *Nb* and *Ta* differ by 0.16%, and the variation in interplanar spacing from the Nb to the Ta regions of a multilayer are sufficiently small that they can be determined by expanding the exponent in the structure factor and keeping the terms up to first order in the displacement of the ith layer from its position in the unmodulated structure. However in most strained layer superlattices this approximation is no longer valid and both the position and composition of each layer in a modulation wavelength must be determined.[1] There will be strong correlations in a refinement between the compositional and positional parameters which compounds the problem of the limited number of reflections which can be observed. A solution to this problem is to combine the high flux and the tunability of synchrotron x-ray sources. The increase in flux of several orders of magnitude makes it possible to measure the intensities of much higher orders of harmonics than can be observed with typical laboratory x-ray sources. In addition by using anomalous dispersion, the data set can easily be doubled or tripled. The atomic scattering factor, f, for each component has dispersion corrections, f′ and f", which vary with the x-ray energy so that $f=f_0+f'+if''$. The dispersion corrections are almost independent of scattering angle whereas f_0 decreases with increasing $\sin\Theta/\Lambda$. The contribution to the intensities of the harmonics arising from the composition modulation varies as the difference in the scattering powers of the components. Near an absorption edge f′ becomes large and negative and f" jumps discontinuously at the edge. By collecting intensity data far from an edge and just below an edge f′ can be varied by tens of electron units relative to f_0 which is equal to the atomic number at zero scattering angle. For a ternary strained layer superlattice such as $(In_x Ga_{1-x} As)_n (InP)_m$ one can collect data below the Ga and As edges and thus probe the structure of each sublattice. An extreme example of the anomalous dispersion effect arises in $Gd_n Y_m$ multilayers where f′ approaches -30 near the Gd L edges. The composition component to the intensity of the harmonics can change sign as $f_{Gd} - f_Y$ goes from $\sim +20 \rightarrow -5$ as the x-ray energy is increased toward the absorption edge. This means that the correlations between the compositional and positional parameters can be reduced as the contribution to the former varies dramatically with x-ray energy whereas the latter contribution is constant.

The interplanar spacing modulation is a useful probe of a number of phenomena. Usually it is used as an indirect measure of the coherency strains built into the superlattice. The amplitude of the interplanar spacing modulation is usually different than simply the percent difference in the spacing of the two pure components, and this is a measure of the strain. The variation in the strain with modulation wavelength is often used as an indication of the coherent-incoherent transition.[1] More recently, measurements of the strain modulation have been used to probe the preferential uptake of hydrogen in transition metal superlattices[13,14] and to measure the magnetostriction as a function of temperature in a superlattice composed of alternating magnetic and non-magnetic components.[15] When hydrogen is absorbed in a transition metal such as Nb the lattice parameter expands and if the hydrogen is absorbed preferentially in one component of a superlattice then the interplanar spacing modulation increases in amplitude. The enhanced solubility of hydrogen in Nb has been observed in both $Nb_n Pd_m$ [13] and $Nb_n Ta_m$ [14] superlattices using this technique.

The temperature dependence of the interplanar spacing modulation in $Gd_n Y_m$ multilayers is an indirect probe of the evolution of magnetic order in these systems.[15] When bulk Gd metal orders ferromagnetically there is an expansion of the lattice along the c axis, resulting from magnetostriction which offsets the normal thermal contraction. In a superlattice of ferromagnetic Gd alternating with non-magnetic Y, the amplitude of the interplanar spacing modulation increases as the Gd regions expand and the Y regions contract with decreasing temperature below the Curie temperature. The intensities of the satellites around several orders of the average reflections along the c axis were measured from $T=13K$ to $333K$ for a superlattice of 21 layers of Gd alternating with 21 layers of Y which were repeated 40 times. The composition and the interplanar spacing modulations were each modelled using ERF functions to describe the smooth variation from Gd to Y over an interfacial region. The results for the best model for the interplanar spacing modulation are shown in fig. 2.[15] Above the Curie temperature the strain interface between the Gd and Y regions is $\sim 2-3$ atomic layers thick, and it is similar in shape to the composition modulation. At low temperatures the strain modulation has more than doubled in amplitude, and it has become substantially more rounded. The effective width of the strain interface has doubled.

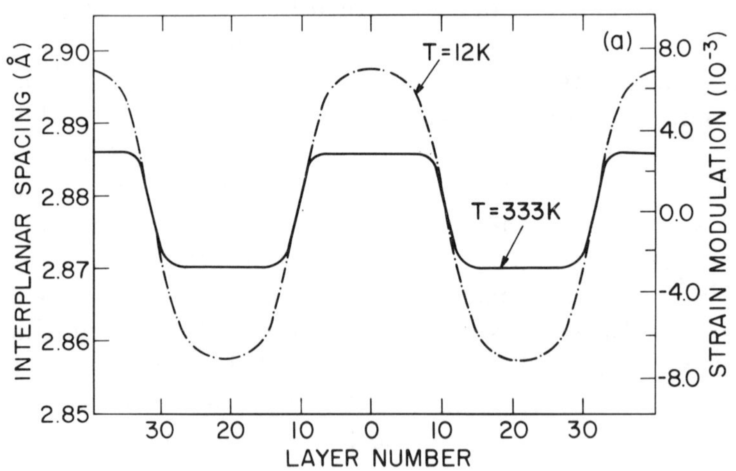

FIG. 2 Interplanar spacing above and below the Curie temperature in $[Gd_{21}\ Y_{21}]$x40.

As the increase in amplitude is magnetostrictive in origin and as the magnetostriction is an indirect measure of the magnetic moment, one speculates that the increase in interfacial width may reflect a reduced moment near the interfaces. This speculation supports the bulk magnetization measurements which show a reduced average magnetic moment for this sample in low applied magnetic fields.[6]

MAGNETIC MOMENT MODULATION

X-ray scattering measurements can be used to directly probe the magnetic moment modulation in a superlattice.[15] The traditional measurement used to determine magnetic structures is neutron scattering,[16,17] but there is a small component of the intensity of scattered x-rays which arises from the electron spin.[18] This component in a ferromagnetically ordered material varies as the projection of the spin structure factor on the normal to the scattering plane, $S_{\perp}(\ell,p)$. The magnetic scattering is very weak relative to the normal scattering from the charge of the electron, but relative changes in the total intensity can be measured by reversing the spin direction and, therefore, the sign of the magnetic component with a small magnetic field. For a fully linearly polarized beam (a typical synchrotron source is $>$80% polarized) the change in intensity or flipping rate is:

$$\frac{\Delta I}{I} = \frac{4\lambda_c}{\lambda} \frac{F''(\ell,p)\left|S_{\perp}(\ell,p)\right| sin\,2\theta}{F'(\ell,p)^2 + F''(\ell,p)^2}$$

where $\lambda_c = h/mc = 0.02426\text{Å}$, λ is the x-ray wavelength, $F'(\ell,p)$ and $F''(\ell,p)$ are the real and imaginary parts of the structure factor.[19] By tuning the x-ray energy just above the Gd absorption edge $F''(\ell,p)$ can be maximized, and consequently, $\Delta I/I$ can be maximized. Typical values of $\Delta I/I$ vary from 10^{-2} to 10^{-4}. In order to verify that the small observed changes are related to the magnetic structure, measurements of $\Delta I/I$ vs λ are made, and the functional form shown to follow that predicted from the changes in $F'(\ell,p)$ and $F''(\ell,p)$ in the vicinity of the absorption edges. As an example of this technique we consider the magnetic scattering from $[Gd_{21}\ Y_{21}]$x40.[15] The determination of the composition and interplanar spacing modulations is described above. To determine the magnetic moment modulation a series of 21 flipping ratios were measured at the Cornell High Energy Synchrotron Source. Several models for the variation in the moment through the Gd regions were tried and a slightly lower R factor was found for a model in which the moment decreased on approaching the interface.[15] This is consistent with the speculation based on the interplanar spacing modulation.

SUMMARY

X-ray scattering using the high flux and the tunability of synchrotron sources is a powerful, non-destructive probe of multilayers. In this paper evidence for a minimum interfacial width in the composition modulation in superlattices is reviewed and the validity of the structural determination assessed. The use of anomalous dispersion to separate the composition and strain modulations is discussed and recent examples of the measurement of the strain modulation are reviewed. Finally, the use of magnetic x-ray scattering to study interfacial magnetism is shown. All of these studies have assumed that the superlattice is perfectly periodic. As this periodicity is controlled by statistical fluctuations in the deposition rates and the opening and closing of shutters, there will not be true one dimensional order. The next challenge in the structure determination of superlattices will be to model these fluctuations.

We thank J. D. Axe, A. Bienenstock, E. M. Gyorgy, R. Kwo, W. P. Lowe, C. J. Majkrzak, I. K. Robinson, J. M. Rowell, and J. M. Vandenberg for helpful discussions.

REF.ERENCES

1. D. B. McWhan, in *Synthetic Modulated Structures,* L. L. Chang and B. C. Glessen EDS (Academic Press Inc. N.Y. 1985), P. 43

2. P. D. Dernier, D. E. Moncton, D. B. McWhan, A. C. Gossard, and W. Wiegmann, *Bull. Am. Phys. Soc.* **22,** 293 (1977)

3. S. M. Durbin, J. E. Cunningham, M. E. Mochel, and C. P. Flynn, *J. Phys.* **F11,** L233 (1981)

4. S. M. Durbin, J. E. Cunningham, and C. P. Flynn, *J. Phys.* **F12,** L75 (1982)

5. G. Hertel, D. B. McWhan, and J. M. Rowell, *Proc. Conf. Superconductivity in d and f Band Metals,* Karlsruhe, Germany 1982, P. 299

6. J. Kwo, E. M. Gyorgy, D. B. McWhan, M. Hong, F. J. DiSalvo, C. Vettier, and J. E. Bower, *Phys. Rev. Letters* **55,** 1402 (1985)

7. G. Allan, *Surface Science Reports,* **1,** 121 (1981)

8. G. Bayreuther, *J. Mag. and Magnetic Mat.* **38,** 273 (1983)

9. The absolute values of the structure factors for the 1st through 7th harmonies normalized to the 1st harmonies are 100, 10, 30.6, 6.3, 6.8, 2.6, 2.3

10. International Tables for X-ray Crystallography, **Vol. IV** (Kynoch Press, Birmingham 1974) p. 288

11. R. W. James, *The Optical Principles of the Diffraction of X-Rays* (Oxborn Press, Woodbridge, Conn 1982) P. 270

12. L. K. Templeton, D. H. Templeton, R. P. Phizackerley, and K. O. Hodgson, *Acta Cryst.* **A38,** 74 (1982)

13. S. Moehlecke, C. F. Majkrzak, and M. Strongin *Phys. Rev.* **B31,** 6804 (1985)

14. P. F. Micele, H. Zabel, and J. E. Cunningham, *Phys. Rev. Letters* **54,** 917 (1985)

15. C. Vettier, D. B. McWhan, E. M. Gyorgy, J. Kwo, and B. M. Buntschuh, B. W. Batterman (to be published)

16. C. F. Majkrzak, J. D. Axe, and P. Böni, *J. Appl. Phys.* **57,** 3657 (1985)

17. G. P. Felcher in *Dynamical Phenomena at Surfaces, Interfaces and Superlattices,* F. Nizzoli, K. H. Rieder, and R. F. Wills EDS (Springer-Verlag, Berlin 1984) P. 316

18. P. M. Platzman and N. Tzoar, *Phys. Rev.* **B2,** 3556 (1970)

19. F. DeBergevin and M. Brunel, *Acta Cryst.* **A37**, 314 (1981)

LAYERED MAGNETIC SUPERLATTICES

Ivan K. Schuller
Argonne National Laboratory, Argonne, Illinois 60439

ABSTRACT

The properties of non-lattice matched layered magnetic superlattices are strongly affected by the layering process. D. C. Magnetization measurements exhibit thin film effects and light scattering observation of the magnon spectra shows the existance of superlattice effects. Nuclear magnetic and ferromagnetic resonance experiments imply the existence of a variety of local fields depending on the distance from the interface.

INTRODUCTION

Metallic multilayers have been used for some time in applications relating to magnetic recording.[1] Magnetic multilayers are also useful in basic studies relating to dimensional[2] and thin film[3] effects. In addition, in some cases, it is possible to produce phenomena which depend on the periodic nature of the stack, i.e.,

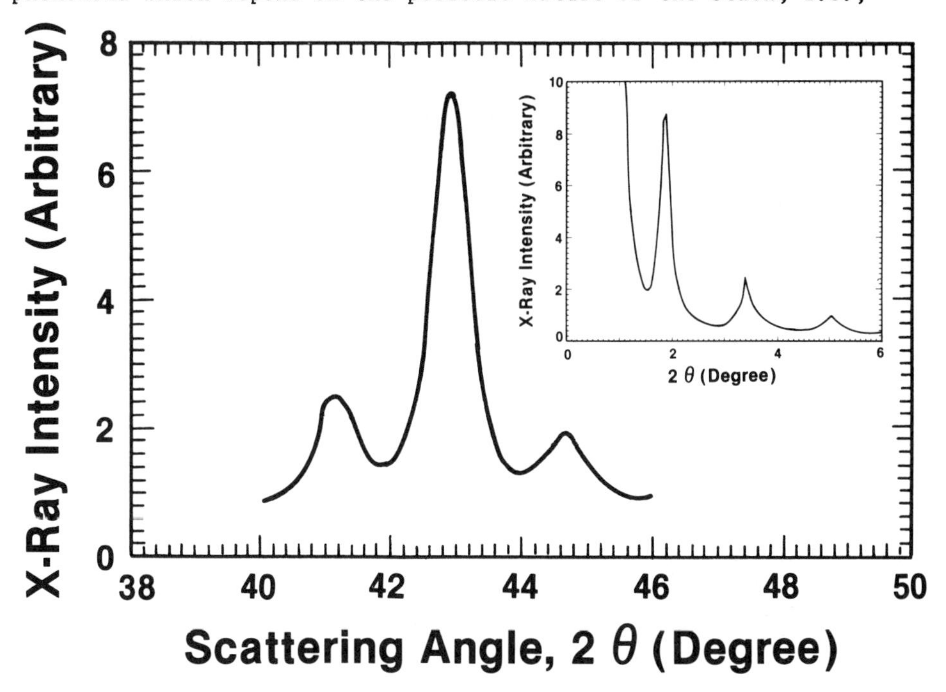

Fig. 1. X-ray diffraction from V/Ni superlattices at high and low (inset) angles.

0094-243X/86/1380093-5$3.00

superlattice effects.[4] Here we summarize our studies in a variety of magnetic superlattices that are formed from elements whose constituents do not form solid solutions in their binary phase diagram and that are not lattice matched.

STRUCTURE AND MAGNETIC PROPERTIES

The superlattices were prepared using a sputtering technique developed earlier.[5] In this method the multilayers are prepared by alternate sputtering of two elements on a heated, single crystal substrate (Al_2O_3, mica, Si, Ge, BaF, MgO, etc.). Structural studies of superlattices are of key importance for the understanding of their physical properties. The importance of the structural studies is emphasized by the fact that not all combinations of elements can be grown on layered form either because of interdiffusion or disorder. Moreover, the reason for layered growth has not been established although some qualitative ideas have been advanced.[6,7]

Figure 1 shows the results of an X-ray diffraction study of V/Ni superlattices.[8,9] The inset shows "small angle" diffraction measurements which directly give the amplitudes of the Fourier transform of the composition modulation. These small angle measurements however give no information on the crystallinity of the layers. Information about the crystallinity can be obtained from scattering at high angles. The existence of the various peaks in this measurement can be related directly to the growth of crystalline layers.[10] In particular, for the bcc/fcc superlattices (V/Ni and Mo/Ni) studied to date the structure consists of bcc(110) planes stacked on fcc(111) planes with random orientation in the plane of the film. Recently, a microcleavage technique has been applied to directly image the layers in transmission electron microscopy measurements.[11]

The magnetic properties are strongly influenced by changes in the layer thicknesses of the magnetic or normal metal constituents.[12] Figure 2 shows the low temperature (~ 5°K) saturation magnetization of Ni/Mo superlattices using a SQUID magnetometer. These results imply that the decrease in the saturation magnetization is mostly due to a decrease in the Ni thickness although there is some indication of coupling across the normal metal (Mo).

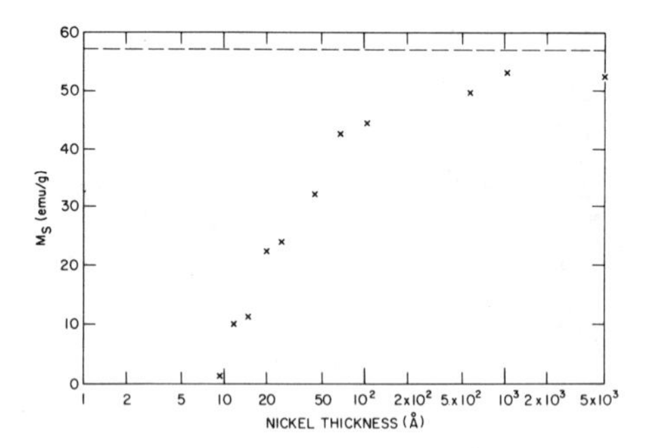

Fig. 2. Saturation magnetization versus nickel thickness of Mo/Ni superlattices.

The Cuire temperature temperature, measured from Arrot plots shows a similar behavior i.e., a strong decrease with Ni layer thickness (Fig. 3). Whether this decrease is due to the existence of dead Ni layers or intermixing at the Ni-Mo interface has not been clarified and is still under study. In any case, it seems that D.C. magnetization changes are mostly due to thin film effects i.e., the changes in the Ni thickness.

The magnons in superlattices exhibit a behavior which is characteristic of the superlattice periodicity. The magnons characteristic of the individual magnetic layers are modified as the layers are coupled via the dipolar interaction across the normal metal Theoretical predictions[13,14] have shown the existence of type of modes: a) a band of modes arising from the interaction of surface-like magnons in each magnetic layer and b) modes arising from the interaction of standing spin waves in each layer. The dependence of the frequency of these modes (ν) on magnetic (d_M) and nonmagnetic (d_N) layer thickness, magnetic field (H), saturation magnetization (M_s) and wave vector (Q) has been theoretically predicted.[13,14] Light scattering measurements[4] have been performed to study experimentally the dependence of ν on d_N, d_M, H, M_s, and Q. Figure 4 shows the dependence of the magnon frequency on magnetic field for a representative set of Mo/Ni superlattices. The solid lines are fits to the theory using the saturation magnetization as the only adjustable parameter. Independent D.C. magnetization[12] measurements of the saturation magnetization were found to be in close agreement (± 15%) with the fits to the light scattering measurements.

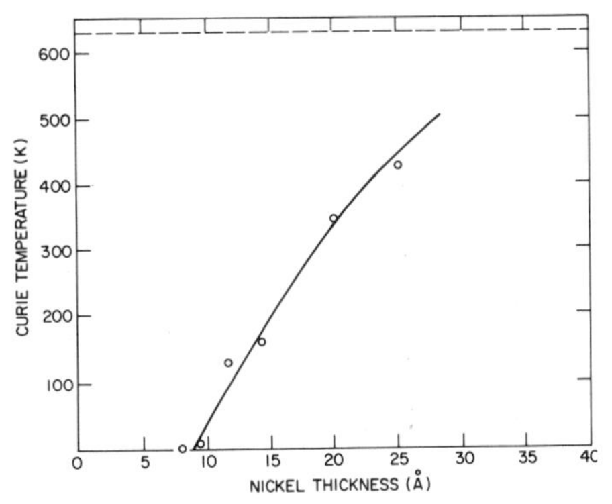

Fig. 3. Curie temperature versus nickel thickness of Mo/Ni superlattices.

The magnetic modulation of Mo/Ni superlattices has been studied using neutron scattering.[15] The results imply that the magnetic modulation is different from the chemical modulation indicating the existence of long range magnetic interactions parallel to the layers. Nuclear magnetic resonance[16] and ferromagnetic resonance[17] measurements show the existence of various local magnetic environments possibly related to the distance from the interface.

In summary, we found that the magnetic properties of metallic superlattices are strongly affected by the layering process. Both thin film and superlattice effects are observable in magnetization, light scattering and/or resonance experiments.

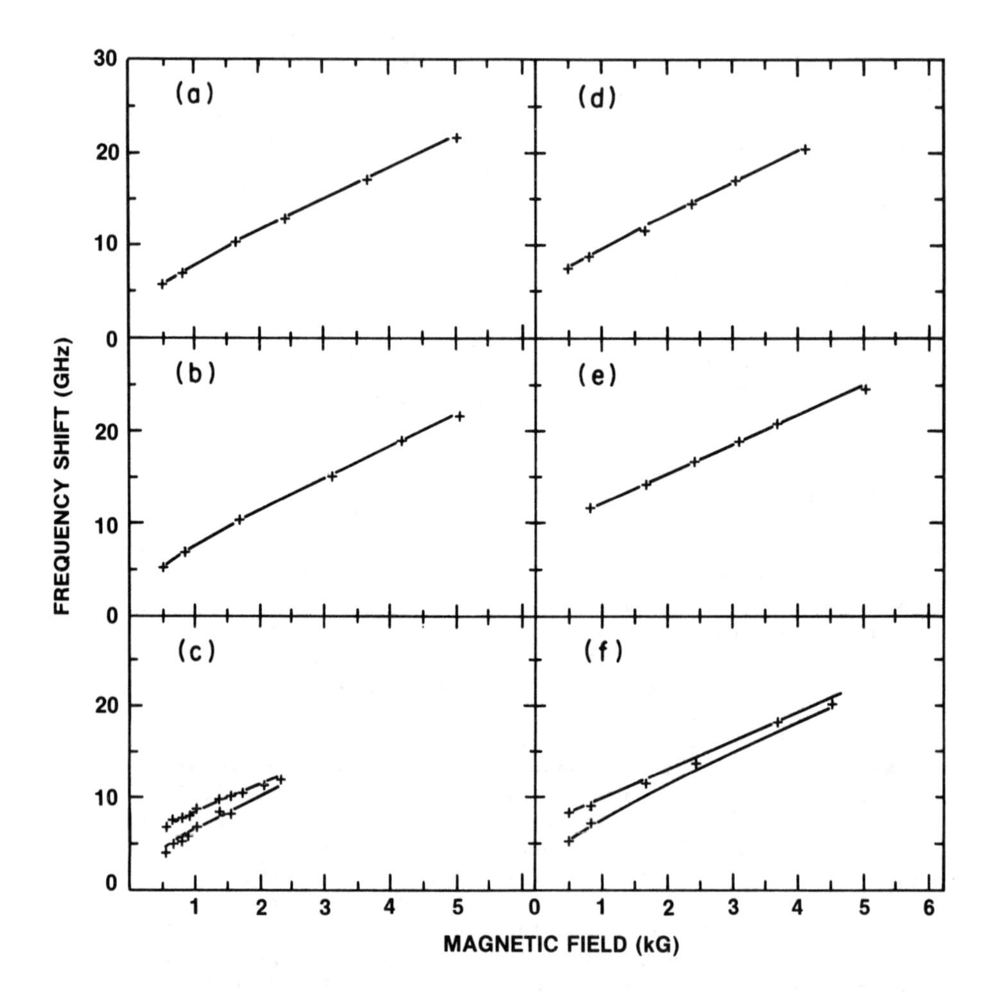

Fig. 4. Field dependence of magnon frequencies in a representative set of Mo/Ni superlattices. The full lines are fits to theory with the saturation magnetization as an adjustable parameter. The samples are as follows: a) d_1 = 100 Å, d_2 = 300 Å; b) d_1 = 100 Å, d_2 = 100 Å; c) d_1 = 138 Å, d_2 = 46 Å; d) d_1 = 250 Å, d_2 = 750 Å; e) d_1 = 5000 Å, d_2 = 5000 Å; f) d_1 = 540 Å, d_2 = 180 Å.

ACKNOWLEDGMENTS

I would like to thank my collaborators J. Cable, G. Felcher, W. Halperin, H. Homma, M. Khan, M. Grimsditch, A. Kueny, M. Pechan, M. Salamon, and M. Yudkowksy in related work. Work supported by ONR Contract No. N000-14-83-F-0031 and the U.S. Department of Energy.

REFERENCES

1. R. M. White, Introduction to Magnetic Recording, IEEE Press, NY, 1985.
2. H. K. Wong, H. W. Yang, B. Y. Jin, Y. H. Zhen, W. Z. Cao, J. B. Ketterson, and J. E. Hilliard, J. Appl. Phys. 55, 2494 (1984).
3. E. M. Gyorgy, J. F. Dillon, D. B. McWhan, L. W. Rupp Jr., L. R. Testardi, and P. J. Flanders, Phys. Rev. Letters 45, 57 (1980).
4. M. Grimsditch, M. Khan, A. Kueny, and I. K. Schuller, Phys. Rev. Letters 51, 498 (1983).
5. I. K. Schuller and C. M. Falco, AIP Conf. Proc. 58, 197 (1979).
6. C. M. Falco, Jour. Phys. 45, C5-499 (1984).
7. R. Ramirez, A. Rahman, and I. K. Schuller, Phys. Rev. B30, 6208 (1984).
8. H. Homma, C. Chun, G.-G. Zheng, and I. K. Schuller (to be published).
9. H. Homma, Y. Lepetre, J. Murduck, I. K. Schuller, and C. F. Majkrzak, SPIE. Proc., Vol. 563 (in press).
10. I. K. Schuller, Phys. Rev. Letters 44, 1597 (1980).
11. Y. Lepetre, I. K. Schuller, G. Rasigni, R. Rivoira, R. Philip and P. Dhez, SPIE Proc., Vol 563 (in press).
12. M. Khan, P. Roach, I. K. Schuller, Thin Solid Films 122, 183 (1985).
13. R. E. Camley, T. S. Rahman, and D. L. Mills, Phys. Rev. B27, 261 (1982).
14. P. Grünberg and K. Mika, Phys. Rev. B 27, 2955 (1983).
15. J. Cable, M. R. Khan, G. P. Felcher, and I. K. Schuller (to be published).
16. M. Yudkowsky, W. P. Halperin, and I. K. Schuller, Phys. Rev. B 31, 1637 (1985).
17. M. J. Pechan, M. B. Salamon and I. K. Schuller, Jour. Appl. Phys. 57, 2543 (1985).

CHAPTER III

STATUS OF U.S. AND JAPANESE SILICON TECHNOLOGY

ADVANCES IN SUBMICRON PROCESS TECHNOLOGIES

A. T. Lowe
Motorola Inc., Phoenix, Az. 85008

ABSTRACT

The present state of process development for submicron CMOS devices will be discussed with emphasis on the challenges common to the scaling of CMOS and bipolar technologies to 0.5 micron dimensions. Significant improvements in circuit density and speed have been achieved as a result of advances in fabrication processes such as direct write e-beam lithography and new metallization techniques. However, at these dimensions, common problems such as substrate auto doping, device isolation, planarization and metal step coverage must be solved.

INTRODUCTION

The introduction of faster, more dense integrated circuits has continued unabated. This is best illustrated by the intensity of the memory market where 256k bit DRAM's prices have eroded by a factor of ten in 1985 and1 Megabit memory chips will soon be introduced by both U.S and Japanese companies. The scaling of devices to obtain ultra dense memories such as the16 Megabit RAM is now in the developmental stages. In contrast to previous generations, where simple scaling factors and old processing techniques often applied, the techniques in fabricating such dense memory chips represent significant changes in process technologies.

To be described here are the advances made in processing submicron geometries and the problems that have yet to be addressed . With present processing capable of fabricating a 1 Megabit memory chip, the emphasis will be on the uniqueness of fabricating a 4 Megabit chip and beyond. The discussion will specifically address CMOS device fabrication, however, many of the challenges in processing such as materials, device isolation and metal interconnects are common to both bipolar and MOS device fabrication.

The submicron regime is defined by the smallest geometry which is usually the gate for MOS devices and the emitter for bipolar devices. For fabrication of a 1, 4, 16 Megabit RAM's, the smallest geometries are 1 to 1.5, 0.75, and 0.50 microns, respectively.

Scaling Issues Lithography drives the process technology in submicron device scaling. There are, however, severe reliability issues near the 0.5 micron level that must be resolved before further scaling of CMOS devices can continue[1]. The first of these short channel effects is the hot carriers which produce substrate current and inject electrons into the gate oxide. Long term effects such as threshold voltage shift due to charge build up in the oxide are measured by accelerated testing which stresses the devices at higher voltages. The effect is cumulative and a figure of merit is to keep the substrate current in the range of 0.1 to 1.0 μamps/micron in order to limit a no greater than 50 millivolt shift over ten years[2]. There are numerous ways to minimize hot carrier injection and the most promising are lightly doped or graded drain structures[3].

A second reliability issue is alpha particle induced soft errors which becomes more severe the denser the structure. This is caused by an alpha particle inducing an electron hole pair at a storage node. It has been shown that lightly doped epi over heavily doped substrate dissipates charge and minimizes this effect[1]. A third reliability issue that is specific to CMOS is latch-up and again is more serious the tighter the spacing between NMOS and PMOS devices. This is minimized with a thin epi structure over a heavily doped substrate as described above and by various other doping schemes which minimizes the beta of the vertical bipolar transistor in the CMOS SCR structure[4].

Submicron Process Flow All submicron process flows can be identified by three main characteristics: low temperatures, low resistances and planarity. The low temperature requirement is best described as short diffusion lengths and involves either low temperatures for long times or high temperatures for short times. The time/temperature budget becomes tighter the more the scaling. This is due to the need to minimize lateral and vertical diffusion of dopants not only during activation, but throughout the process. If a boron doped epi is used, diffusion and auto doping can occur at any process step above 800° C. With junction depths on the order of 0.1 to 0.2 micron, and with precise control of well or sidewall dopant profiles needed, any exposures to temperatures above 900° C must be for very short times or eliminated entirely. Therefore, processes such as glass reflow, typically at temperatures of 1000° C or more, must be avoided. The uses of rapid annealing techniques[5,6] (flash lamp, e-beam, etc.) and low temperature depositions (low pressure, plasma enhanced[7],etc.) are a necessity in the submicron flow.

The second essential element in the submicron regime is low resistance. The resistance of submicron lines and contact openings causes significant reduction in circuit speed. The most effective way of lowering this resistance is with salicides, such as PtSi, TiSi, or WSi [8,9,10]. This is a selectively grown or deposited and self-aligned process on the gate and in the source/drain region. Salicides reduce the sheet resistance of the polysilicon runs and contact resistance without increasing the doping level in the source/drain. This is consistent with the need to minimize hot carriers with a lightly doped drain structure. It also reduces the resistance (source series resistance) between the edge of the contact and the gate.

A parameter which becomes critical when fabricating submicron geometries is planarity of the structure. Improved planarity or semi-planarity is essential for two reasons. The first is the need for resist planarity when exposing either optically or with direct write e-beam. Resist thins over edges and thickens in corners making any step undesirable.

Also, change in topography increases the difficulty in focusing the exposure tool. This is solved with multilevel resists[11,12] such as a trilevel structure of thick planarizing resist under a hard mask under a thin exposure resist or a bilevel structure which is a thick resist under a thin resist. A second planarizing step is required prior to metal etch. The etch of metal over steps leaves picket fences which results in shorting problems.

The undesirable feature of planarization at any step is the problem of registering to a planarized alignment key. The solution is an extra clear out step to expose that key.

In addition to the above characteristics, the submicron flow utilizes processing that has become fairly standard in the industry such as an all ion implanted dopant process (no diffusions) and all reactive ion etched to obtain the straight side walls necessary for small geometries. A generic submicron CMOS process flow is shown in fig. 1 and a cross section of the completed structure is illustrated schematically in fig. 2.

SUBMICRON LITHOGRAPHY

All fabrication of scaled devices is keyed off the development of the patterning techniques. Advances in obtaining fine line geometries with various lithography techniques have been significant[13]. Due in part to the advances in its use as a mask maker, the use of direct write e-beam lithography is maturing as a manufacturing process. At the present time, however, applications are limited due to low throughput and large capitol expense (over 3 million dollars each). Therefore it is best suited for custom or semi-custom chip fabrication. Other techniques such as X-ray[14] and I-line[15] steppers will provide much higher throughputs in the future. Near term however, e-beam provides the best developmental tool for testing submicron processes. There are now commercial suppliers of direct write e-beam equipment and there has been significant advances in throughput and much improved overlay accuracy. Previous systems have a throughput of 1 to 2 wafers per hour and the new e-beam equipment, such as the AEBLE 150, has a throughput of 10 to 20 wafers per hour depending of the level written. Also, overlay accuracy is now within 0.1 to 0.15 microns. With the throughput low, the best approach has been hybrid technology which mixes optical exposures for gross geometries and e-beam exposures for the smallest geometries and only at critical (gate, etc.) levels[16].

The e-beam process is now less resist limited because of much improved resist sensitivities which increases throughput. These new resists have the contrast necessary to resolve submicron geometries, are capable of withstanding the harsh environments of reactive ion etch, and have the necessary process latitude for use in a manufacturing mode[17,18]. The new positive resists now have a sensitivity of 1E-5 $\mu C/cm^2$ (typically, 2 to 4 wafers/hour) and a contrast of 1.5 to 3. The new negative resists have a sensitivity of 1E-6 $\mu C/cm^2$ (typically,10 to 20 wafers/hour) and a contrast range of 0.8 to 2.5.

STARTING MATERIALS

As discussed earlier, it is advantageous to build submicron CMOS in a thin epitaxial silicon layer[19,20,21]. Either boron doped or arsenic or antimony doped epi over heavily doped substrate can be used in fabrication of CMOS. There is evidence of a higher number of defects in n/n+ than in p/p^+ material[22]. It also appears that the p/p^+ provides a better latch up free CMOS device structure[23]. However, the disadvantage of p/p^+

epi is that auto doping occurs during growth and subsequent processing steps[24]. This necessitates low temperature processing for p/p^+ material. Advances in low pressure, low temperature epi growth techniques have helped minimize this effect.

The quality of thin epi has improved with numerous techniques for gettering defects in the epi[25,26]. Temperature cycling for long periods of time precipitates oxygen defects in an antimony doped substrate which getters epi defects[27]. The levels of oxygen contamination that can be tolerated for submicron geometries remains a concern. Part of the problem is the difficulty in measuring oxygen levels in the heavily doped substrates now in use[28].

DEVICE ISOLATION

Fully scaled submicron designs require narrow isolation regions between devices. Numerous techniques have been demonstrated[29] that are compatible with these designs. Examples of such isolation schemes include SEG (selective epi growth) which is the selective growth of single crystal silicon over openings in an oxide layer[29,30] and SOI (silicon on insulator) which provides for total isolation between devices with a buried oxygen layer[31]. The formation of a deep trench which is then refilled with a dielectric results in an isolation scheme which provides minimal oxide encroachment and lends itself to planarization and latch up suppression[32,33]. Additionally, trenches provide a third dimension which can be utilized for the formation of side wall capacitors for memory devices. An example of such a poly filled deep trench is shown in fig. 3.

Trenches compatible with submicron geometries have been demonstrated on a small scale but the challenge remains in reproducibly etching and filling large numbers of trenches. Fills which coat conformally such as thermal oxide place severe requirements on the sidewall etch, while a less conformal coat such as CVD polysilicon allows for a slightly sidewall. Because of this, the fill of choice has been polysilicon.

GATE TECHNOLOGY

Using direct write e-beam, devices are being fabricated (with low throughput) with gate lengths of 0.5 micron and reasonable device characteristics have been observed with gate lengths down to 0.25 micron. An example of a 0.25 micron gate structure that has been e-beam written into a trilevel resist structure is shown in fig. 4.

Silicided gates are used to reduce the sheet resistance of the poly gate runs thus improving circuit speed. The choice at present, however, is to keep the poly runs short and having multiple layers of poly rather than use silicides. There is only a limited use of silicides in the present generation of products for a variety of reasons. An example of the types of problems encountered is the stresses in thicker TiSi films which result in cracking.

As mentioned, for submicron CMOS, siliciding the gates alone is not sufficient[8]. The use of a salicide is a necessity. This selective process with PtSi grown on both the gate and in the source/drain is shown in fig. 5.

METALLIZATION

The choices for interlevel dielectrics which isolate gate and metal interconnect and metal to metal interconnect are numerous[34] and at present there is no clear cut choice. To reactive ion etch the metal layer, improved planarity is a necessity and this planarity must be obtained with low temperature processing. This suggests the use of a spun-on material such as polyimide[35] or a spin-on glass[36], usually in combination with a deposited glass. An option which also shows promise for low temperature planarization is bias sputtered quartz[37].

After planarization, metal coverage of submicron contact openings is extremely difficult. With an aspect ratio (depth of contact hole to width of contact hole) of one or greater, standard physical deposition techniques do not provide sufficient step coverage. An aspect ratio of 1.5 is not uncommon for submicron geometries. The options recently developed include a metal plug such as CVD tungsten, or a bias sputtered metal[38,39].

As early as 1974[40], resputtering of Al by RF biasing of substrate was demonstrated. Recent studies have shown[41] that sputtered Al with this substrate bias will not only provide sufficient step coverage, but will fill contacts and actually planarize the metal. This was shown to occur when the resputtering rate of Al was above 50%. The fill has been successfully demonstrated for aspect ratios up to 3. This technique appears to have a potential technique for filling submicron vias. D.C. bias sputtered aluminum on an MRC 662 illustrates this planarizing ability in fig. 6.

Challenges which remain for metallization include the identification of the proper Al alloy system (Al/Si, Al/Ti, etc) and several problems associated with bias sputtering such as the characterization of radiation effects of RF biasing on submicron devices and the ability to bias refractories (TiW, etc.).

CONCLUSIONS

The lower limits for scaling MOS devices have been revised downward continually. At present, a manufacturable circuit built around a 0.25 micron gate appears to be a reality. Within a few years, 0.1 micron geometries may be a possibility. Limitations of scaling due to reliability problems such as hot carriers are solvable within the realm of the 0.25 micron regime. The pace for fabricating manufacturable products will be established by the lithography development. Significant progress is being made in direct e-beam lithography and X-ray steppers. Other process technologies such as isolation, gate, and metallization have also advanced to the point of enabling the fabrication of an entire submicron circuit.

ACKNOWLEDGEMENTS

The author would like to express his appreciation for useful discussions with the following individuals: C. Lund, L. D. Sikes, J. Smith, M. Hodel, and numerous other colleagues in the Process Technology Labs of Motorola Semiconductor Products Division.

REFERENCES

1 M. H. Woods, Proceedings of the Third International Symposium on VLSI Science & Technol. (The Electrochem. Soc., Inc., 1985), p. 41
2 E. Takeda and N. Suzuki, IEEE Electron Device Lett., Vol. EDL-4, 111 (1983)
3 S. Meguro, S. Ikeda, K. Nagasawa, A. Koike, T. Yasui, Y. Skai, and T. Hayashida, IEDM Tech. Dig., 59 (1984).
4 Y. Taur, W. H. Chang, and R.H. Dennard, IEDM Tech. Dig., 398 (1984).
5 S. R. Wilson, W. M. Paulson, R. B. Gregory, Solid State Technol., 185 (1985).
6 C. Russo, Proceedings of the Fifth International Conference on Ion Implant Equip. & Tech., (Vermont,1984), 298.
7 B. Gorowitz, T. B. Gorazyca, and R. J. Saia, Solid State Technol., 197 (1985).
8 A. L. Butler and D. J. Foster, IEEE Trans. Electron Devices, Vol. ED-32, 150 (1985).
9 M. E. Alpirin, T. C. Holloway, R. A. Haken, C. D. Gomeyer, R. V. Karnaugh, and W. D. Parmantie, IEEE Trans. Electron Devices, Vol. ED-32, 141 (1985)
10 T. Yachi and S. Suyama, J. Vac. Sci.,. Technology, B3(4), 992 (1985).
11 E. Reichmanis, G. Smolinsky, and C. W. Wilkins, Solid State Technol., 130 (1985).
12 P. Burggraaf, Semicond. Int., 88 (1985).
13 A. N. Broers, Solid State Technol., 119 (1985).
14 C. R. Fencil, G. P. Hughes, Proceedings of SPIE, Vol. 333,100 (1984).
15S. Lee, S. Grillo, and V. Miller, SPIE Optical Microlithography IV, Vol. 538, 17 (1985).
16 J. N. Helbert, P. A. Seese, A. J. Gonzales, and C. C. Walker, Optical Engineering, Vol. 22, 185 (1983).
17 S. Fok and G. H. Hong, Proceedings of the Kodak Microelectronics Seminar, 1983.
18 P. H. Singer, Semiconductor Int. 68 (1985).
19 D. J. Foster, IEE Colloquium on Design & Fab of Devices for Improved Reliability, Dig. No. 27, 1 (1985).
20G. J. Hu and R. H. Bruce, IEEE Electron Device Intl., Vol. EDL-5, 211 (1984).
21 B. L. Gingerich, J. M. Hermsen, J. C. Lee, and J. E. Schroeder, IEEE Trans. Mech. Sci., Vol. NS-31, 1332 (1984).
22F. Secco d'Aragona, J. W. Rose, and P. L. Fejes, Proceedings of the Third International Symposium on VLSI Science & Technol., Vol 85-5, 106 (1985).
23 Y. Taur, W. H. Chang, and R. H. Dennard, IEDM Tech. Digest, 398 (1984).
24 M. W. M. Graef, B. J. H. Leunissen, and H. H. C. de Moor, J. Electrochem.,
Soc., 1942 (1985).
25 J. O. Borland, M. Kuo, J. Shibley, B. Roberts, R. Schindler, T. Dalrymple, Proceedings of the 1984 Custom Integrated Circuits Conf., 176 (1984).

26 S. Kishino, Proceedings of the Third International Symposium on VLSI Science & Technol., Vol. 85-5, 399 (1985).
27 F. Secco d'Aragona, to be published.
28 F. Shimura, W. Dyson, J. W. Moody, and R. S. Hockett, Proceedings of the Third International VLSI Science & Technol., Vol. 85-5, 507 (1985).
29 J. O. Borland and C.I. Drowley, Solid State Technol., 141 (1985).
30 S. Nagao, K. Higashitani, Y. Akasaka, and H. Nakata, IEDM Tech. Dig., Sec. 26.6, 593 (1984).
31 H. Shichyo, Conf. Solid State Devices & Material, 265 (1984).
32 T. Yamaguchi, S. Morimoto, H. K. Park, and G. C. Eiden, IEEE Trans. Electron Devices, Vol. ED-32, 184 (1985).
33 H. P. Vyas, R. S. L. Lutze, and J. S. T. Huang, IEEE Trans. Electron Devices, Vol. ED-32, 926 (1985).
34 H. Okabayashi, 1984 Symposium on VLSI Technol., 20 (1984).
35 D. L. Bergeron, J. P. Kent, and K. E. Morrett, Annual Proceedings Reliability Physics, Vol. 22, 229 (1984).
36 S. L. Chang, K. Y. Tsao, M. A. Meneshian, and H. A. Waggener, Ext. Abs. Electrochem. Soc., Vol. 85-1, 323 (1985).
37 K. Eguchi, H. Sakurai, H. Harada, and T. Kashiwagi, J. Vac. Soc. Japan, Vol. 27, 759 (1984).
38 T. Mogami, M. Morimoto, and H. Okabayashi, Ext. Abs. of the 16th Conference on Solid State Devices and Material, 46 (1984).
39 J. F. Smith, Solid State Tech, 135 (1984).
40 J. L. Vossen, G. L. Schnable, and W Kern, J. Vac. Sci. & Technol., Vol. 11, No. 1, 60 (1974).
41 Y. Homma and S. Tsunekawa, J. Electrochem. Society, Vol. 132, No. 6, 1466 (1985).

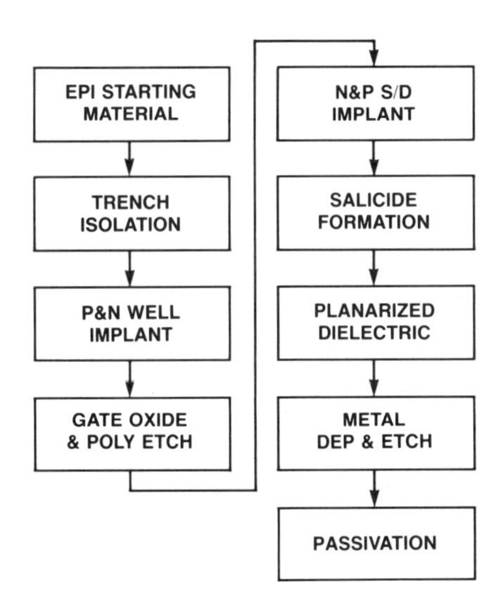

FIG. 1 CMOS PROCESS FLOW

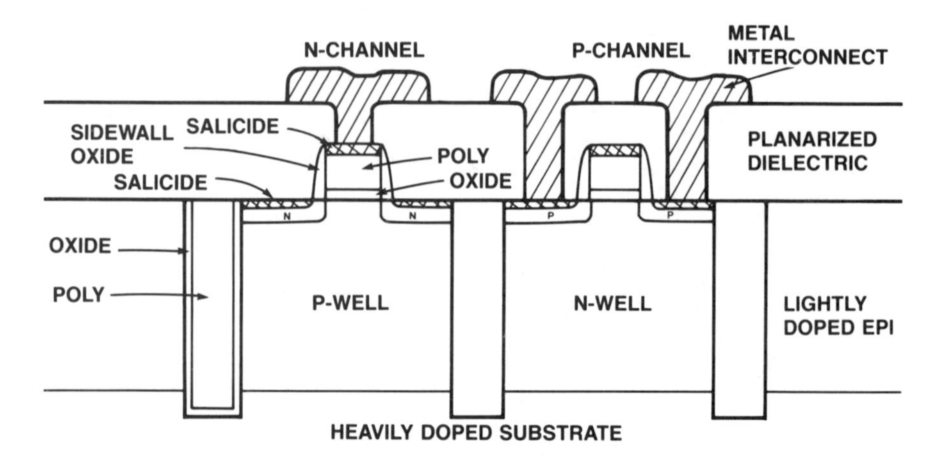

FIG. 2 SUBMICRON CMOS CROSS SECTION

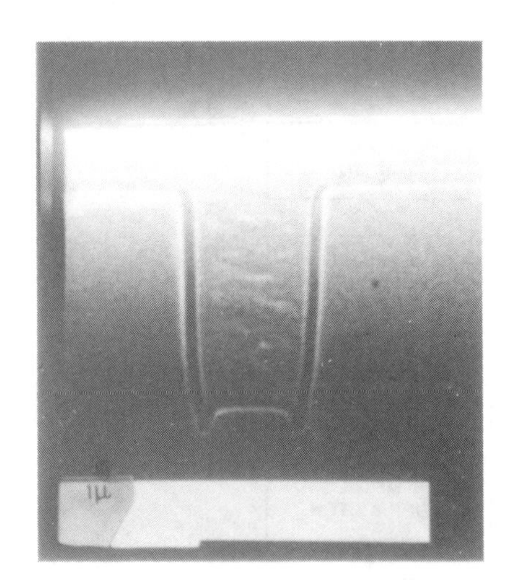

FIG. 3 POLY REFILLED TRENCH

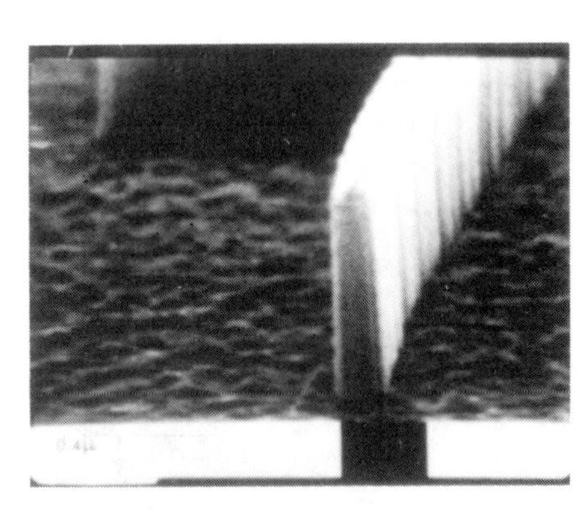

FIG. 4 0.25 MICRON TRILEVEL STRUCTURE

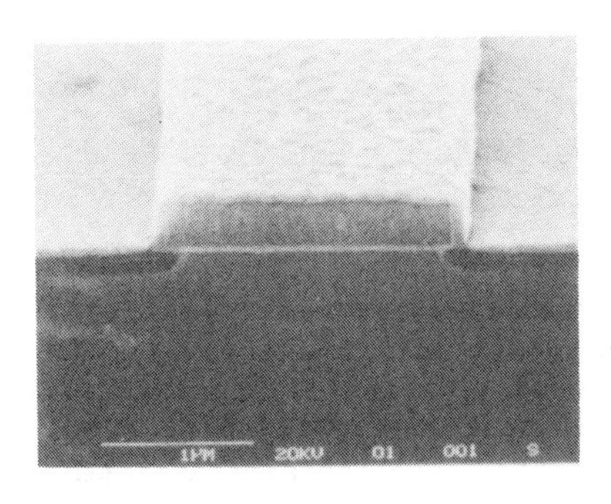

FIG. 5 PtSi GATE & SOURCE/DRAIN

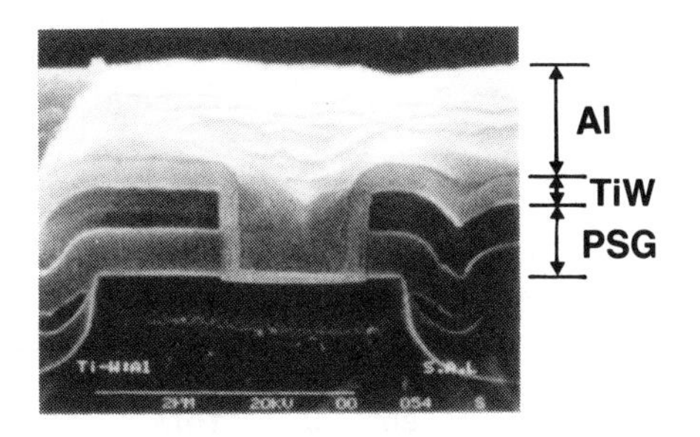

FIG. 6 MRC FOCUSED CATHODE/D.C. BIAS SUBSTRATE

DEVICE AND MATERIAL REQUIREMENTS IN VERY LARGE SCALE INTEGRATED CIRCUITS

Shojiro Asai
Hitachi Central Research Laboratory,
Kokubunji, Tokyo 185, Japan

ABSTRACT

An attempt is made to relate increasing performance/cost of VLSI chips to what it imposes on the constituent devices and materials. The requirements are discussed in terms of physical properties as quantitatively as possible. This highlights problem areas where further R and D efforts are needed for the next several years. Possible approaches to solve the problems are discussed and compared where alternatives exist.

INTRODUCTION

Progress in silicon integrated circuits over the past decade is undoubtedly among the most remarkable in pace in the history of human technological achievements. Integration of increased number of components on a chip, as shown in Fig.1, has made the systems built with ICs more reliable, less costly, and faster. As a result, the data processing speed of the fastest computers has increased by a factor of 10 over the decade. The performance of smaller systems like micro-computers has improved even more rapidly, since their architecture is centered on the design and use of the fastest-evolving parts--VLSIs.

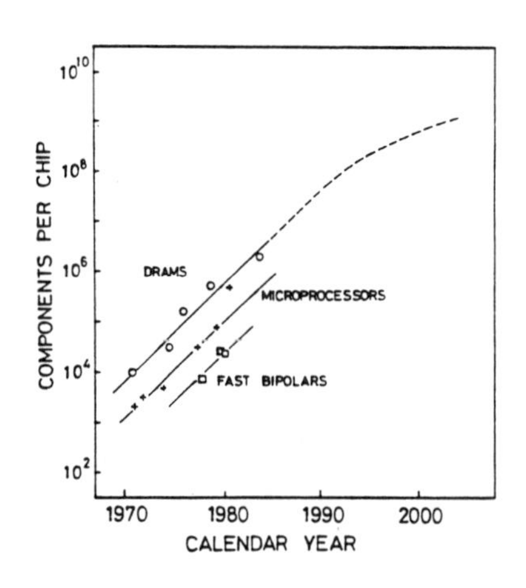

Fig.1 Number of components per chip versus calendar year.

At the present time, micro-processors with several hundreds of thousands of transistors are on the market. 256Kbit MOS dynamic random access memories(DRAMs) with more than a half million devices are being mass-manufactured. 1 Mbit DRAMs with four times more components are just around the corner. Technology employed in fabricating these devices is highly sophisticated. Photolithography with resolutions down to 1 micron, polycide gate, 20nm-thick gate oxide are now industry standard. Most high-density LSI circuits are being built with CMOS FETs with carefully tailored impurity distribution to maintain 5V operation. Double-level metal interconnects are being used more extensively.

The level of process complexity of the present-day VLSIs has become quite high over the past 15 years in terms of linewidths, structure, mask count, variety of materials used, and types of equipment employed.

The evolution in process technology reflects the needs for higher density and lower power-delay product. In the next several years, further development in materials and device structures is needed to realize VLSIs with multi-million transistors in submicron dimensions[1]. The requirements for devices and materials here will still be based on density and/or speed-power consideration[2]. Another, rapidly growing, concern is in the reliability-related aspects of VLSIs[3]. Alpha-particle-induced soft-errors, hot-carrier or electrical-stress-related wearouts, and electromigration are cases in point. These problems, which always pose practical limits to integration, are of paramount importance, and call for extensive and intensive investigation. We shall address these problem areas and specify relevant physical properties of devices and materials. In the subsequent sections, implications of density- and speed-power-related requirements will be discussed in terms of device and material properties. Another section to follow deals with reliability-related requirements. The last section will discuss basic research required for the understanding of the crucial issues, and tight control in manufacture to contain the problems within manageable tolerances.

DENSITY REQUIREMENTS FOR VLSIS

It is the high-packing density of components on a chip that has made VLSIs the center of electronic innovation, making it possible to provide data storage and processing at constantly decreasing costs. This is most typically exemplified by the size of memory cells which has been reduced greatly over the past 15 years[4], as shown in Fig.2.

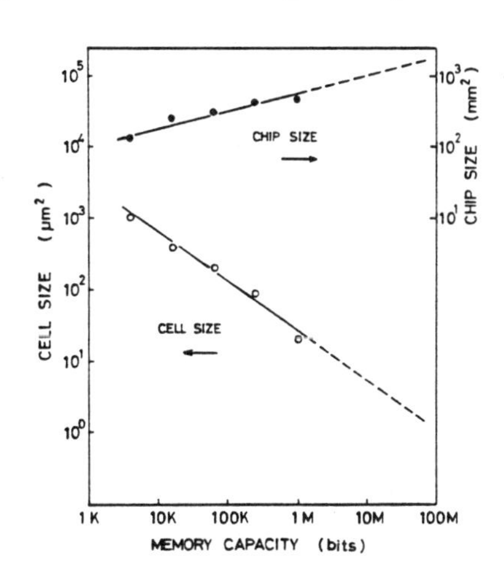

Fig.2 DRAM cell size and chip size.

Of vital importance in reducing the feature size of VLSIs is patterning technology, which consists of lithography and etching. Mask printing machines have evolved from contact printers, 1X projection printers and then to 5X and 10X reduction printers. Resolution capability down to 1μm is now achieved routinely. Optical printing of 0.5μm feature sizes is being seriously discussed, but will require innovations both in process technology and equipment. The latter includes the use of shorter wavelengths of light: from the Hg g-line (436nm) to i-line (346nm) and even the wavelengths of excimer lasers (249 to 190nm) and the relevant improvement in optics. Process technology improvement includes the use of tri-level[5] or double-level resist and contrast-enhancement layer[6] materials. The theoretical limit for optical resolution is given by:

$$l = \alpha\lambda / NA \qquad (1)$$

where λ is the wavelength and NA is the numerical aperture of the lens. The use of the techniques mentioned above improves α from Lord Rayleigh's value of 0.61 by approximately 10-15 %, which is significant in the submicron range.

As the linewidths are decreased, the thickness of MOS gate oxide has been decreased proportionately. In the ideal, constant-electric-field scaling, the supply voltage is to be reduced by the same factor. However, from the ease-of-use consideration in the established systems environment, the 5V supply voltage has been used for digital VLSIs for the last three generations, 3μm, 2μm and 1.3μm. This has resulted in the increase of electric field in the silicon as well as in the oxide. The figure-of-merit of the gate dielectric in high-density MIS structures is the dielectric strength, Emax, where Emax is the breakdown voltage, as compared in Fig. 3. Growth of thin SiO_2 films with improved quality is thus a subject of great concern. Materials with dielectric strength higher than SiO_2 are being extensively looked for. These include SiON, silicon oxynitride and Ta_2O_5. Thin films of Ta_2O_5 sputtered onto Si have proven integrity which is even superior to SiO_2[7]. These films will be useful for capacitances in memory devices with enhanced alpha-particle immunity, as well as for transistor gate dielectrics.

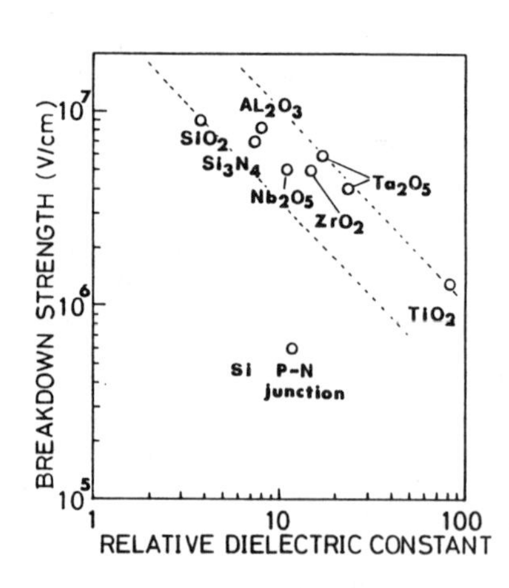

Fig.3 Relation between dielectric constant and breakdown field of various dielectrics.

The principle of scaling meets a difficulty in the interconnects; it is not practical to reduce the thicknesses of Al or poly-Si wiring and inter-layer dielectrics. Consequently, the aspect ratio or ratio of the thickness of the layer to the linewidth has kept increasing. For the 1.3μm-technology, the aspect ratio is close to 1. It will reach even 2 or 3 in the submicron technology. What is needed here is an etching technique with homogeneous, selective, aspect-ratio-independent, and high etch-rates. The interaction between gases, ions, and substrate surfaces is being studied intensively to build machines and find gas materials which meet the requirements[8]. Aside from the linewidth reduction, what contributes most to VLSI density innovation is the use of 3-dimensional structures. The first, but firm step was taken when a double-level poly-Si gate was used to achieve a high-density dynamic RAM cell, where the second-level poly was heavily doped and used as a transfer gate. Soon after that, a static RAM cell using a lightly doped, second-level poly-Si as the flip-flop load was proposed, which now is an industry standard. The stacked capacitor cell[9] and corrugated capacitor cell[10] are both 3-dimensional structures proposed for megabit dynamic RAMs where the density is of utmost importance. With the advent of technology which can deal with film deposition, etching and doping in three dimensions, more and more use will be made of three-dimensional structures.

SPEED-POWER REQUIREMENTS FOR VLSIS

The delay time in an LSIs is expressed as[11],

$$t_d = t_{do} + \Delta t_d C_{in} f_o + \Delta t_d C_w, \qquad (2)$$

where t_{do} is the intrinsic delay of the circuit with no loading capacitances, C_{in} the gate input capacitance, f_o the fanout, and C_w the wiring capacitance. Contributions of the three terms in (2) are of the same order. In LSIs containing a large number of transistors, one has to be particularly careful about the wiring delay. Since C_w is directly proportional to the wire length,

$$C_w \propto l_w \propto A^{1/2}, \qquad (3)$$

reducing the chip area A by means of multi-level wiring is important for the reduction of delay time also. How effective adding more levels of wiring is can be quantitatively evaluated in the following way. The total length of wiring required to provide interconnection between logic gates has been empirically shown to be[12],

$$l_T = (4/3)(A/B)^{1/2}(B^{1/6} - 1)N_p, \qquad (4)$$

where B is the number of blocks into which logic functions are modularlized, and N_p is the total number of signal pins these blocks have. Both B and N_p are proportional to the total number of gates, G, on the chip. With n levels of wiring,

$$nA = Kp^2G + 2l_Tp, \qquad (5)$$

where p is the pitch of wiring. Empirical assumptions are made here that a gate occupies an area of Kp^2 and that 50% of wiring is utilized. Coupled equations (4) and (5) are easily solved to yield A as a function of G. The results are shown in Fig.4 with n as a parameter. If the total floor space, nA, is dominated by wiring and the number of gates is extremely large, the solution reduces to,

$$A^{1/2} \simeq pG^{2/3}/n \qquad (6)$$

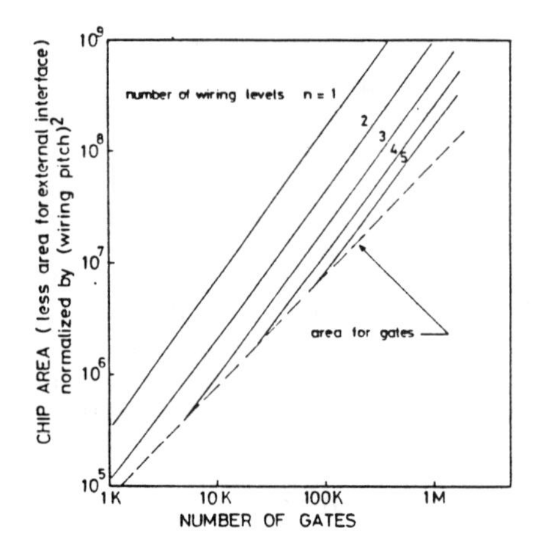

Fig.4 VLSI chip area as a function of number of gates.

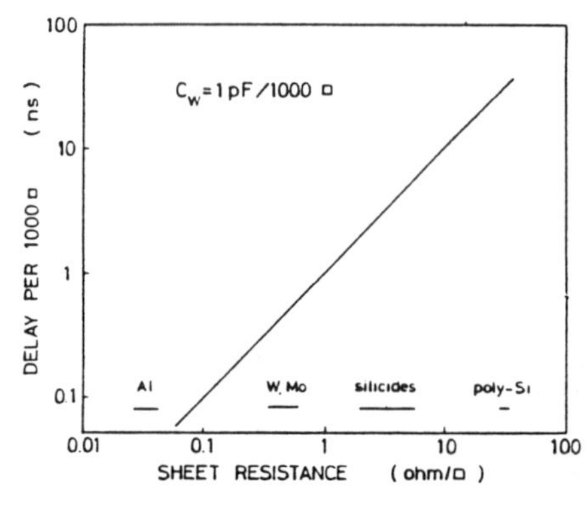

Fig.5 RC delay time for various MOS gate materials.

which is a pertinent representation of the Rent's rule. Equation (6) implies that the wiring delay, $\Delta t_d C_w$, is inversely proportional to the number of wiring levels. This will justify up to at least 4 levels of interconnects in some applications. Multi-level wiring is where various new materials and new processes are called for. Inter-level insulators, both inorganic and organic[13], and interconnect metals, both aluminum-based and refractory[14], are being extensively investigated. Damage-free, and self-planarizing processes are badly needed. Bias-sputtering of SiO_2[15] and Al[16] has been reported as a self-planarizing process. Spin-on polyimides[13] and selective metal CVD[17] are attractive from damage considerations.

In many VLSI memory cell arrays, the switching MOS gate is directly used as the word line. In this case, the RC delay due to the gate resistance can make a significant contribution to the access time. Doped poly-Si gate has a rather high sheet resistance of about 30Ω/□. If we assume a floating capacitance of 1pF/1000□, the line delay can amount to 30ns/1000□, as shown in Fig.5. This can be very serious for the design of a memory VLSI with an access time below 100ns. Polycide gates[18], or refractory metal silicides (WSi_2, $MoSi_2$, $TiSi_2$ and $TaSi_2$) on top of poly-Si, have proven to be viable with a sheet resistance of 2-5Ω/□, suitable for high-density dynamic memories. The next step of lowering the gate resistance is targeted at below 0.5Ω/□, which is the realm of pure refractory metals, W or Mo. Deposition by sputtering, as well as by non-selective CVD of refractory metals is being studied. Wet-hydrogen annealing[19], which results in oxidation of Si but reduction of W, is a very promising technique for realizing pure W gate.

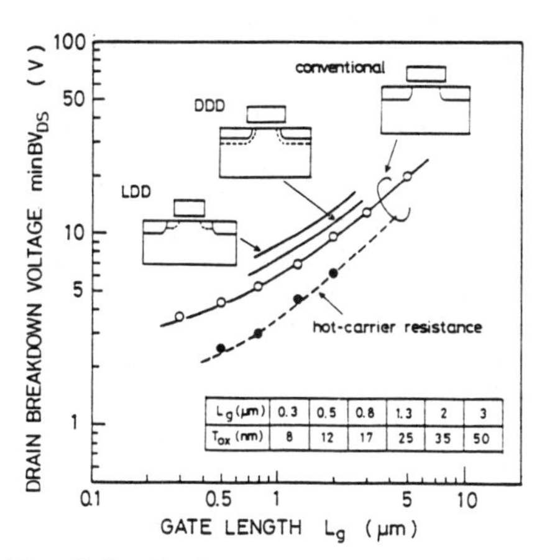

Fig.6 Drain breakdown voltage and hot-carrier resistance versus MOS gate length.

The speed-power characteristics of LSIs depend on the power supply voltage, which is limited by the breakdown voltage of transistors. The breakdown voltage, $minBV_{DS}$, for MOSFETs is plotted in Fig.6 against gate length. The gate oxide thickness is properly scaled in this experiment. As-P double-diffused drain (DDD) and lightly doped drain (LDD) have been proposed to lower the electric field near the drain which gives rise to avalanche multiplication. However, the use of 5V power supply looks almost prohibitive for submicron MOSFETs even when sophisticated structures like LDD[20] are used. Scaling the dimensions by a factor of $1/\lambda$ and the supply voltage by $1/\alpha$ will change the power dissipation by λ/α^3, and the intrinsic gate delay will be decreased by α/λ^2. Since the inverse load-driving capability Δt_d in (2) is

$$\Delta t_d = 1/g + R_W , \tag{7}$$

where R_W is the wiring resistance and g is the conductance of the circuit driver,

$$g = I/V \propto \lambda/\alpha . \tag{8}$$

Therefore, a reduction in supply voltage could cancel the effect of decreased sizes on the speed. Moreover, in eq.(7), the second term can become relatively large in the scaled-down device, since R_W increases as a result of scaling. This causes the delay time of a circuit to depend on the length of wiring it contains, which gives rise to a racing hazard. It is therefore very important to maintain the breakdown voltage and conductance of the device as high as possible. MOSFET structures, both n- and p-channel, need further study in this regard.

While MOS is superior to BIPOLAR in packing density, the opposite is true in speed, or load-driving capability. In fact, a typical bipolar transistor having an emitter size $L_E \times W_E$ of $1 \times 1.5\mu m^2$ has a conductance, 40mS at 1mA, more than two orders of magnitude higher than that of a MOS, 0.1mS, having the same gate area Lg x Wg of $1 \times 1.5\mu m^2$. Although a bipolar transconductance,

$$g = S_E J/kT \propto 1/\lambda^2 \tag{9}$$

decrease on the second power of linear dimensions, BIPOLAR's edge on MOS in drivability will not diminish down to below 0.1μm. High-speed bipolar transistors, therefore, still remains to be important research subjects. Various self-aligned structures to minimize parasitics as well as mask counts have been proposed. ECL ring oscillators with delay times per gate below 100ps have been reported[21,22]. Such devices will enable high-speed memories with an access time near 1ns, which compares well with GaAs memories. Another derivative from the above discussion is that BICMOS, a combination of BIPOLAR and CMOS circuits built on a chip, is a very attractive approach toward high speed and high density. Especially, genuine-BIPOLAR-based BICMOS, not the free bipolar that comes with CMOS, will open up application in an area which has been hard to reach with either pure ECL or pure CMOS.

RELIABILITY REQUIREMENTS FOR VLSIS

All the topics discussed in previous sections are more or less related to the reliability of VLSIs, which is discussed in this section. Dielectric robustness is studied in long-term stress tests such as constant-voltage TDDB(time dependent dielectric breakdown) or constant current-stress tests. The degradation mechanism of SiO_2 has not yet been well identified, but discussed in terms of injected carriers, avalanche multiplication, and trapping. The time to failure, t_{BD}, has been related to the total injected charge, Q_{BD}[23],

$$Q_{BD} = \int_0^{t_{BD}} J\ dt = Jt_{BD}. \qquad (10)$$

The lifetime of oxide has been related to the electric field E in the oxide as[23]

$$t_{BD} \propto \exp(+B/E). \qquad (11)$$

This indicates how using thin oxides under increased electric field is compromised with device reliability.

Degradation in MOSFET conductance and threshold voltage caused by hot-carrier injection puts a practical limit to VLSI operating voltage. The hot-carrier lifetime is also determined by a similar equation,

$$Q_{HC} = \int_0^{t_{HC}} J_G dt = J_G t_{HC}, \qquad (12)$$

where J_G is the injected current. The relation between the t_{HC} and the operating voltage has been obtained as[24]

$$t_{HC} \propto \exp(+C/V). \qquad (13)$$

Both equations (11) and (13) reflects the mechanisms of current injection into the oxide, Fowler-Nordheim tunneling and energetic hot-carrier emission. Oxide degradation should be studied more intensively to identify the traps involved in degradation mechanisms. A closer examination of the hot-carrier-induced MOSFET degradation will then be possible by correlating the local build-up of trapped charge near the drain with better understood charge build-up in a MOS diode.

Electromigration of materials along the paths of current in VLSIs causes formation of metal voids, eventually resulting in open circuits. The electromigration lifetime of a metal line is given by[3],

$$t_{EM} = AJ^{-n} \exp(D/kT), \qquad (14)$$

where A is a material- and geometry-dependent constant, n is an empirical exponent between 2 and 3, and J is the current density. It is readily recognized that electromigration is aggravated by the use of smaller linewidth, which is a must for higher packing density. Standard MOS VLSIs with Al-Si (Al with a few % Si) interconnects are designed with a maximum allowable current density which is set around 1×10^{5} A/cm^{2}. Al-Cu-Si (Al with a few % Cu and Si) has been used for BIPOLAR LSIs, where current density is higher, but will be used for future MOSVLSIs also. Sandwiches of Al with refractory metals, Al/X/Al, where X is W, Ta or Ti, has an electromigration lifetime more than a hundred times longer than Al-Si. Use of Al-Cu alloys and Al-refractory sandwiches also alleviates the stress-induced open failures, which is another reliability problem in wiring. A multi-level structure favorable from reliability viewpoints is obviously like the one shown in Fig. 7(b), in contrast to the conventional structure shown in Fig. 7(a). It consists of intermetal and metal levels, each being perfectly planarized. Bias-sputtered SiO_2 and Al may not be completely free from plasma-induced damage, but are self-planarizing. Polyimide provides an intermetal layer, self-planarizing, damage-free, and sufficiently moisture-resistant. The increase in the variety of materials used in VLSI structures, however, can be a potential source of reliability problems. Figure 8 shows where some of these problems can arise. Reaction, interdiffusion, and precipitation can take place at the interfaces between materials.

Alpha-particle-induced upset in digital semiconductor circuits is a soft error, or a temporary failure[25]. It can happen to any semiconductor circuit when noise charge due to ionization by an alpha-particle is collected in excess of the signal charge at the node. There are two, device-oriented and material-oriented, ways of coping with this problem. Three-dimensional structures and thin, robust dielectrics are used to enhance signal storage. The use of buried layers in the substrate as a potential barrier against minority carriers is another structural approach. The material-oriented approach is to eliminate the sources of alpha particles: U or Th impurities in various materials of VLSI including wiring, passivation, and packaging. Purification of gate,

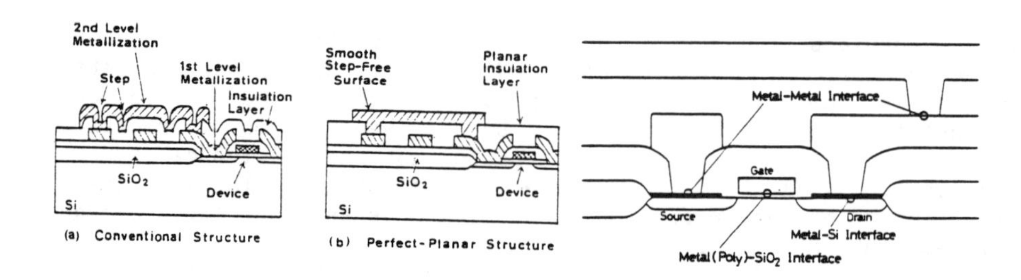

Fig.7 Conventional(a) and ideal (b) two-level MOS interconnect schemes.

Fig.8 Interfacial problem areas in MOS VLSI.

interconnect, and insulating materials down to the sub-ppb level is strongly needed.

DISCUSSIONS

An important problem which has not been discussed in previous sections is the problem of power dissipation. The thermal resistance of VLSI packaging puts a practical limit on the number of components in a VLSI. Let the allowable power dissipation from a VLSI be P=10W.

$$P = Gp \times duty = Gp\, t_d/T \qquad (15)$$

where G is the total number of gates, p is the power dissipation per gate, t_d is the delay per gate, and T is the average switching period of a gate in the VLSI circuit. The power-delay product,

$$pt_d = [pdp], \qquad (16)$$

depends on types of devices (ECL, CMOS, I^2L, etc.). Assuming CMOS technology,

$$[pdp] \simeq 10^{-14} J, \quad t_d = 0.1 \sim 1 ns. \qquad (17)$$

The ratio,

$$k \equiv T/t_d \qquad (18)$$

has been discussed by Masaki et al[11]., and is determined to be on the order of 100 for CMOS logic circuits. The total number of gates which can be integrated in a VLSI is then,

$$G = \frac{P}{[pdp]/td} k = 10^7 \sim 10^8 \qquad (19)$$

This means that there still is a plenty of rooms left for increasing the number of components on a VLSI chip. Figure 9 shows the power dissipation of various VLSI chips. The extrapolated power for a chip with 10^7 to 10^8 transistors agrees well with the above discussion. This level of integration will be realized by using CMOS technology with linewidths near a quarter of a micron. The basic components will be transistors with gate lengths or emitter widths in that range, which have been proven to work at least on a discrete basis.

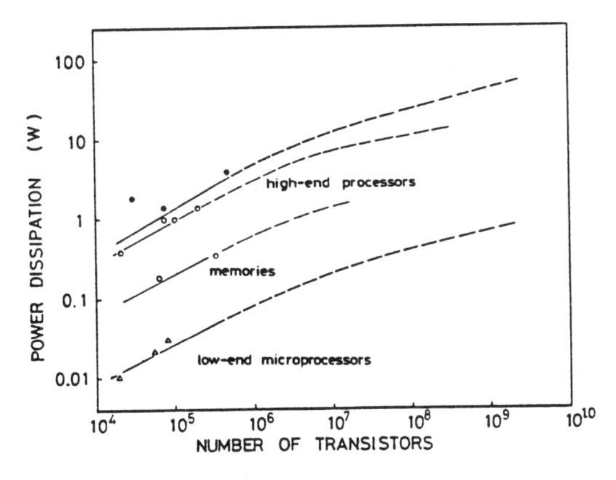

Fig.9 Power dissipation of VLSI chips as a function of number of transistors.

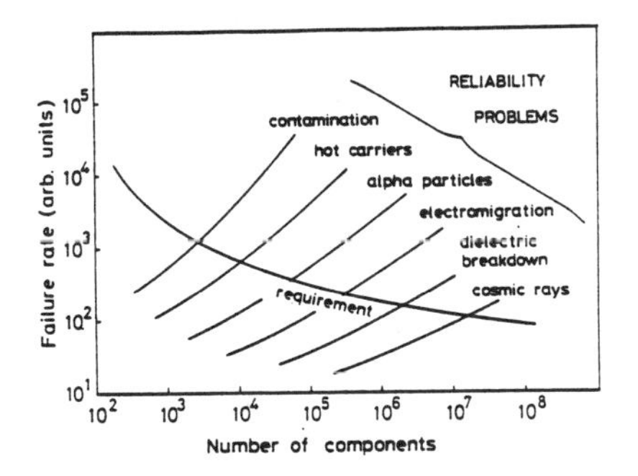

Fig.10 Reliability problems limiting the level of integration.

In order to realize VLSIs of the above scale, however, we need, first of all, extremely tight control on parameters which determines the VLSI structure. Lines and spaces, which will probably be delineated by X-ray lithography, will have to be controlled within several tens of nm. Doping will have to be controlled not only vertically but also horizontally. Control on topography, not just control on thicknesses, will have to be practiced by advanced deposition and etching technology. Secondly, finer features also calls for more strict control of defects. Random defects, such as photo-(or X-ray-) lithographic defects, pinholes in insulator films, and foreign particles arising from processing equipment, will remain to be critical factors dominating the yields of future LSI chips. Generation, motion, and localization of crystal imperfections which give rise to excessive leakage current of junctions, have to be controlled by improving crystal growth (including epitaxial) processes and carefully designing succeeding high temperature processes[26]. In this context, contamination control against heavy metal impurities such as Fe requires re-examination of processing equipment. Thirdly, reliability requirements will become even more demanding for future VLSIs. New materials, processes, and structures such as those discussed in the previous chapter are strongly needed. Material purity must be controlled in units of ppb. Mechanical stress must be more strictly controlled not only during processing but until the chip is packaged avoiding stress-induced reliability problems[27]. The reliability requirements of gate oxides and inter-level insulators have become so stringent that one has to be careful about the damage in the dielectrics caused by radiation which wafers undergo during processing itself. The susceptibility of oxides to radiation seems to depend on the growth technique and to be related to H or OH content of the oxides.

Reliability problems are manifold. They never can be totally eliminated. At one time, a problem appears and sets a practical limit to the number of devices on a chip from the failure-rate viewpoints. At another time, we will be faced with yet another problem, as shown in Fig. 10. A same problem can come back over and over again. To overcome these problems and build VLSIs with more than 10 million gates, all the topics discussed here need further basic research on the one hand, and careful design compromise on the other.

REFERENCES

1. J. D. Meindl, Tech.Digest, 1983 IEEE IEDM, Washington D.C. Dec. 1983, p.8.
2. S. Asai, Extended Abstract, 16th Conf. Solid state devices and Materials, Kobe, Aug. 1984, p.221.
3. M. H. Woods, Proc. 3rd Int. Symp. VLSI Sci. Technol., Toronto, May 1985, p.41.
4. S. Asai, Tech. Digest, 1984 IEEE IEDM, San Francisco, Dec. 1984, p.6.
5. J. Moran et al., Proc. 15th Symp. Electron, Ion, and Photon Beam Technol., p.1620, 1980
6. B. E. Griffing and R. E. West, IEEE Electron Dev. Let., EDL-4, p.14(1983).
7. Y. Nishioka et al., Ext. Abs., 165th Electron. Soc. Mtg., Cincinnati, 1984, p.190
8. Y. Miyake et al., J. Appl. Phys., 53, 3214(1982).
9. M. Koyanagi et al.,Tech.Digest, 1978 IEEE IEDM, Washington D.C. Dec. 1978, p.348.
10. H. Sunami et al.,IEEE Trans. Electron Devices, ED-31, p.746, (June 1984).
11. A. Masaki and T. Chiba, IEEE Trans. Electron Devices, ED-29, p.751 (April 1982).

12. M. Yamada et al.,1977 Nat. Conf. Rec., IECE Japan, paper6.99 (March 1977).
13. T. Nishida et al.,1985 IEEE Reliability Physics Symp., Orlando, March 1985, paper 5.4.
14. J. P. Roland et al., HP Journal, 34, 30 (1983)
15. T. Y. Ting et al., J. Vac. Sci. Technol.,15,1105(1978).
16. Y. Homma and S. Tsunekawa, J. Electrochem. Soc., 131, 123(1984).
17. T. Moriya et al., Tech.Digest, 1983 IEEE IEDM, Washington D.C. Dec. 1983, p.550.
18. S. P. Murarka, J. Vac. Sci. Technol.,17, 775(1980).
19. N. Kobayashi et al., Digest Tech. Papers, 1983 Symp. VLSI Technol., Maui, Hawaii, Sept. 1983, p.94.
20. G. Baccarani et al., IEEE Trans. Electron Dev., ED-31, 452 (April 1984).
21. Y. Kobayashi et al., Digest Tech. Papers, 1985 Symp. VLSI Technolo., Kobe, May 1985, p.40
22. T. Nakamura et al., Digest Tech.Papers, IEEE ISSCC, San Francisco, 1984, p. 152
23. I.-C. Chen et al., IEEE J. Solid State Circuits, SC-20, p.333(1985).
24. E. Takeda, Digest Tech. Papers, 1985 Symp. VLSI Technol., Kobe, May 1985, p.2.
25. T. C. May, Proc. 29TH Electron. Comp. Conf., Cherry Hill, May 1979, p.247.
26. H. Iwai et al., IEEE Trans. Electron Dev., ED-31, 1149(1984).
27. K. Arimoto et al., Digest Tech. Papers, 1985 Symp. VLSI Technol., Kobe, May 1985, p. 92.

After completion of the manuscript, the following paper came to the author's attention. It is a comprehensive review of the topics discussed here; A. Reisman, Proc. IEEE, Vol. 71,550(1983).

ION IMPLANTATION DAMAGE AND RAPID THERMAL PROCESSES IN SEMICONDUCTORS

J. Narayan,
Materials Engineering Department, North Carolina State University, Raleigh, N.C. 27650
Microelectronics Center of North Carolina, Research Triangle Park, N.C. 27709

O.S. Oen*, S.J. Pennycook*
Solid State Division, Oak Ridge National Laboratory, Oak Ridge, TN 37831

ABSTRACT

The characteristics of ion implantation damage, which influence the subsequent annealing behavior, are determined by both implantation and substrate variables. Fundamentals of ion-solid interactions and a model for damage accumulation and amorphization are presented. Ion channeling effects that change damage and dopant profiles were calculated as a function of ion implantation and substrate variables using MARLOWE, a well-known computer program. These calculations have been compared with experimental results on dopant profiles and residual damage. The crystalline to amorphous phase transition has been shown to occur at a critical damage energy of 12 eV/atom for silicon, in the absence of annealing effects. The conventional method of removing ion implantation damage involves isothermal annealing treatments in a furnace at about 1000°C for 30 minutes. However, as device dimensions become smaller, shallower junctions are needed that must be processed at reduced thermal budgets. Recently, more advanced methods for removing have been developed, which are based upon rapid heating or annealing at high temperatures for a short period of time. Almost a "complete" annealing of displacement damage is possible for shallow implants, provided that loop coalescence does not lead to formation of a crossgrid of dislocations. For deep implants, the free surface cannot provide an effective sink for removal of all the defects as in the case of shallow implants. Dopant profile broadening can be controlled by transient annealing to less than 500 Å in layers having excellent electrical properties. Annealing mechanisms associated with amorphous layers involve solid-phase-epitaxial growth, whereas annealing of dislocation loops occurs by the glide and climb of dislocation loops.

---

*Part of this research was sponsored by the Division of Material Sciences, U.S. Department of Energy under contract DE-AC05-840R21400 with Martin Marietta Energy Systems Incorporated.

0094-243X/86/1380122-16$3.00

INTRODUCTION

Ion implantation offers many advantages for controlled doping of semiconductor devices. For doping purposes, the defects (ion implantation damage) must be removed as efficiently as possible in order to recover electrical activation and carrier mobility.[1] As dimensions of semiconductor devices get smaller in VLSI and ULSI circuits, shallower junctions of the order of 1000 Å depth are needed. The formation of shallow junctions requires low energy ion implantation and a damage annealing cycle with a minimum thermal budget. The typical ion energies and species are 10 keV B^+ for p-type doping and 80 keV As^+ for n-type doping. The incident ions exhibit enhanced channeling under low energy ion implantations. Although, in general, ion channeling should be avoided in forming shallow junctions, it can be used to manipulate dopant profiles. The annealing of ion implantation damage is a critical function of its nature. In this review, we cover fundamentals of ion implantation damage in semiconductors, process of amorphization, factors influencing straggling damage, solid phase epitaxial growth of amorphous layers, annealing of straggling damage, annealed microstructures, and dopant profile control.

FUNDAMENTALS OF ION IMPLANTATION DAMAGE

The nature of residual ion implantation is a strong function of the incident ion and the substrate variables. The light ions produce simple defects that cluster in the form of dislocation loops, whereas heavy ions directly produce amorphous cascades.[2,3] Figure 1 shows an example of amorphous cascades produced by 100 keV, $^{208}Bi^+$ ions. These specimens were implanted to a dose of 1.0×10^{12} cm^{-2} at 7° from [110] surface normal along the [110] axis. The collision cascades appear as black spots in the image with an average size of 50 Å. It is interesting to note that some of the cascades contain subcascades that are bunched around a primary cascade. The number density of cascades was determined to be $5.0 \pm 1.0 \times 10^{11}$ cm^{-2}. A high resolution image of a cascade is shown in Figure 2 in which the central region is amorphous with the absence of <110> chains of atoms. The outer regions of the amorphous cascades contain significant amounts of disorder in the form of disrupted <110> chains of atoms. From the studies on the nature of residual damage as a function of ion implantation and substrate variables, the following model of the process of amorphization was derived. Above a critical value of deposited damage energy, a first order crystalline to amorphous phase transition is observed. In the case of silicon ion implanted at 4 K, the critical damage energy of amorphization[3] was determined to be 6.0×10^{23} eV cm^{-3} or 12 eV/atom. At higher temperatures, where annealing effects become important, higher implant doses are required for amorphization; however, the critical value remains constant. Figure 3 is a high resolution cross-section electron micrograph from a sample implanted above a critical dose in which the crystalline-amorphous boundary is clearly shown. This boundary is atomically sharp, which is consistent with first order crystalline to amorphous phase transition. It should be pointed out that if the residual damage is in the form of dislocation loops, then the crystalline to amorphous transition occurs when dislocation cores are close to overlapping.

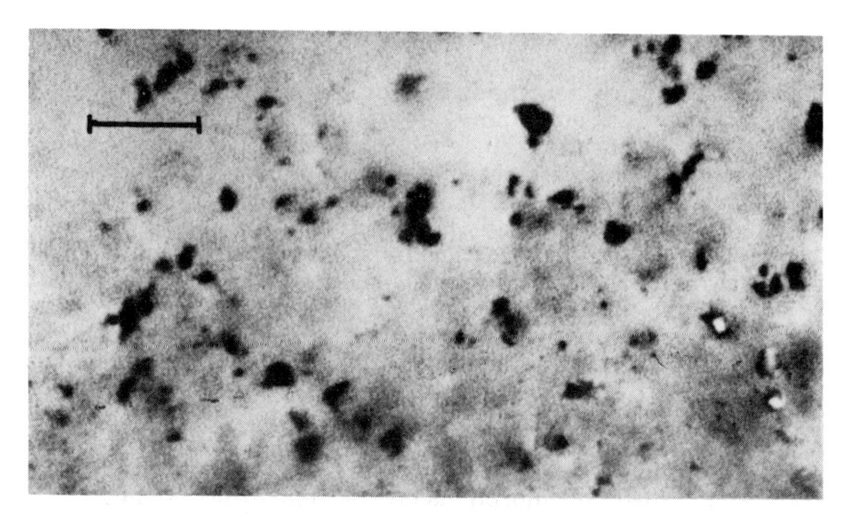

Figure 1. Bright-field electron micrograph of damage regions in (110) silicon produced by implanting 100 keV 208 Bi^+ ions to a dose of 1.0×10^{12} cm^{-2}. The length of the marker is 40 nm.

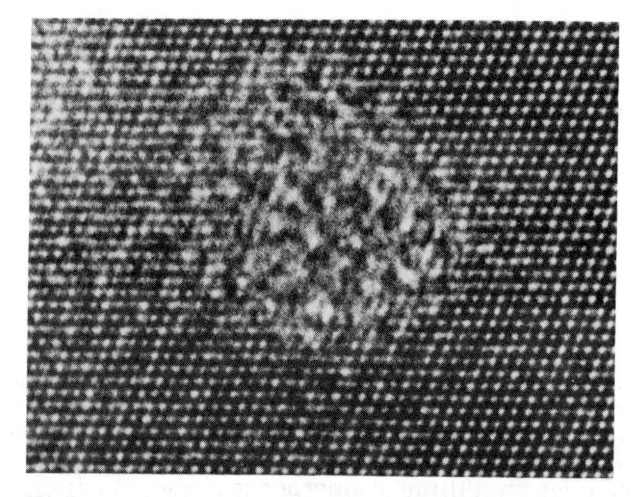

Figure 2. High-resolution electron micrograph (plane-view) showing <110> chains of atoms around damaged regions. The amorphous structure in the central regions of the cascades is clearly shown.

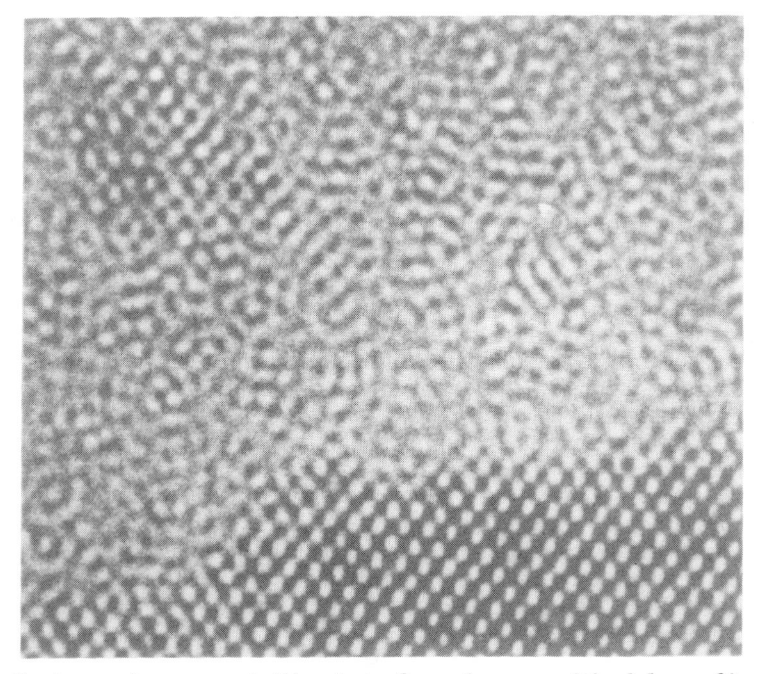

Figure 3. Amorphous-crystalline interface above a critical dose of ion implantation. The micrograph shows a sharp interface with undulations due to ion range straggling.

DAMAGE AND DOPANT PROFILES AND CHANNELING EFFECTS

The formation of shallow junctions requires low energy ion implantation and a thermal annealing procedure that results in dopant profile broadening of less than 500 Å. The dopant and displacement damage profiles are very sensitive to the incident ion directions with respect to planes and axes in crystalline solids, the nuclear charge of the incident ion (Z_1) and the substrate atoms (Z_2), and the energy of the incident ion (E). There is a critical angle of channeling $\Psi_c \alpha (Z_1 Z_2/E)^{1/4}$ below which significant deviations from a Gaussian profile corresponding to a random solid are observed.[4] Table 1 summarizes the calculated values of Ψ_c for different ions. The ion channeling is also very sensitive to the presence of overlayers on the surface. If the structure of the overlayer on the surface is random, then the incident beam is dechanneled in this layer, thus producing fewer channeling effects in the crystalline lattice below this layer. During ion implantation, the accumulated damage can also affect the channeling of subsequent ions; this is particularly true in the case of high dose implants. The channeling effects on the damage and range profiles were calculated using the computer code MARLOWE.[5] In these calculations, the effects of ion implantation (ion mass, energy) and substrate (mass, orientation, amorphous overlayer) variables on damage and range profiles were examined. The program follows the incident ion, collision by collision, and the target atoms that are set into motion. The fate of target atoms is followed until their energy drops below 15 eV, which is the displacement threshold. The MARLOWE program uses binary collision approximation to construct trajectories of energetic particles moving in crystalline or amorphous material. In these calculations, the Molierè approximation to the Thomas-Fermi interatomic potential[6] with a Firsov screening distance[7] is used to describe binary elastic collisions. The (nonlocal) electronic slowing down theory of Lindhard et al.[4] is used to account for the inelastic energy losses. It is assumed that only the elastic collisions result in displaced atoms and thereby stable interstitial-vacancy pair production. The interstitial-vacancy (Frenkel) pairs included here are those whose separation distance is greater than or equal to the second neighbor distance. It is assumed that the pairs closer than second neighbor distance recombine. The net effect of incorporating atom transport is to produce a damage profile at deeper target depths than those deduced from the energy loss profile of the incident ion alone. The channeling effects are produced because the ions are constrained to move in the more open avenues of the crystal, which greatly reduces the probability for violent collisions that produce cascade damage. The MARLOWE program can model random or structure-less targets by rotating the target crystal randomly in three dimensions about a lattice before each collision. This procedure preserves target density, but destroys directional correlation. The program also includes thermal vibrations of the target atoms in calculations.

$Bi^+ \rightarrow Si(100)$

TABLE 1

Displacement Damage and Average Penetration Depth as a Function of Beam Direction

Energy (keV)	Beam direction θ (degrees)	Beam direction Φ (degrees)	Divergence of beam (degrees)	Total number of displaced atoms	Average pen. depth of incident ion (Å)
100	7	0	0	1954	955
100	3	15	0.25	1129	3878
100	0	0	0.50	491	5665
100	random run		---	1916	453

Figure 4 shows the calculated number of displaced atoms as a function of target depth for a 100 keV Bi^+ ion incident onto a random target and a crystalline target. For the case of crystalline target, the incident beam direction was chosen to be 7° from the [110] surface normal. Since this angle is larger than the critical axial channeling angle of 4.5°, estimated from Lindhard's channeling theory,[8] channeling effects are significantly reduced. The damage profiles for random and crystalline (7°) targets are quite similar. For example, the total number of displacements agree to within two percent of each other. However, there is a slight penetration tail in the case of the crystal, indicating that a few ions are channeled, which produces damage at greater depths than the Gaussian profile.

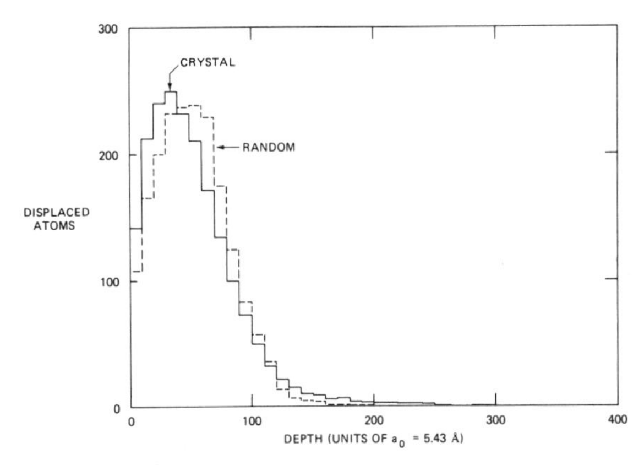

Figure 4. Number of displaced Si atoms as a function of penetration depth in a Si target produced by a 100 keV Bi ion. The curve labeled random is for a structureless target and that labeled crystal is for a crystalline target. For the crystalline target, the polar angle of the beam direction is 7° from the [110] crystal normal and the azimuthal angle is 0° from the [110]. The integrated average number of displacements per incident ion is 1954 for the crystal case and 1916 for the random case.

The damage profiles at different angles from the [110] surface normal are shown in Figure 5. For curve 3, $\theta = 7°$ and $\phi = 0°$, for curve 2, $\theta = 3°$ and $\phi = 15°$, and for curve 3, $\theta = 0°$, and $\phi = 0°$, where ϕ is the azimuthal angle (measured from [110]). Table 1 summarizes the total number of displaced atoms and the average penetration depth of incident ions as a function of ion implantation and substrate variables. The channeling reduces the accumulated damage as much as 75% and increases the average penetration depth by approximately an order of mgnitude. The damage peak is reduced by an order of magnitude with deep penetrating tails. The experimental results showed about a factor of 10 decrease in the number density of damaged regions. It should be noted that the damage is produced even when the beam is incident within the critical channeling angle. There are several reasons for this. Firstly, some ions never become channeled since they undergo large deflection angles in collision with the surface atoms. Secondly, dechanneling mechanisms, such as thermal vibrations, cause some channeled ions to dechannel before they slow down to rest. Thirdly, for systems of large projectile to target mass ratios, it is possible for the projectile to produce displacements and still remain channeled. For instance, a 10 keV Bi^+ ion undergoes a laboratory deflection of less than one degree in transferring an energy of 15 eV (displacement threshold) to a Si atom. Both the second and third reasons enable the creation of damage at much greater target depths in a crystalline than in a random target because of the greater range of channeled ions.

$B^+ \rightarrow Si(100)$

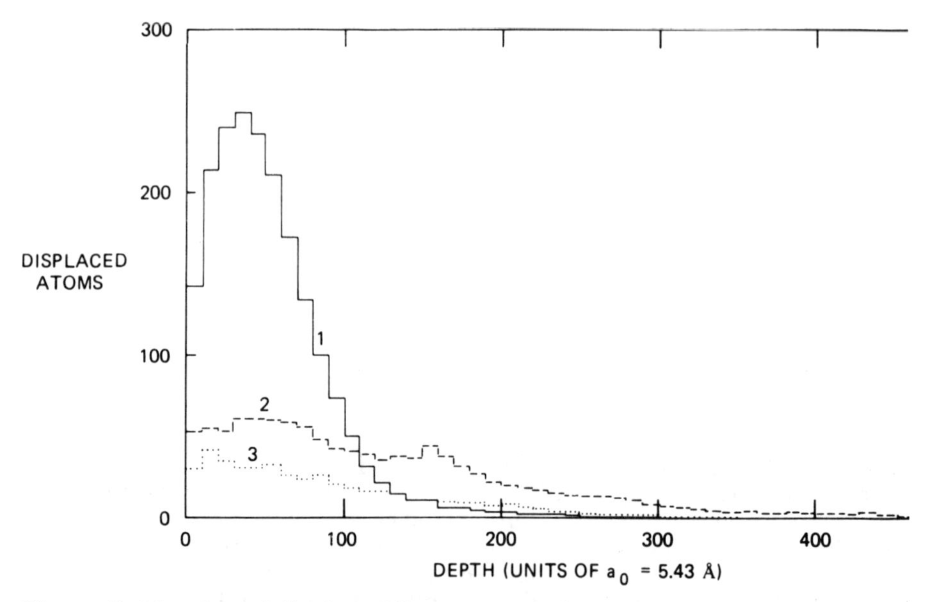

Figure 5. Number of displaced Si atoms as a function of penetration depth in a crystalline Si target produced by a 100 keV Bi ion. The three cases represent three different incident directions of the Bi beam onto the (100) Si target. For curve 1, the polar angle $\theta = 7^o$ and the azimuthal angle (measured from [110] $\phi = 0^o$; for curve 2, $\theta = 3^o$, $\phi = 15^o$; for curve 3, $\theta = 0^o$ and $\phi = 0^o$. For the latter two cases, channeling reduces the damage near the surface and increases the damage at greater depths. The average number of total displaced atoms per incident ion in the three cases is 1) 1954, 2) 1129, 3) 491.

In the following, we present displacement damage and dopant profiles for 10 keV, $^{11}B^+$ ions incident upon a Si(100) single crystal. The 7 degree profile is close to being Gaussian except for the channeling tail as shown in Figure 6. The critical angle using Lindhard's theory[8] was estimated to be 4.1^o. The 3° degree profile shows a significant reduction in damage peak and integrated damage. The damage peak shifts into the deeper regions. Further reduction in damage and deeper penetration of damage profile is observed at zero degrees. It is interesting to note that the damage is quite uniform in the case of a zero degree tilt angle.

The effect of the presence of a random surface layer on damage and dopant profiles was investigated and the results for 10 keV, $^{11}B^+$ ions in (100) Si are shown in Figure 7. For each of the three cases, a beam divergence of 1° was used. In each of the three, cases 3 nm thick random SiO_2 is assumed. The underlying substrate for 0 degree and 7 degree profiles is crystalline and for amorphous target calculation the underlying substrate is assumed to be random silicon. The 7 degree and amorphous target profiles are fairly similar except for the channeling tails in the crystalline substrate. The zero degree profile shows a considerable reduction in channeling effects due to the presence of 3 nm thick SiO_2. As the thickness of the amorphous overlayer is increased, the channeling effects decrease. At a thickness of 22 nm amorphous overlayer, the channeling effects are almost eliminated. Figure 8 shows the fraction of projectiles stopped as a function of depth, corresponding to Figure 7. In the tail region, there is a significant fraction of projectiles that do not produce displacement damage. Figure 8 also includes the fraction of incident ions R_N, which is reflected during the ion-solid interaction

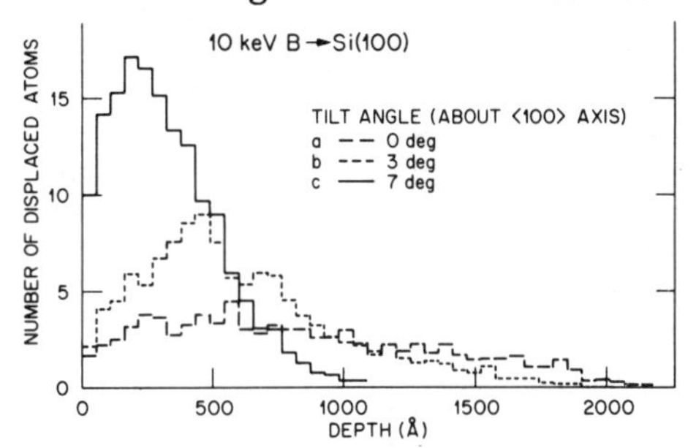

Figure 6. Number of displaced atoms as a function of depth. The profiles are for 10 keV $B^+ \rightarrow$ Si (100) at 0, 3, and 7 degrees about the <100> axis.

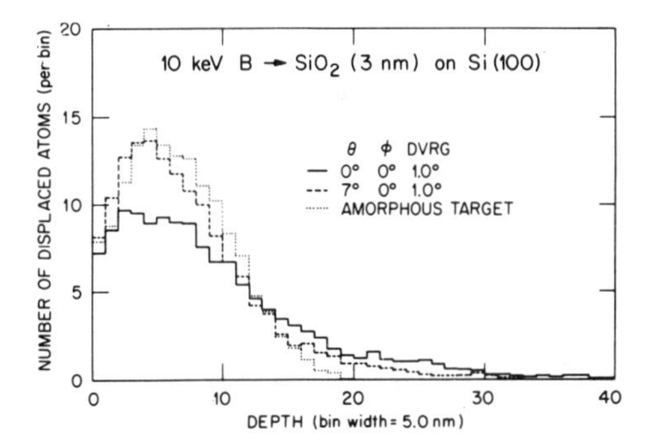

Figure 7. Effect of the surface overlayer (3 nm thick SiO_2) 10 keV B^+ damage profile. The profiles at are 0 and 7 degrees from the <100> axis including the profile in amorphous silicon with 3 nm thick, amorphous SiO_2,

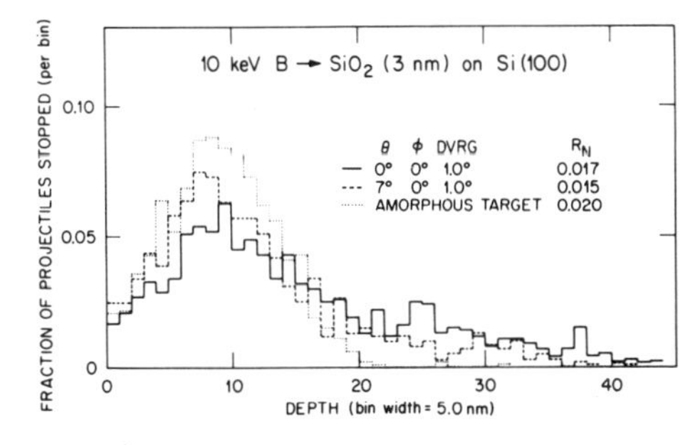

Figure 8. Fraction of the projectiles (boron atoms) as a function of depth corresponding to the specimen in Figure 7. R_N is the fraction of B reflected from the target.

Figure 9 shows results for 35 keV, $^{11}B^+$ ions incident upon (100) Si single crystals. A comparison between these profiles and the previous results (Figure 6) shows that the channeling effects at higher energies (35 keV) are considerably less than those at 10 keV. This is also reflected in the critical angle, which is 3.0 degrees at 35 keV compared to 4.1 degrees at 10 keV (Table 2). Figure 10 illustrates the variation of azimuthal angles on damage production. Here curve a) is the same as curve c) in Figure 9. It should be noted that the peak in the damage curve b) is considerably smaller than curve a); it shows an enhancement in the damage tail. The integrated number of displaced atoms in case b) is 78% that of case a). The reason for the difference in the two cases in Figure 10 is planar channeling in case b) arising from the relatively large open (022) channels. Figure 11 shows experimental results (obtained using SIMS, secondary ion mass spectrometry techniques) on channeling for 35 keV, $^{11}B^+$ ions in (100) Si single crystals at 0, 3, and 7 degrees from the [001] axis. These specimens retained about 3 nm thick native oxide during ion implantation. Since the critical angle for channeling for 35 keV, $^{11}B^+$ ion is 3.0 degrees, the differences between 7 and 3 degree profiles are small. However, the channeling effects are considerably pronounced at zero degrees despite the presence of 3 nm thick native oxide.

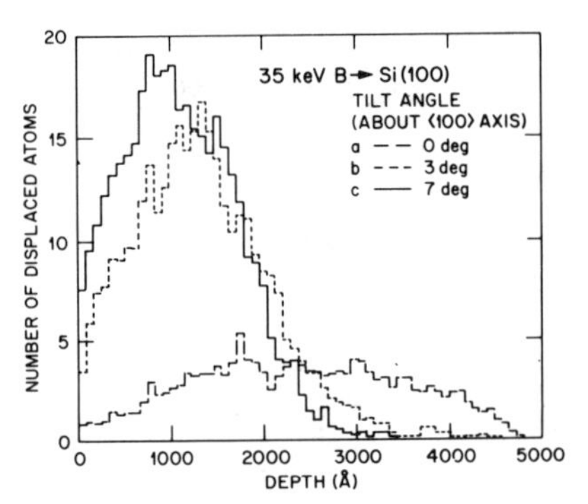

Figure 9. Number of displaced atoms as a function of depth for 35 keV B^+ → Si (100) at 0, 3, 7 degrees about the <100> axis.

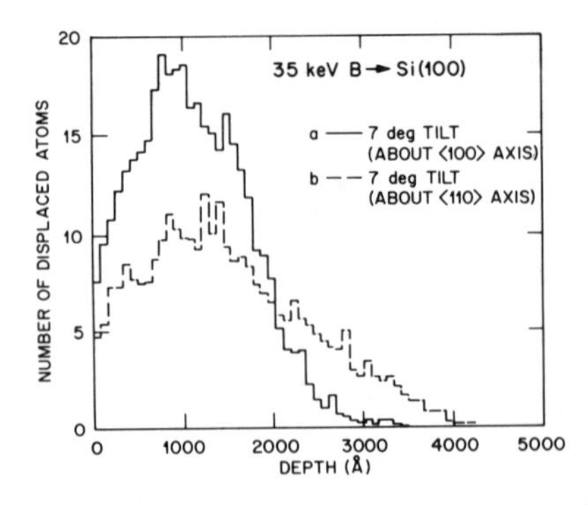

Figure 10. Number of displaced atoms as a function of depth for 35 keV B^+ → Si (100) at 7 degrees from the <100> and <110> axes.

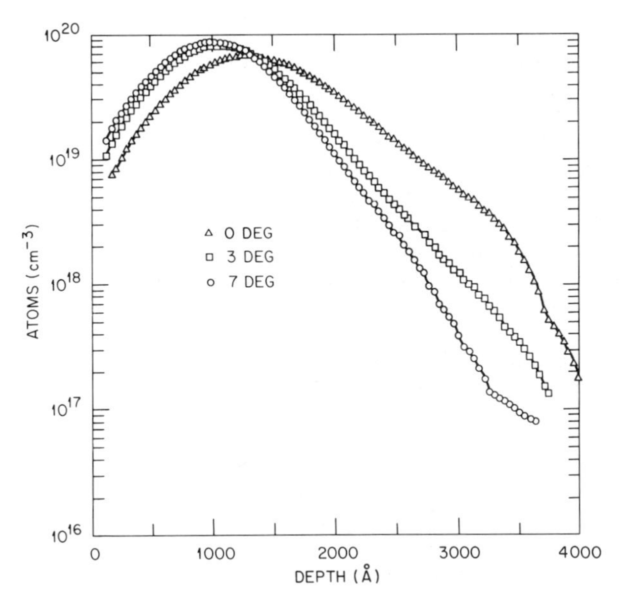

Figure 11. Boron concentration as a function of depth as measured by SIMS for 35 keV B^+ (1.0×10^{15} cm^{-2}) into (100) Si at 0, 3, and 7 degrees about the <100> axis.

TABLE 2

Ion Channeling in Si and GaAs

Ion (Z_1)	Energy (E) (keV)	Target (Z_2)	ϕ_c Critical angle for channeling (degree)
Bi^+ (83)	100	Si (14)	4.6
As^+ (33)	35	Si	4.7
As^+ (33)	80	Si	3.9
B^+ (5)	5	Si	4.8
B^+ (5)	10	Si	4.1
B^+ (5)	35	Si	3.0
Si (14)	50	GaAs	4.3

SOLID PHASE EPITAXIAL GROWTH

If the amphorous layers are formed with the underlying crystalline substrate, solid phase epitaxial (SPE) growth provides an efficient mechanism of annealing. The straggling damage below the amorphous layer is mostly in the form of dislocation loops. The annealing of loops occurs primarily via a dislocation climb and glide mechanism. The SPE growth process is thermally activated over a wide temperature range and is well described by an Arrhenius expression of the form

$$V = V_0 \exp(-E_a/KT)$$

where V is the SPE rate and E_a is the activation energy. For the self ion implanted amorphous layer, the value of V_o and E_a were determined to be 3.07×10^8 cm s^{-1} and 2.68 ± 0.05 eV, respectively.[9,10] The SPE growth kinetics have been extensively studied in silicon below 600°C using Transmission Electron Microscopy (TEM) and Rutherford backscattering (RBS) techniques. At higher temperatures, beam heating techniques coupled with laser-based diagnostic methods are needed to monitor the rate of solid phase crystallization and the growth temperature in real time. These time-resolved reflectivity measurements[10] now allow us to explore the SPE growth techniques from 1 to 10^8 As^{-1} and temperatures in the range of 500 - 1400°C.

From cross-section TEM studies, a complete detail of SPE growth can be obtained. Figure 12 illustrates SPE growth as a function of time at 525°C in a 200 keV, Sb^+ implanted specimen to a dose of 6.0×10^{15} cm^{-2}. The As implanted specimens contained an amorphous layer ~ 1580 A thick, followed by a band of dislocation loops. The amorphous layer above the dislocation band grows toward the surface with the substrate acting as a seed for crystal growth. The dislocation loops in the underlying dislocation band exhibited only a slight coarsening at these temperatures of annealing. The thickness of regrown layers was determined directly from the cross-section electron micrographs. The regrown regions were devoid of defects and the crystalline-amorphous interface remained planar during growth with no indication of the presence of a transition layer.

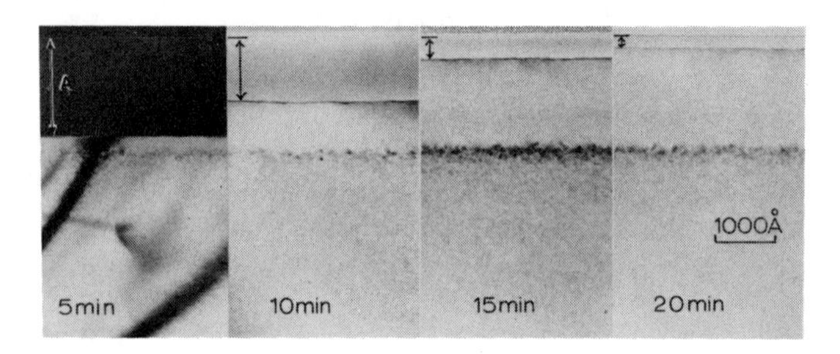

Figure 12. Cross-section TEM micrographs showing SPE growth in Sb^+ implanted (200 keV Sb^+, 6.0×10^{15} cm^{-2}) specimens of (100) Si after annealing at 525°C.

The time resolved reflectivity technique is quite powerful for a quantitative determination of SPE rate, but it does not provide direct information on dopant precipitation and interface structure. Therefore, it becomes necessary to combine TRR measurements with TEM studies. Figure 13 shows TRR and cross-section TEM results for specimens with arsenic doses of 2.0×10^{15}, 1.0×10^{16}, and 2.0×10^{16} cm^{-2} (220 keV As^+) and recrystallized at 640°C. In 2.0×10^{15} cm^{-2} specimens, the SPE growth is fairly "defect free" except for the underlying dislocation band. The SPE rate versus depth curve shows a maximum near the average projected range of the ions. The increase in SPE rate with dopant concentration is a normal phenomenon. The SPE rate enhancement relating to the intrinsic rate generally occurs when dopant concentration exceeds ~ 10^{20} cm^{-3}. In the case of 1.0×10^{16} cm^{-2} specimens, the arsenic precipitates are formed after about 1000 A "defect-free" SPE growth. When precipitation occurs, the growth rate is slowed down considerably and then increases again due to an increase in dopant concentration. For 2.0 x 10^{16} cm^{-2}, the precipitation of arsenic starts after about ~ 800 A "defect-free" SPE growth. The growth rate is slowed down consider-ably due to enhanced precipitation of dopants. The growth rate recovers and then it slows down again near the surface. The precipitation of dopants can be minimized by raising the substrate temperature in such a way that the interface passes before the dopants are able to cluster.

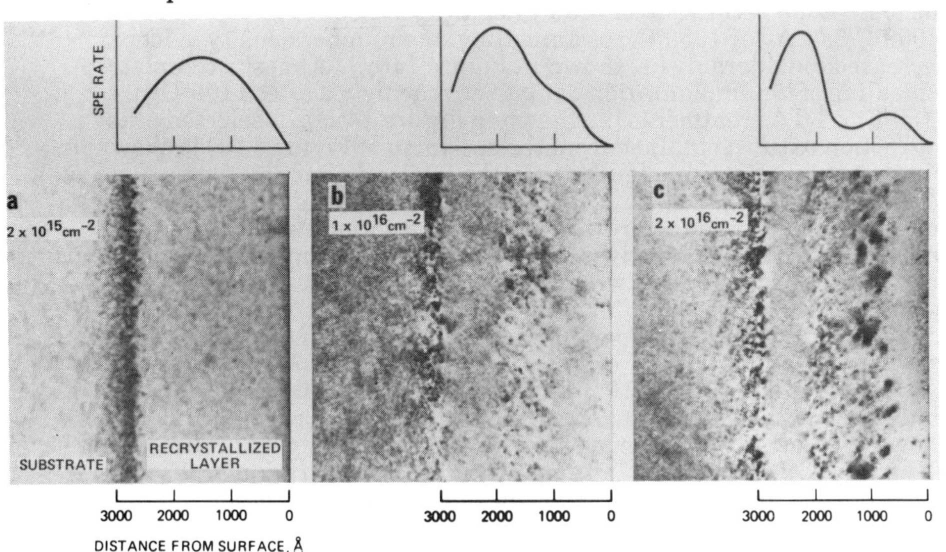

Figure 13. Cross-section TEM micrographs of SPE grown layers at 640°C in silicon (100) specimens implanted with 220 keV As^+ at three different doses: (a) 2.0 x 10^{15} cm^{-2}, (b) 1.0 x 10^{16} cm^{-2}, and (c) 2.0 x 10^{16} cm^{-2}. SPE growth rate versus depth data obtained from transient reflectivity techniques are shown for the corresponding specimens.[10]

RAPID THERMAL ANNEALING OF ION IMPLANTED SEMICONDUCTORS

From the above studies on SPE growth, it can be shown that amorphous layer of the order of 1000 A can be recrystallized in less than a millisecond above 1000°C. The SPE grown layers contain a certain concentration of trapped point defects and dislocation loops in the underlying dislocation band. The annealing of these defects requires times of the order of a few seconds at ~ 1100°C. Thus, annealing around ~ 1100°C for a few seconds that accomplishes SPE growth and removal of remaining defects is known as rapid thermal annealing.[11]

Taking silicon as an example, optimum RTA involves a temperature and a time of about 1100°C and 5 seconds, respectively. Thermal diffusion distances in silicon for these times and temperatures are of the order of 1 cm. Thus wafers, which are usually 0.05 cm thick, are heated uniformly. At these temperatures, amorphous layers are annealed by solid phase epitaxial growth and annealing of dislocation loops involves glide and climb mechanisms. Since glide and climb of dislocations are very sensitive to the proximity of the free surface, the depth of the dislocation band plays an important role in efficient removal of displacement damage.

Figure 14 shows cross-section TEM micrographs from an arsenic implanted specimens (dose = 1.0 x 10^{16} cm^{-2}) after RTA treatments at 1050 and 1100°C. The As implanted specimens contained about a 1300 Å thick amorphous layer followed by a 250 Å wide dislocation band. The micrograph in Figure 14(a) shows the presence of large dislocation loops in the top ~ 2000 Å thick layer after annealing at 1050°C for 5 seconds (1050°C/5s). After 1050°C/10s annealing, the number density of loops decreased considerably [as shown in Figure 14(b)]. Almost a complete annealing of ion implantation damage was achieved after 1100°C/10s or 1150°C/5s RTA treatments [as shown in Figure 14(c)]. A selected area diffraction pattern obtained from the specimen of Figure 14(c) is shown in Figure 14(d) indicating regularity in the lattice diffraction spots and thus confirming a "complete" annealing of ion implantion damage. The arsenic concentration profiles before and after RTA treatments of arsenic implanted specimens are given in Figure 15. The As implanted specimens contain a Gaussian profile where arsenic atoms are in random nonsubstitutional sites. After 1050°C/5s annealing, a very little broadening (~ 100 Å) of the profile is observed. It is interesting to note that a very high fraction (> 95%) of the arsenic is in substitutional (electrically active) lattice sites. After 1050°C/10s annealing, the dopant profile is flattened out in the top 1000 A region, followed by an exponential fall as shown in Figure 15(b). The average profile broadening was estimated to be about 300 Å, and the total and substitutional concentration profiles of dopants overlap each other, indicating again a high (> 96%) substitutionality.

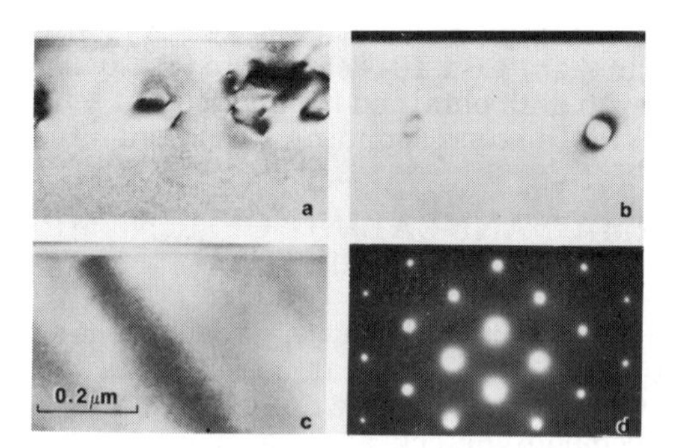

Figure 14. Cross-section TEM micrographs from arsenic implanted (100 keV, 1.0 x 10^{16} cm^{-2}) (100) Si specimens after the following RTA treatments: (a) 1050°C for 5s, (b) 1050°C for 10s, (c) 1100°C for 10s, and (d) SAD diffraction pattern from the specimens of (c).

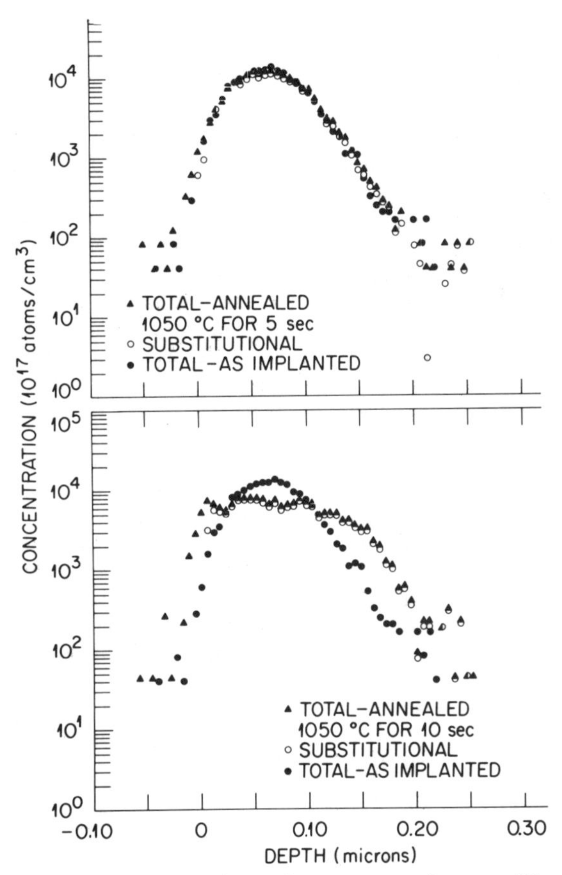

Figure 15. Substitutional and total concentration profiles before and after rapid thermal annealing treatments: (a) 1050°C/5s and (b) 1050°C/10s.

One of the problems during SPE growth and rapid thermal annealing is the formation of hairpin dislocations. These dislocations are formed primarily by two mechanisms: (1) original amorphous-crystalline interface intersecting dislocation loops in the underlying dislocation band create hairpin dislocations that continue to grow with the crystallizing interface; (2) isolated microcrystallites present near the amorphous-crystalline interface, if misoriented with respect to the underlying substrate, generate dislocations that continue growing with the interface. The formation of hairpin dislocations can be minimized by reducing the number density of dislocation loops in the underlying dislocation band, by minimizing the number of microcrystallites and by smoothening the original amorphous-crystalline interface. The number density of dislocation loops and isolated microcrystallites can be reduced by manipulating ion implantation and substrate variables (lowering dose and dose rate, substrate temperature, substrate and incident ion mass). The number density of isolated microcrystallites is also reduced by low temperature thermal annealing treatment.

In conclusion, we have shown the influence of ion implantation and substrate variables in the retained damage. The ion channeling effects on damage and dopant profiles were studied as a function of ion implantation and substrate variables. The channeling effects produce deep penetrating tails. These effects can be minimized by substrate orientation or by creating a random/amorphous layer on the surface. The damage accumulated in the form of complex defects or clusters thereof. Above a critical damage energy, we obtained crystalline to amorphous phase transition. This value was determined to be 6.0 x 10^{23} eV cm^{-3} or 12 eV/atom, in the case of silicon. During ion implantation, amorphous layers are generally followed by a band of dislocation loops. Amorphous layers anneal by solid phase epitaxial growth, while annealing of dislocation loops involves glide and climb of dislocation loops. If the loop is located within twice its diameter from the free surface, it glides out of the surface and is annealed out. During SPE growth, dopants start precipitating out (above a certain concentration) ahead of the crystallizing itnerface. This can be minimized by controlling the interface kinetics. Rapid thermal annealing includes time for SPE growth and additional times for removal of point defects and their clusters. During RTA treatments, times and temperatures of annealing can be manipulated to minimize dopant profile broadening and, at the same time, obtain efficient removal or "almost a complete" annealing of damage. Due to the above desirable characteristics, RTA is expected to play a major role in the fabrication of advanced VLSI and ULSI devices. A main thrust of the paper is that RTA is ready for VLSI and ULSI processing. However, a fundamental understanding of the nature of ion implantation damage and annealing phenomena are required to obtain efficient removal of damage and the concomitant electrical activation of dopants.

REFERENCES

1. J. Narayan and O.W. Holland, J. Electrochem. Soc. 131, 2651, (1984); A. Lietola, J.F. Gibbons, T.J. Magee, J. Peng, and J.D. Hong, Appl. Phys. Lett. 35, 532 (1979).
2. J. Narayan, O.S. Oen, D. Fathy, and O.W. Holland, Mat. Lett. 3, 67, (1985).
3. J. Narayan, D. Fathy, O.S. Oen, and O.W. Holland, J. Vac. Sci. Technol. A2(3), 1303 (1984); J.R. Dennis and E.B. Hale, J. Appl. Phys. 49, 1119 (1978); O.S. Oen, Nucl. Inst. Meth. (in press).
4. J. Lindhard, M. Scharff, and H.E. Schiott, K. Dan. Vidensk. Selsk., Mat.-Fys. Medd. 33, No. 14 (1963).
5. M.T. Robinson, User's guide to MARLOWE (version 12), 1984 (unpublished), describes a version of the program which is available from the National Energy Software Center, Argonne National Laboratory, Argonne, Illinois 60439, and from the Radiation Shielding Information Center, Oak Ridge National Laboratory Center, P.O. Box X, Oak Ridge, TN 37831.
6. G. Molierè, Z Naturforsch, 2a, 133 (1947).
7. O.B. Firsov, Zh Eksperim, Teor. Fig. 33, 696, (1957) [Sov. Phys. JETP 6, 534, (1958)].
8. J. Lindhard, K. Dan. Vidensk. Selsk., Mat.-Fys. Medd. 34, No. 14 (1965).
9. J. Narayan, O.W. Holland, and B.R. Appleton, J. Vac. Sci. Technol. B1 (4), 871, (1983).
10. G.L. Olson, J.A. Roth, L.D. Hess, and J. Narayan, p. 73 in Layered Structures and Interface Kinetics, ed. by S. Furukawa, KTK Scientific Publishers, Tokyo, 1985.
11. J. Narayan and O.W. Holland, J. Appl. Phys. 56, 2913, (1984).

CURRENT LOW TEMPERATURE PROCESSING OF VLSI IN JAPAN

K. Maeda

Applied Materials Japan Inc., Tokyo, Japan

ABSTRACT

The current trend of VLSI device technologies in Japan, especially of low temperature processing, is introduced. As the density of semiconductor devices is rapidly getting higher and the dimension of the pattern is rapidly getting finer year by year, various kinds of development regarding low temperature processing have been made in Japan as well as in the U.S. In this paper, recent topics of CVD processing at low temperature, planarization, and application of RTA(Rapid Thermal Anneal), which seem to be especially necessary for the advanced devices, are discussed.

By introducing the present status of 4M DRAM development and the future aspect of other devices following that in Japan, it is made clear what kinds of technical problems are left unsolved.

INTRODUCTION

VLSI technology has made such remarkable progress that advanced integration and processing of high density devices are under way as well as minimizing device dimensions. Not only in the U. S. but also in Japan, 256K DRAM has technically and economically arrived at the period of maturity. 1M DRAM has almost completed the stages of development, pilot production and mass production. The device structure and the processing parameters of 4M DRAM seem to be nearly established. Major device makers and research organizations are active in research and development of 4M DRAM in Japan. It seems that development and technology in Japan have the following aspect. First, all the alternatives are considered, and the safest one is always chosen. In addition, process always takes precedence over design. That is, no IC design which necessities excessive or complicated process specifications is adopted.

Minimizing dimensions is necessary for devices to increase their density, not only horizontally but also vertically. Problems in reliability and yield arise as a result of roughness on the surfaces of devices caused by three dimensional metallization and the complex device structures themselves. Also, the realization of highly accurate film thickness, precise control of diffusion of impurity and formation of shallower junction depth are strictly pursued, especially in minimizing the vertical dimension. The diameter of the silicon wafer inevitably increases to 8".

The most important problem in processing technology to be solved in the nearest future is to lower the temperature of the whole process. The pursuit of techniques for lowering the temperature seems to extend the boundaries of VLSI device technology, because VLSI devices are produced by subjecting silicon wafers to repeated heat cycles. Some of the techniques currently being explored to lower the processing temperature for VLSI production in Japan are discussed below.

VLSI DEVELOPMENT IN JAPAN

Fig. 1 [1)] shows the chronology of MOS memory device development in Japan. The times when pilot production could be commenced are marked in the figure, and introduction of 4M DRAM is foreseen at the end of the 1980s. The design rule for the existing 1M DRAM is about 1.2μm and for 4M DRAM it is expected to be about 0.7∿0.8μm, because there is a very clear view of fine patterning technology of 0.7∿0.8μm from the point of view of its process and mass production facilities. Each device maker in Japan aims at establishment of the technology of 0.5μm for 16M DRAM as its next target. Of course, not only fine patterning technology but also many peripheral technical problems must be solved. As the density of devices has gotten higher, the device technology for CMOS has taken the place of NMOS, because CMOS requires less power consumption. This development requires new technological breakthroughs.

Table 1 [2)] shows a forecast for changes of CMOS device parameters. As the minimum pattern size decreases, the gate oxide thickness gets still thinner, and PN junction depth drastically becomes shallower. Some specific designs, such as trench structure also become essential.

The future direction of VLSI processing can be summarized as finer patterns, shallower junction depth, larger wafer diameters and elimination of defects. As mentioned above, lower temperature processing is an extremely important factor in achieving these results.

Fig. 2 [3)] is an illustration of the cell structure of 4M DRAM which was composed into CMOS. An epitaxial substrate to suppress latch-up effect, polycide electrodes, trench isolation and trench capacitor processes are applied. The process parameters are summarized in Table 2. [3)] Two key factors to increasing device densities are the minimization of patterns and considerations in design structure. In this case, the most important goal is to obtain smaller cell size for memory devices. It is also essential not only to reduce the area of the capacitors but also to solve the problems of hot carriers, lower resistance of gate electrodes, and multilayer metallization processes. Every device maker in Japan is highly motivated to solve these problems.

LOW TEMPERATURE REQUIREMENT

The use of low temperature processes is vital to achieve higher density and higher integration of VLSI devices. In formation of shallower junctions, redistribution of impurities in the heat treatment cycles must be prevented, and wafers of larger diameter must be free of distortion and their flatness must be preserved in order to improve the precision of the whole process. Processes which require lower temperature are oxidation, epitaxy, doping, CVD and reflow. The concept of lower temperature processing must also be introduced into metallization, and dry etching. The specific minumum temperature must be considered for each different process. Processing should be taken into consideration when discussing the range of these temperatures. Assembling procedures must also be considered. If the lower temperature processing is achieved, devices with higher density will be obtained by improving devices' characteristics increasing yield, and establishing higher reliability. In all the present approach to lower temperature processing, it is aimed to change the temperature range shown in Tabel 3 to even lower values.

LOW TEMPERATURE PROCESSING FOR VLSI

1. Oxidation

Thermally-grown SiO_2 is used for gate dielectrics. The thickness in present advanced devices is about 150∿250Å; but this is becomming ever thinner. The oxide film for capacitors in memory cells is still getting thinner. A temperature between about 850°C and 1100°C is applied in the existing process, but it is preferable to lower the temperature to decrease thermal effect on the substrate in which impurity has already distributed. Problems may also arise such as deterioration of film quality. Many considerations on physical properties and limitation for thin SiO_2 film have been made in Japan.[4) 5)] With regard to lower temperature oxidation process itself, the high pressure oxidation method has been in practical use in Japan too, and it is applied to formation of thick SiO_2 film.

The relation between high density of VLSI and the thickness of its dielectric film is shown in Table 3.[6)] To form thin SiO_2 film at a low temperature, microwave discharged oxygen plasma can be used,[7) 8)] with which films of several hundred Å in thickness with excellent interface properties are obtained at a temperature between 300∿700°C. Experiments to form thin silicon nitride film by direct nitridation of silicon[9)] instead of SiO_2, have been tried. but the treating temperature is too high. Experiments of nitridation of silicon in plasma[10) 11)] or nitridation of SiO_2 have been performed instead. The former has demonstrated good nitride films at temperatures between 600°C and 900°C.

Fig. 4 is a summarized picture of a nitridation apparatus and an example of deposition data.[10] It is also reported that direct nitridation was made with an excimer laser at 460°C.[12] As film for capacitors of memory cells requires a high dielectric constant, the low temperature formation of silicon nitride indicates the possibility of future applications for this purpose. These approaches have just begun, and some time is required until they are put into practical use.

2. Epitaxy

Epitaxy is the essential process for VLSI devices using CMOS structure to prevent latch-up effect and to increase yield. In order to prevent redistribution of buried impurity, autodoping, and formation of distortion, epitaxy at a low temperature is needed for practical use. In the conventional pyrolytic decomposition method or hydrogen reduction method, the optimum temperature which gives good crystallinity and properties is 1000∿1150°C. It is also now possible to make epitaxial growth at a temperature below 1000°C by using disilane(Si_2H_6).[13] Some trials to lower the temperature have been carried out using a combination ultra-violet rays and laser beams. Silicon epitaxy is now under development by MBE with approaches for practical use of it in the future.[14),15)]

3. Junction Formation

The process required for ion implant shallow junction formations must be stable with high current and a low energy level. Since the junction depths for p-channel and n-channel in 4M DRAM are both less than 0.2μm, extreme care should be taken for the anneal after ion implantation. For anneal of ion implanted layers, what is called furnace anneal and RTA(Rapid Thermal Anneal) are available. RTA seems to have been steadily introduced into device lines. RTA is a process which heats the surfaces of substrates rapidly with radiation from lamps or carbon heaters, in which the effect on redistribution of impurity is mitigated in time. Some development has been carried out to form junction, which has lower sheet resistance, by using an ion implantation in combination with RTA. Many developments aimed at CMOS devices are under way today. The recent paper [16] on the implantation of boron ions and anneal effect says that the anneal temperature must be lowered to about 800°C in order to attain a junction depth of about 0.2∿0.4μm.

Attempts to obtain a range of low contact resistance by ion implantation and RTA with silicide have also been made.[17),18)] One of the examples of the process is shown in Fig. 5.[17] In each of these examples titanium silicide is used. After the junction and silicide contact are formed, an aluminum electrode is easily formed. The electrode structure displays high reliability.

The above technology is a combination of ion implantation, RTA, and silicide technology. It is expected to contrribute significant improvements to the performance of VLSI.

Another method of technical application of RTA is solid-to-solid diffusion from SOG(Spin-On-Glass). When boron glass was deposited, a shallow junction depth of about 0.1∿0.2μm was obtained by RTA for 1∿2 seconds at 1200°C.[19]

It is a method to obtain a very shallow junction depth at extremely low cost. However, there are some problems in the stability, the reproductivity, and the purity of materials using this method.

4. Chemical Vapor Deposition

As shown in Table 3, film formation by CVD has been examined over a wide range of temperatures. The formation of interlayer dielectric and passivation films at a temperature below 400°C is of particular interest. Low temperature experimentation is being examined for plasma CVD and photo CVD technologies. Photo CVD is a new process aimed at the formation of dielectric films such as SiNx and SiO_2 under near room temperature by exciting SiH_4, NH_3, or O_2 with ultraviolet rays or laser irradiation and sometimes with the assistance of mercury sensitizing reaction. The photo CVD film is believed to have advantages over the plasma CVD film; such as, less radiation damage, better step coverage, and better film quality. The research and development of the photo CVD technology by Japanese device manufactures, research organizations and universities is extensive.

Table 4[20] shows a comparison of photo CVD and plasma CVD. From the table, it can be assumed that the deposited films are almost equal in physical and chemical properties. Fig. 6[21] shows the deposition rate, the refractive index, and the infrared spectra of the SiNx film deposited by the photo CVD. A practical problem lies in the low growth rate. The photo CVD of dielectric film is being developed by NEC.[20], Mitsubishi[21],[22] and Hitachi[23].

The selective deposition method of tungsten through the scanning of laser beam and CVD in combination was reported by Hitachi[24] and NEC.[25] This technology is attracting attention for its new potential in CVD. Fig. 7 shows the result.

Another approach to lower the temperature for CVD is the ECR (Electron Cyclotron Resonance) plasma CVD method.[26] In this mehtod, silicon nitride is deposited on substrates by the interaction between N_2 plasma generated in the ECR ion source and SiH_4. In this way high density films can be deposited without heating substrates. Fig. 8 shows the principle of the system and deposition data. This method has just beendeveloped, and many applications are now under consideration. The application to lift-off process taking advantage of its being anistropic deposition is its only application at the present time.

5. Planarization

Planarization is essential to VLSI devices to increase reliability and yield. It is an important key to smoothing out surface uneveness, which becomes worse as the aspect ratio of devices gets higher and processes get more complicated. It is also essential to increase the resolution of the pattern formation by projection mask alignment. There are several methods for planarization. For example:

Glass reflow anneal
Etch back planarization
Bias sputtered quartz
Lift-off process
Spin-on-glass coating

Reflow can be most easily applied between polysilicon gates and Al layers. PSG, depending on the concentration of phosphorus, the temperature must be over 950°C. Many investigations using various kinds of glass other than PSG have been made. Practical use of BPSG has just been achieved in Japan.[27] Another method to reflow glass of the $ZnO-B_2O_3-SiO_2$ system at 600∿800°C is also under consideration.[28] About 800°C is thought to be the lowest temperature for reflow of BPSG, but is not low enough to meet the requirements for VLSI. A further concern is that the concentration of phosphorus and boron must not be higher than the present levels. On the other hand, a report was made that reflow of PSG gave good results under high pressures of up to 10Kg/cm^2 and at about 850°C even though the concentration of phosphorus was low. Etch back is a process which has already been put into practical use, but because it requires a RIE system, the process is complicated. Lowering the temperature for reflow is required as a low cost method for planarization.

Surface planarization using SOG has been adopted in Japan as the easiest method. Fig. 9[30] illustrates an example of this process. Although SOG is used in a subsidiary way, in most cases it cannot achieve a perfect planarized surface. SOG is extremely useful in improving device reliability and yield. Application of RTA to reflow is also under development.

CONCLUSION

The present condition of the VLSI process in Japan, especially low temperature processing developments,were discussed. Some of these technologies have been applied in specialized usages, and some are under development. Futhermore, some processes will be needed in the future rather than at present. As the target temperature is considered to be about 800°C, oxidation, anneal(RTA) and reflow must be studied intensely. It is also forecasted that in the future all silicon processes must be performed at temperatures lower than 800°C.

To summarize, future low temperature technology goals are as follows:

Final method for low temperature reflow
Practical use of RTA
Establishment of metallization technology for shallow junction
Development of low temperature CVD technology to improve reliability of multilayer interconnection structure

The application of low temperature epitaxy and laser technology are also considered to be technologies inevitably necessary for next step together with the ion beam technology. These technology developments are being aggressively pursued in Japan.

References

1) Y.Akasaka : Semiconductor World, P.53, May, 1985

2) Y.Nishi : Japan Semiconductor Technology News, Vol.3,No.6 p.19, Dec., 1984

3) Y.Mizokami : ibid., Vol.3, No.6, p.19, Dec.,1984

4) K.Yamabe : Extended Abstract of the 17th Conference on Solid State Devices and Materials, Tokyo, 1985, p.261

5) K.Yoshikawa, Y.Nagakubo and K.Kanzaki : Extended Abstract of the 16th Conference on Solid State Devices and Materials, Kobe, 1984, p.475

6) N.Hashimoto, H.Sunami, Y.Horiike and S.Takasu : Nikkei Micro Devices, p.146, July, 1985

7) S.Kimura, E.Murakami, K.Miyake, T.Warabisako and H.Sunami . Extended Abstract of the 16th Conference on Solid State Devices and Materials, Kobe, 1984, p.467

8) S.Kimura, E.Murakami, T.Warabisako, K.Miyake and S.Sunami: Extended Abstract of the 17th Conference on Solid State Devices, Tokyo, 1985, p.271

9) T.Ito, H.Ishikawa and Y.Fukukawa : Proc. of the 12th Confernce on Solid State Devices, Tokyo, 1981, p.33

10) M.Hirayama, T.Matsukawa, H.Arima, Y.Ohno, N.Tsubouchi and H.Nakata : J.Electrochem. Soc., Vol.131, No.3, p.663, 1984

11) T.Ito, T.Nozaki and H.Ishikawa : J.Electrochem. Soc., Vol.127 No.9, p.2053, 1980

12) T.Ito, T.Sugii and H.Ishikawa : Extended Abstract of 16th Conference on Solid State Devices and Materials, Kobe, 1984 p.433

13) T.Yamazaki, T.Ito and H.Ishikawa : Proc. of Jap. Appl. Phys. Soc. Fall Meeting, p.461, 1984

14) A.Ishizaka, P.A.Callen and Y.Shiraki : Extended Abstract of the 16th Conference on Solid State Devices and Materials, Kobe 1984, p.39

15) H.Yamada and Y.Torii : Extended Abstract of the 17th Confernce on Solid State Devices and Materials, Tokyo, 1985, p.305

16) K.Yamada, S.Naritsuka, M.Manoh, K.Shinozaki, T.Yuasa, Y.Nomura, M.Mihara and M.Ishii : Extended Abstract of the 15th Solid State Device Conference, Tokyo, 1983, p.157

17) N.Matsuaki, K.Ohyu, T.Suzuki, N.Kobayashi, N.Hashimoto and Y.Wada : Extended Abstract of the 17th Conference on Solid State Devices and Materials, Tokyo, 1985, p.325

18) T.Okamoto, K.Tsukamoto, M.Shimizu, Y.Mashiko and T.Matsukawa: ibid., p.321

19) M.Yoshida, M.Inoue : Semiconductor World, p.59,April, 1985

20) K.Hamano, Y.Numasawa and K.Yamazaki : Proc. of Dry Process Symposium, Tokyo, 1983, p.85

21) H.Ito, M.Hatanaka, H.Abe and H.Nakata : ibid., p.91

22) H.Ito, H.Hatanaka, Y.Akasaka and H.Nakata : Extended Abstract of the 16th Conference on Solid State Devices and Materials, Kobe, 1984, p.437

23) H.Okuhara and A.Shintani : Extended Abstract of the 17th Conference on Solid State Devices and Materials, Tokyo, 1985 p.173

24) A.Shintani, T.Tsuzuku, E.Nishitani and M.Nakatani : ibid,p.189

25) H.Uesugi, H.Yokoyama and S.Kishida : Jap. J. Appl. Phys., Vol.22, No.4, P-L210, 1983

26) S.Matsuo and M.Kiuchi : Jap. J. Appl. Phys., Vol.22, No.4, P-L210, 1983

27) H.Matsui : Proc. SEMI Technology Symposium, Tokyo 1984, P-W-4

28) Y.Misawa : J.Electrochem. Soc., Vol.131, No.8, p-1862, 1983

29) M.Inoue : Private Communication

30) J.Kanamori : Semiconductor World, p.129, Oct., 1984

Fig. 1.
VLSI trend in Japan[1)]

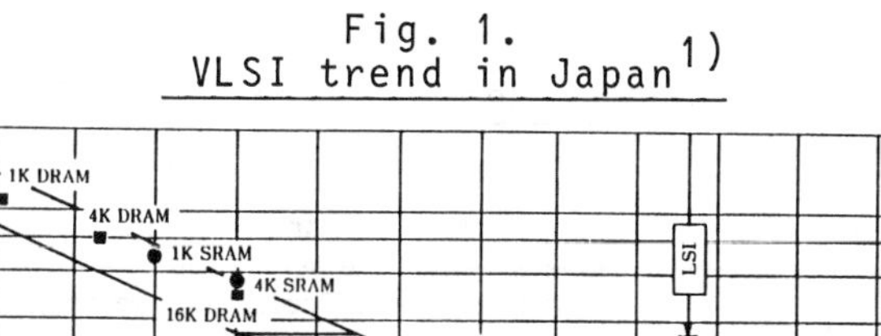

Table 1.
CMOS device parameters[2)]

Parameters		Poly-3	Poly-2	Poly-1.2	Poly-0.8	Poly-0.5
Devices	DRAMs	64K*	256K*	1M	4M	16M
	SRAMs	16K	64K	256K	1M	4M
Gate length —	NMOS (μm)	3	2	1.2	0.8	0.5
	PMOS (μm)	3.5	2.2	1.5	0.8	0.5
Gate oxide thickness (nm)		70	45	25	15	10
Capacitor oxide thickness (nm)		40	25	15	–	–
N^+ junction depth (μm)		0.4	0.25	0.2	0.2	0.15
P^+ junction depth (μm)		0.8	0.5	0.35	0.2	0.15
Well depth (μm)		9	5	3	2	1.5

* Mainly NMOS; no mark indicates CMOS

Fig. 2.
Submicron CMOS structure[3)]

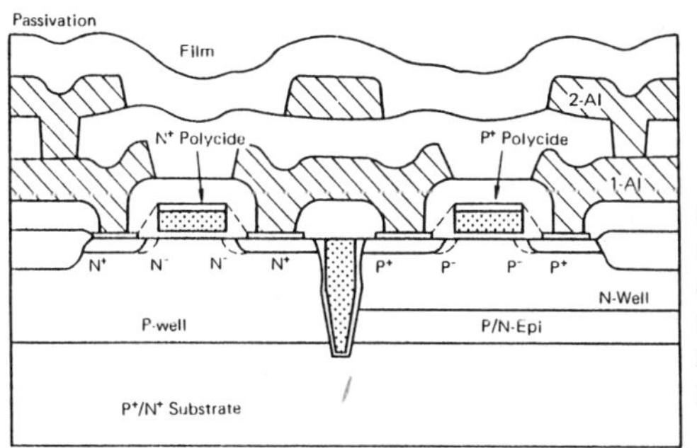

Table 2.
4M DRAM process parameters[3)]

1. 0.7μm processes
   1) ○Transistors — Lg = 0.7μm; tox = 200Å SiO_2; Xj = 0.15μm
      ○Cs — tox = 100Å SiO_2 equiv. trench cell
   2) Measures against hot carriers; sidewall/LDD
   3) Trench isolation
   4) Trench cell; cell area 10μm^2
2. N-well CMOS (peripheral circuits)
3. Refractory metal gate; 2-level Al metallization; salicide
4. Planarization; via fill
5. 850°C process; RTA
6. Large diameter wafer processes; 8-inch diam.
7. Pattern forming
   1) 0.7μm patterning; superhigh-resolution optical stepper
   2) Multilevel resists
   3) Taper controlled etching

Table 3.

Wafer process temperature for VLSI

Process step	RT 200 600 1000 1400
Oxidation	Thermal Oxidation xxxxxxxxxx High Pressure Oxidation xxxxxxxxxx
Epitaxy	Vapor phase Epitaxy xxxxx
CVD	High Temp. CVD xxxx Med.Temp.CVD xxxx xxxxx Low Temp. CVD
PVD	xxxxxxx Sputtering xxxxx Sintering
Reflow	PSG Reflow xxxxx
Etching	xxxx RIE
Impurity Doping	Diffusion xxxxxxxxxx xxxx Ion Implantation Anneal xxxxxxxxxx

Fig. 3.

Trend of VLSI cell area and oxide thichness[6)]

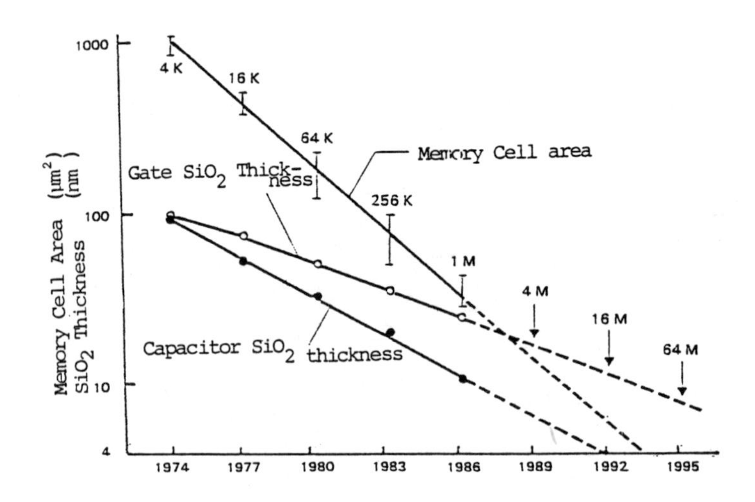

Fig. 4.

Plasma anodic nitridation[10)]

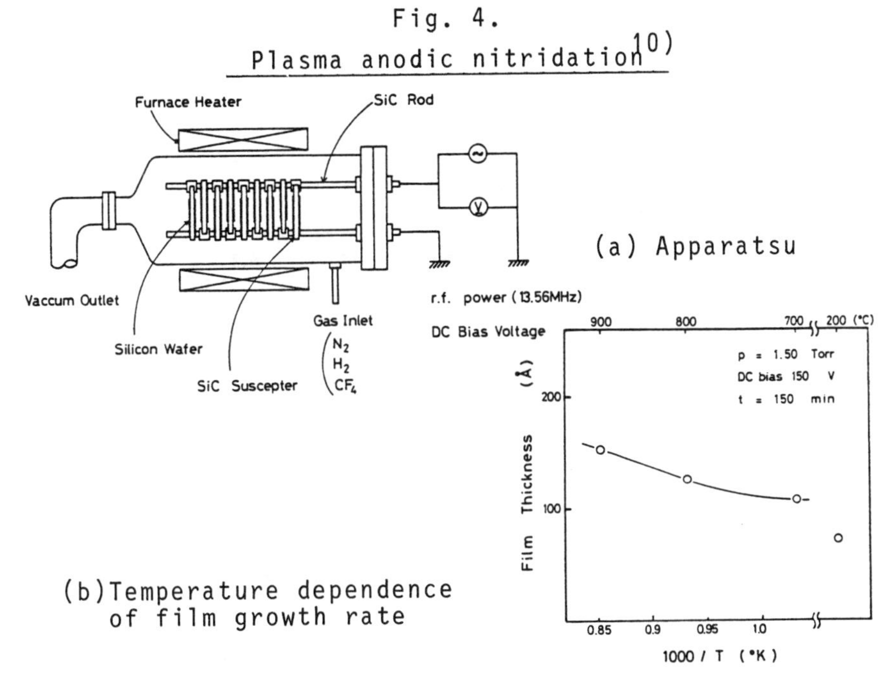

(a) Apparatsu

(b) Temperature dependence of film growth rate

Fig. 5.

Metal silicide contact formation[17)] by rapid thermal anneal(RTA)

(a) Process sequence

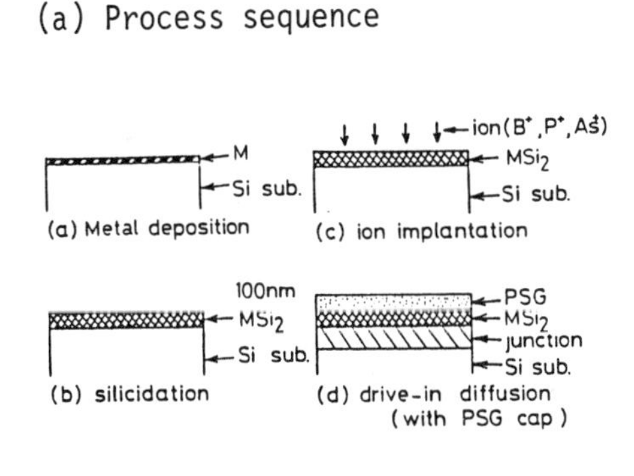

(b) Sheet resistance changes during silicidation and annealing

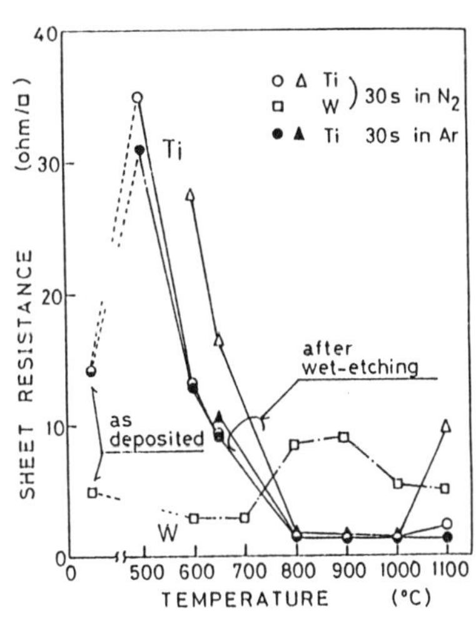

Table 4.
Comparison of photo CVD and plasma CVD[20]

	Si-N ab. coeff.(cm^{-1})	H-content (cm^{-3})	Si/N	Ref. index	Etching rate (nm/sec)	Deposition temp (°C)
Hg-sensitized Photo CVD SiN	$8.5x10^3$	$1.8x10^{22}$	0.65	1.85	2.5	300
Direct Photo CVD SiN	$8.7x10^3$	$1.5x10^{22}$	0.69	1.87	3.8	250
Plasma CVD SiN	$9.7x10^3$	$1.7x10^{22}$	0.75	1.98	0.1	300

Fig. 6.
Photo CVD silicon nitride film[21]

(a) Temperature dependence of deposition rate and refractive Index

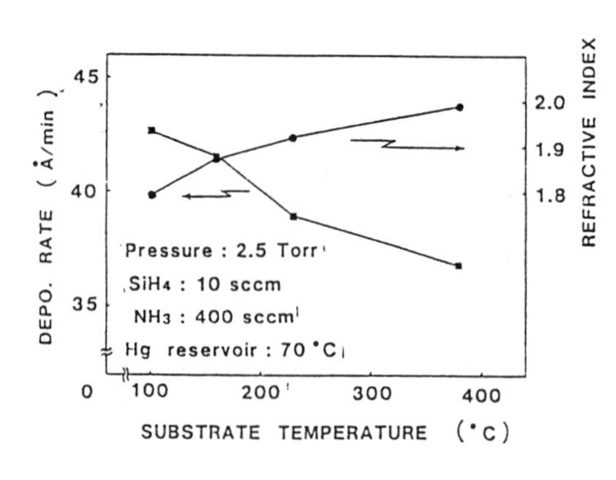

(b) Infrared spectroscopy of photo CVD silicon nitride film

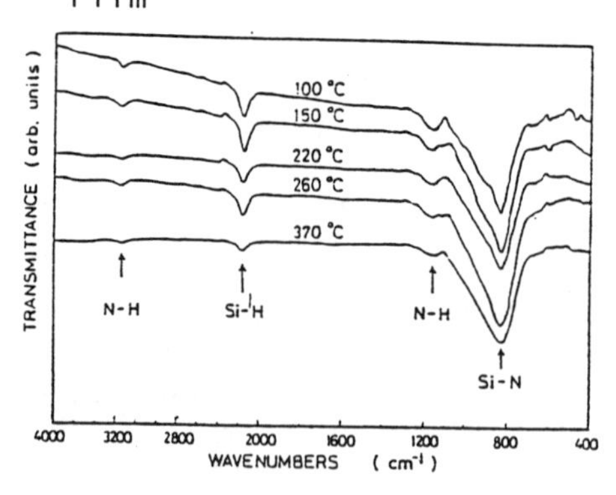

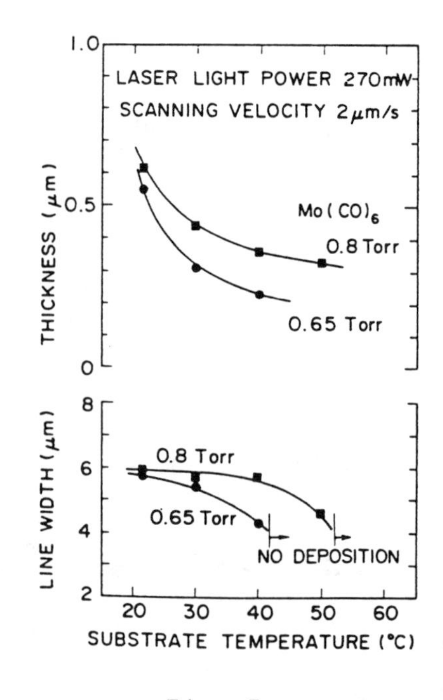

Fig. 7.
Mo line width and thickness dependence on substrate temperature[25]

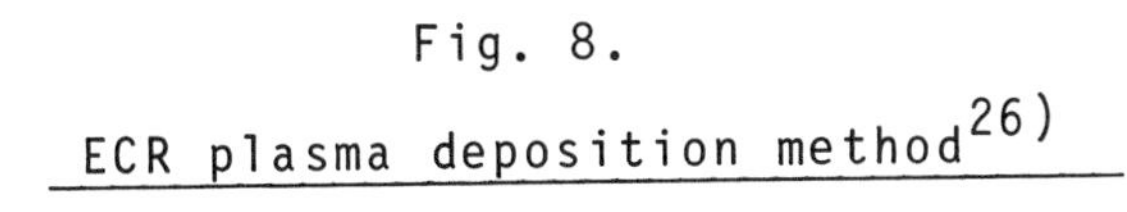

Fig. 8.

ECR plasma deposition method[26)]

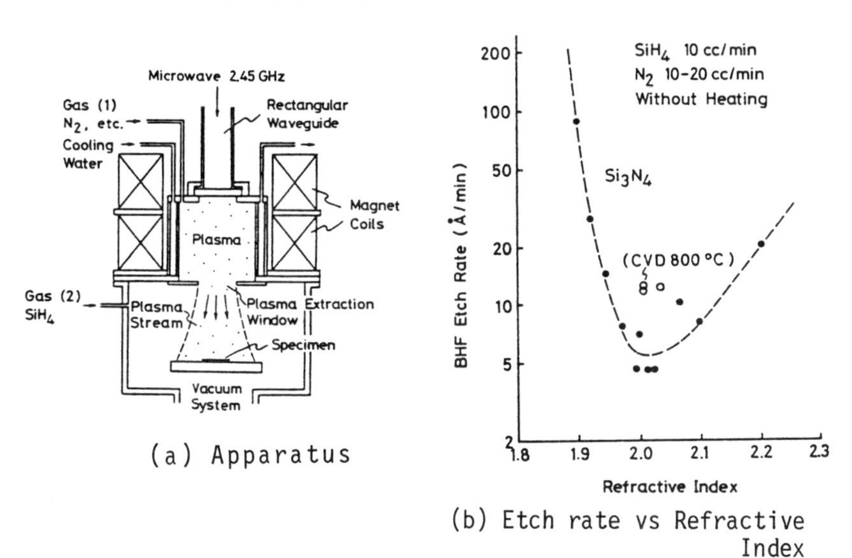

(a) Apparatus

(b) Etch rate vs Refractive Index

Fig. 9.

Planarization by Spin-On-Glass (SOG)[30)]

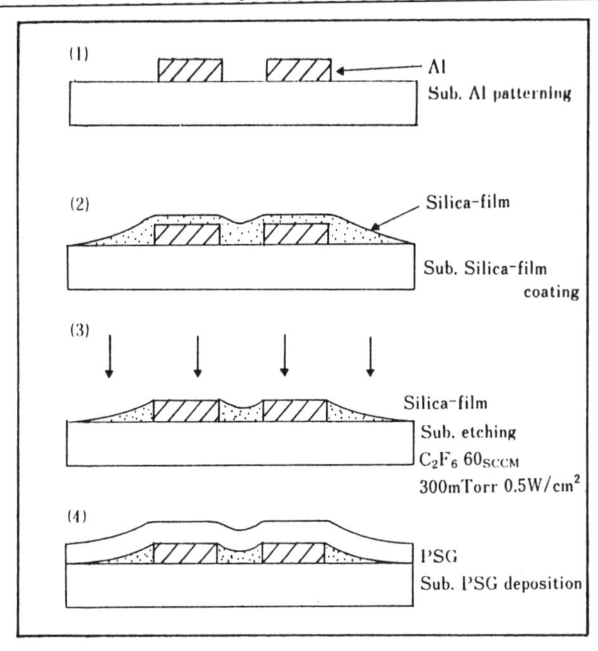

CHAPTER IV

STATUS OF U.S. AND JAPANESE III-V COMPOUND SEMICONDUCTOR TECHNOLOGY

ORGANIZATION OF THE JAPANESE EFFORT ON NON-SILICON BASED OPTO- AND MICRO-ELECTRONICS

W. E. Spicer, Stanford University
Stanford, CA 94305

I. INTRODUCTION

The organization of Japanese technology and science (and the philosophies which underline this organization) are sufficiently different from those in the United States that we must attempt to understand them in order to put a study like this in perspective. Since this is intended for a U. S. audience, emphasis will be placed on areas where the Japanese system differs from that in this country.

The last two decades have seen profound reorganization of government funding for Japanese electronics and related areas. (This pattern is being applied in many technologies, not just electronics.) Since all details of Japanese funding and organization are not available to the author, the purpose here will be to provide a broad overview with increased detail when we approach the area of interest for the reports which follow.

II. THE OVERALL ORGANIZATION OF JAPANESE ELECTRONICS

The overall organization of Japanese electronics and related areas is shown in Figure 1 which is divided into four parts. The Ministry of Education Science and Culture (MESC), the Ministry of International Trade and Industry (MITI), and Ministry of Communications (MC) are government agencies; whereas, the electronics industry is lumped together to make up the fourth group. In recent years it has appeared that the MITI sector has been the fastest growing. However, the work and facilities supported by MESC cannot be ignored. Of particular interest for this report is the MESC funding for research of particular importance for national strength. Through this means, strong funding for areas such as 3-5's is made available for the academic community. Thus, when an area is considered of national importance, funding is not only coupled into the industrial sector by MITI as outlined above but also into University research by special funds from MESC. This should be reflected by university research results mentioned in this report.

Another function of the MESC is the support of large national facilities. For example, a facility strongly used in electronics work is the $60-100M synchrotron facility called the Photon Factory built in Tsuhuba City, the science city near Tokyo. This facility is just coming into operation and is being used by universities, the Optoelectronics Joint Laboratory, industrial laboraties, and NTT (see Figure 1).

Since the Japanese system of technical and scientific education differs from that in the United States, a few words are necessary concerning this. Here I will concentrate on the national universi-

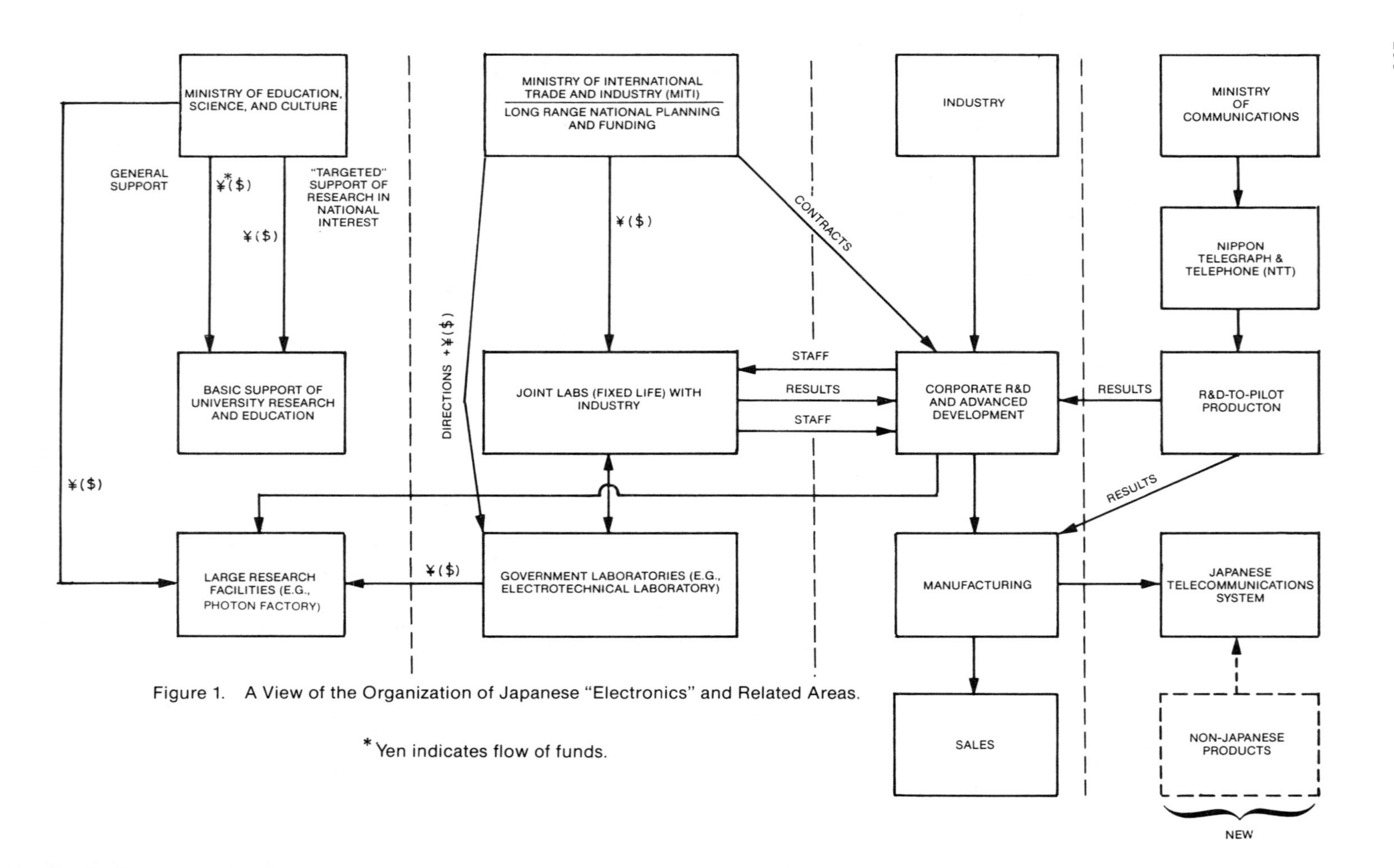

Figure 1. A View of the Organization of Japanese "Electronics" and Related Areas.

*Yen indicates flow of funds.

ties since they are usually considered the leading Japanese universities. Partially because of the highly competitive exams which are the sole factor determining whether or not a student will be admitted to a given national university (e.g. the University of Tokyo), the Japanese high school students work much harder than their counterparts in the U.S. and, thus, come much better prepared to the university. Beginning in their freshman university year, the students concentrate much more strongly on their area of technical or scientific interest than do their American counterparts. As a result at the end of the junior year the Japanese student is still far ahead in his technical or scientific area of speciality. The senior year is spent as a member of a research group containing also master and doctoral candidates. Consequently upon graduation the Japanese undergraduate is in a very advanced position. Many of the best are recruited directly into industry, and a certain fraction of these obtain their doctoral degree (D.Sc.) on the basis of research and publication in industry without further work at the university. This provides a rather unique "bonding" between the D.Sc. and his company. The other path to the D.Sc. is essentially identical to that used in the U.S. However, whichever route is taken, the degree is the same - a Doctor of Science.

The "industrial" column in Figure 1 will not be discussed in any detail as it is quite similar to its U.S. counterparts. However, the reader's attention is directed to the various arrows between the industrial sector and the government and NTT sectors. In general these are quite different from anything in the U. S. although some parallels exist. To understand these cross linkages and their implications is to understand much of the difference in the Japanese and U. S. organization of research and development leading to improved and/or new products.

III. NIPPON TELEGRAPH AND TELEPHONE (NTT)

Let me turn to the last column in Figure 1, Nippon Telephone and Telegraph, (NTT) and the Ministry of Communications (MC). Historically NTT has played much the same role in Japan as ATT in the USA; however, it has been more closely tied to the government through the MC. As of April 1, 1985, NTT was being divested in somewhat the same way as has happened to ATT. Since most of the material presented in this report predates that divestment, it has little effect on this report. However, the future effects may be quite significant. One parallel action, the opening of the Japanese telecommunication industry (the operating branch of NTT) to imported equipment, has, to date, been something of a disappointment to non-Japanese companies. One reason for this will be apparent as we describe NTT.

Despite many similarities, two historic differences between ATT and NTT should be noted. First, the work of NTT's research laboratories has been much more directed toward applications than that of Bell Labs with the broad fundamental component of Bell Laboratory largely missing. Second, and most strikingly, it has no manufacturing and is equivalent to the old ATT without Western Electric, i.e.

it has the R & D plus operating components. In contrast to the operation of ATT, the results from the NTT R & D efforts have been fed into private Japanese electronic manufacturing companies such as NEC, Toshiba, Hitachi, etc. and manufacturing contracts are let to these companies to supply the equipment needed by NTT. The specifications for the equipment was dependent on NTT work in a detailed way. This is probably one reason that non-Japanese companies have found so much difficulty with the specification presently required for equipment to be bought by the operations arm of NTT.

It appears that NTT has the largest electronics R & D effort in Japan; however, it has a surprising lack of visibility in the United States because of its lack of manufacturing and, sales in the U.S. NTT appears to be making a concentrated effort to present more of its work at United States meetings, open up its laboratories more for United States visitors, and, in general, build better relations with the United States.

NTT is organized into bureaus: Engineering, Plant Engineering, Data Communication, for example, are within its R & D Bureau; others are outside of its R & D Bureau. It has just completed the new Atsugi laboratory which is the fourth laboratory under the R & D Bureau. It consists of several buildings which I estimate have over 100,000 square meters of floor space. When I visited it in September, 1983, I was told that it had a research staff of 400 members plus support. It was not clear whether or not it was yet fully staffed. The research emphasized at Atsugi are: 1) R & D on Si VLSI, 2) GaAs and Josephson VLSI, and 3) optoelectronic semiconductor devices based on 3-5 materials. The paralled with the MITI thrust mentioned below is striking. One difference from U. S. work was close contact between the Si and GaAs work with frequent transfer of personnel from GaAs to Si and vice versa.

Particular pride was taken in 50,000 square feet of continuously connected Class 10 clean room space. I was shown (in Sept. 1983) a pilot production line for IC's such as the Si 256K RAM completely set up in this space. Each processing step was being automated. The wafers were transported in an enclosed cassette system via an overhead trolley system between the approximately twenty-three processing stations. The wafers were never exposed to the Class 10 atmosphere. Each processing station was enclosed and the cassettes transferred into them through air locks. By using computer and remote control, the ultimate objective was to remove all personnel from the processing room thus achieving, with the cassette transport, better than Class 1 processing environment.

I also noticed in the same clean space as the the pilot line, two MBE machines for GaAs-based research - the latest model Ruber plus a Japanese made unit. I estimate that I saw only one-third to one-fourth of the clean room space.

As an example of the use of National facilities [supported by the Ministry of Education, Science, and Culture (MESC) - see Figure 1], NTT is mounting a strong research program at the large (approximately $100M) synchrotron radiation facility, named the Photon Factory. NTT has constructed one beam line with three experimental stations. One station is being used intensively for X-ray lithography (IBM has a comparable effort at the National Synchrotron Radiation Light Source, Brookhaven). A second station is being used to study the effects on materials of simultaneous exposure to high intensity pulses of laser and synchrotron radiation (ultra-violet to X-ray wavelength) - a pioneering area with obvious applications to semiconductor processing. The third NTT experimental station has not been put into operation. Hitachi has also built a beam line and Fujitsu will build one. A number of other industrial or government laboratories, e.g. the Joint Optoelectronics Laboratory, are collaborating with groups (usually University) who have access to other beam lines at the photon factory. Thus not only NTT but many industrial laboratories and government facilities are making use of MESC's Photon Factor. The situation is very similar to that of the two largest synchrotron facilities in this country which are supported by the Department of Energy.

From contacts with NTT, I was impressed with the magnitude and focus of the electronics effort and the quality of the scientists and engineers at NTT. The only United States analogy I can think of is if all the talent at Bell Laboratories were sharply and successfully focused for the last twenty years into problems dealing directly with IC, opto-communication, and closely related areas.

IV. MITI

A. Introduction

The emergence of MITI (The Ministry of International Trade and Industry), see Figure 1, and its role as the dominant force spearheading the Japanese attempts to "leap frog" into dominance in new technological areas is critical to understand. As indicated in Figure 1, one mechanism developed by MITI to do this is the "joint laboratories". With guidance from senior members of the Japanese Technical and scientific communities, MITI allots large amounts of funds and organizes "joint laboratories" with a predefined and relatively short lifetime to develop the generic technology and scientific knowledge necessary for such "leap frogging"; however, as will be described below, the reduction of this knowledge to practice and the introduction of new systems is left to individual Japanese companies who operate in a competitive mode. It is important to note that MITI funding is restricted to industry and government laboratories. While its domain does not extend to the Universities, MITI supported facilities or laboratories may use large MESC facilities such as the Photon Factory. This seems to provide one of the few places where MITI and MESC efforts strongly interact.

In the electronics area, the first MITI "joint laboratory" in electronics (The VLSI Cooperative Laboratory) was formed about a decade ago and was disbanded after its predetermined life expired around the turn of the decade. The success of this effort can perhaps best be judged by the growth of the Japanese integrated circuit (IC) industry in the last decade.

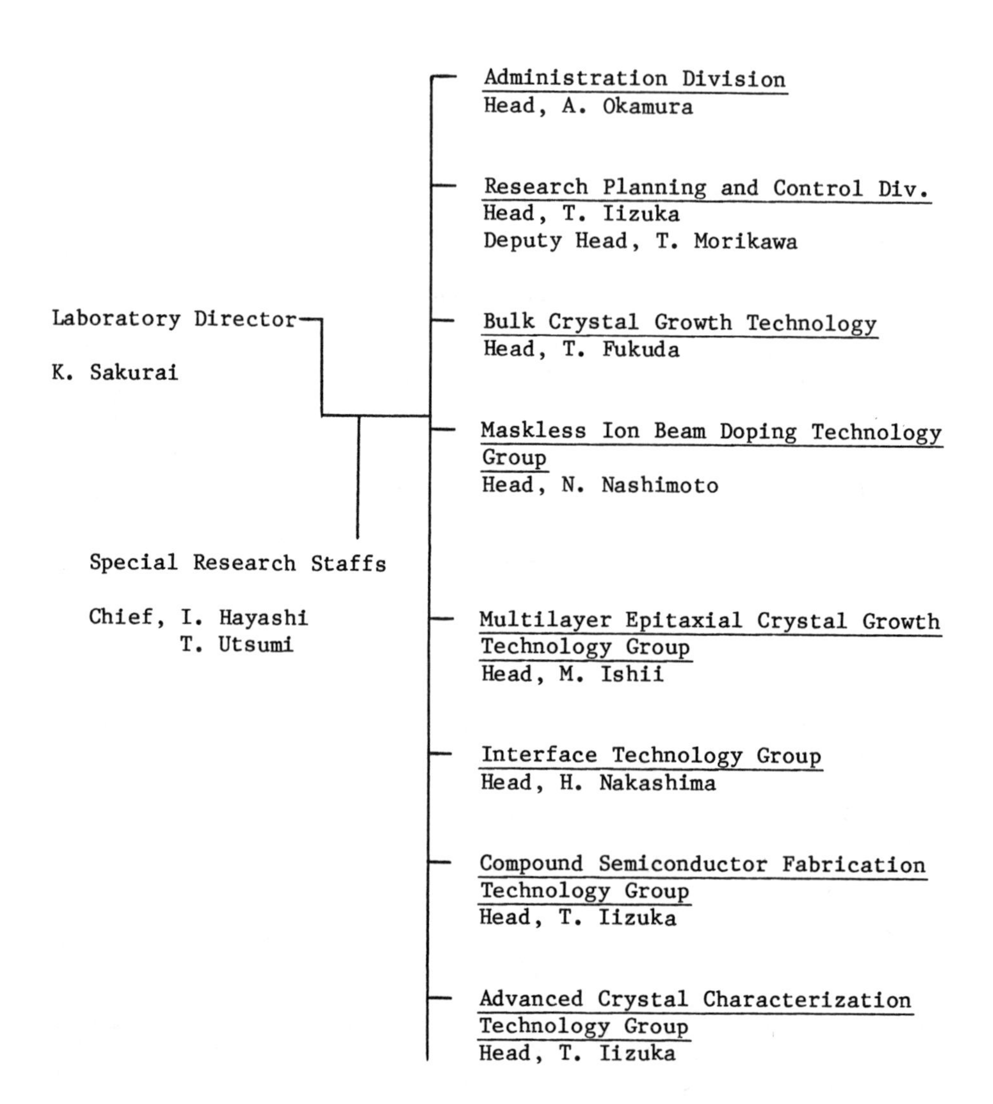

Fig. 2. Organization of Optoelectronics Joint Research Laboratory.

OKI ELECTRIC INDUSTRY CO., LTD.

OPTOELECTRONICS INDUSTRY AND TECHNOLOGY DEVELOPMENT ASSOCIATION

SHIMADZUI SEISKUCHO, LTD.

SUMITOMO ELECTRIC INDUSTRIES LTD.

TOSHIBA CORPORATION

NIPPON SHEET GLASS CO., LTD.

NIPPON ELECTRIC CO., LTD.

HITACHI, LTD.

FUJITSU LTD.

FUJI ELECTRIC COMPONENTS

RESEARCH AND DEVELOPMENT LTD.

FUJIKURA CABLE WORKS, LTD.

FURAKAWA ELECTRIC CO., LTD.

MATSUSHITA ELECTRIC INDUSTRIAL CO., LTD.

MITSUBISHI ELECTRIC CORPORATION

YOKOGAWA ELECTRIC WORKS, LTD.

- Each company provides professionals to make up research staff of approximately 50 for Joint Lab.
- Laboratory to be dissolved in 1986 and staff returned to companies.

Fig. 3. Member companies of MITI Joint Optoelectronics Laboratory.

B. The Optoelectronics Joint Research Laboratory of MITI

To illustrate how such joint laboratories operate and to focus on the area of this report, let us examine the Optoelectronics Joint Research Laboratory. By opto-electronics one means a very wide range of technology in which optical as well as electrical functions are carried out. It includes, for example, optical communications and fast computing. In some cases classical electrical functions may be replaced by optical functions. In this laboratory, there is considerable emphasis on high frequencies (giga Hertz). This plus the fact that Si is not a useable light emitter, automatically brings concentration on 3-5 compound semiconductors, e.g. GaAs. Thus, all the work of this laboratory lies in the area of this report. The funding for this laboratory is $75 M.

Figure 2 indicates the original organization of that laboratory. It is especially important to note the six research groups into which the laboratory was organized since this gives a good overview of the areas of concentration. Note two things: First, the organization is strongly built around materials growth and processing; second, it goes from the most basic materials aspect - the growth of "ideal" large single crystals - to such advanced topics as Maskless Ion Beam Doping and finally to Fabrication Technology. Again it is important to emphasize that the titles of the various groups indicate emphasis on the generic problems of the technology being developed with detailed development of systems to be manufactured left up to the individual companies.

By examining the way in which the MITI joint labs are staffed, one can get the best insight into the joint government - private industry partnership represented by these laboratories. Figure 3 gives a list of the companies which are members of the Opto-electronics Joint Laboratory. The research staff of approximately fifty professionals comes from the permanent staffs of these companies, and they will return to their parent company either during the life of the Joint Lab or when it is dissolved in about 1986. The optoelectronics thrust started in 1979; however, the Joint Laboratory was still receiving large influxes of new equipment (e.g. another MBE machine) when I visited it in September, 1983.

Perhaps the important points to emphasize about the joint labs are:

1) These laboratories are funded by the government,
2) The research staff comes from and <u>returns to</u> the companies which are members of the laboratories,
3) Each research staff member has responsibility to transfer back to his company the results obtained by the laboratory. The companies may then use this knowledge for rapid development of new product lines.
4) The Joint Laboratories are dissolved after a fixed period (perhaps eight years).

Thus, the Joint Laboratories represent something almost unique in peace time. They most closely resemble laboratories such as the Radiation Laboratory at M.I.T. which was formed during World War II and was dissolved at the end of that war. Questions have been raised in

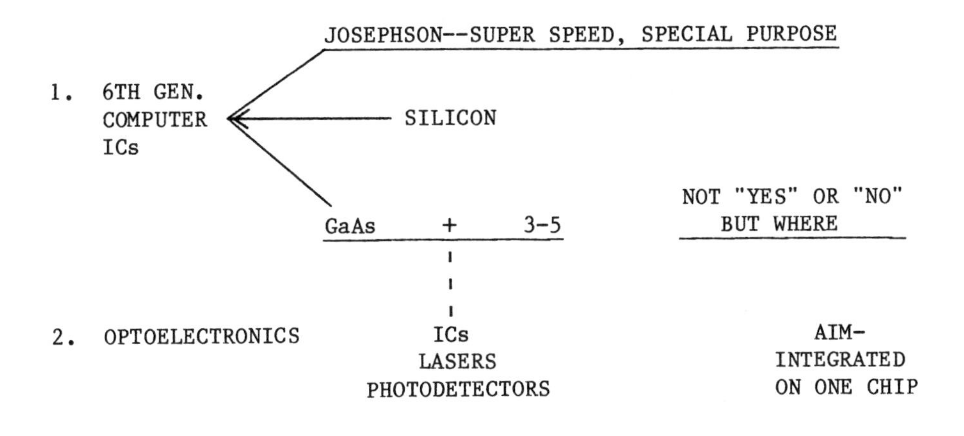

OBJECTIVES AND PROJECTIONS

OPTOELECTRONICS--TO DEVELOP INTO OEIC (OPTOELECTRONIC INTEGRATED CIRCUITS)

JUNE 1983--NO EXISTING OEIC, BUT JAPANESE BELIEVE IT WILL BE DEVELOPED AND GROW TO BECOME KEY ELEMENT IN FUTURE ADVANCEMENT OF OPTICAL COMMUNICATIONS, INFORMATION PROCESSING, AND SENSING SYSTEMS

MITI JOINT OPTOELECTRONICS LABORATORY $75M 1979-1986

SIZE JAPANESE OPTOELECTRONICS INDUSTRY

1978	$0.1B	1983	$1.8B (E)	1990	$ 8.4B (E)
1982	1.1B	1985	4.2B (E)	2000	50.0B (E)

(E) = ESTIMATE

Fig. 4. Organization of Japanese advanced optoelectronics and microelectronics.

recent years about U.S. Government Labs which may have outlived their mission (see, e.g. The Packard Report). The Japanese Joint Laboratories may serve as one example of an alternative approach.

In the above we have concentrated on the Joint Optoelectronics Laboratory as an example of one part of the Japanese strategy to move forward quickly and efficiently in electronics. Figure 1 attempts to put the joint laboratories into perspective within the whole Japanese structure in electronics. We have concentrated on the Joint Laboratory since it is perhaps the most innovative part of this structure.

C. The MITI Optoelectronics and Its 6th Generation Supercomputer Programs

As Figure 4 shows, R and D on 3-5 semiconductors is also a part of MITI Supercomputer Program. In this section we will first assess the overall effort in the optoelectronics area and then attempt to take into account that in the Supercomputer Program.

The Joint Optoelectronics Laboratory is one part of the overall MITI thrust in optoelectronics which started in 1978 and will be finished in 1986. At that time MITI will have invested $75 M in that program. It is not clear how much the industrial funding will add to this in the overall research and development phase which extends beyond the time when the Joint Laboratory closes and the industrial staff move back to their respective laboratories; however, a factor of four increase in funding over MITI's original investment is probably not unrealistic. It should not be assumed that the work in industrial laboratories starts when the joint laboratory closes; rather, it should be considered a continuous process stimulated by information flow between the joint laboratory and, the member companies. Remember that each member of the companies staff at a joint lab has the responsibility of keeping communications flowing with his own company. Because of lower salaries and longer working hours, the actual productivity per dollar may be as much as twice that in the United States. The inflated value of the dollar with respect to the yen is another factor. This reduction in costs is somewhat offset by capital equipment expenditures which seem to be considerably higher per researcher than in many U.S. laboratories. I found the Joint Optoelectronics Laboratory to be perhaps the best equipped per staff member that I have ever seen. Moreover, as we all know, the increase in capital equipment/R & D staff member, should lead to higher productivity.

There is another very large MITI thrust area which includes work covered by this report - the supercomputer thrust. In addition to the work in computer science and architecture, strong efforts are being made in Si, 3-5 and Josephson junction hardware under this MITI program; I do not have any actual numbers on the 3-5 expenditures but would suspect that they were comparable to that in the Optoelectronics area. It is to be expected that there is a strong element of mutual support between these areas, particularly after information is transmitted back to the companies involved.

V. ONE EXAMPLE: A BRIEF GLIMPSE OF THE DEVELOPMENT OF 3-5 SEMICONDUCTOR INTERFACE

In the preceding part of this paper, we have tried to provide an outline of the organization of Japanese technology and science with emphasis on work on 3-5 compound semiconductors. In this section we will attempt to illustrate how this system works by examining the development of surface and interface work on 3-5 semiconductors in Japan.

A decision based on studies made by the Japanese appears to have been made in the early 1970's that a concentrated Japanese effort was needed in 3-5 surfaces and interfaces (with the emphasis on interfaces). The first result I know of was a five year University program funded by the Ministry of Education, Science and Culture (MESC) in the mid 70's. This effort was capped by the International Conference on Solid Thin Films and Surfaces held in Tokyo, July 5-8, 1978. The Proceedings of the conference were published as Vol. 86 of Surface Science, in 1979. This conference was followed by a two-day post-conference symposium, "Basic and Applied Studies on Surfaces and Interfaces between Semiconductors and Other Materials Related to Electronic Devices" where some thirty leading foreign and Japanese scientists discussed critical problems in a seminar setting.

This phase of university work ending in 1978 could be looked upon as part of a determined effort by the Japanese Universities to bring themselves up to the state-of-the-art in surface and interface experimental techniques. This program also greatly increased overall knowledge relative to surface analysis and interface studies (particularly high vacuum technology) as well as other surface thin film studies and provided the trained professionals necessary to transfer such work to Japanese industry and government laboratories. MBE was particularly emphasized. At about the same time as the start of this "thrust", approximately $60M was committed by MESC for the building of the Photon Factory, a massive synchrotron radiation facility. On a smaller scale such facilities in the West have proved valuable in surface and interface studies as well as in other phases of science and technology important for 3-5 semiconductor interface studies. Thus, the possibilities of such studies were probably among the motivations for the investment in the Photon Factory which is now being used for such studies.

Incidentally, other areas of interest for this study are also involved in the Photon Factory. One example is the use of synchrotron radiation for X-ray lithography, an area which - except for work at IBM - is being largely ignored in this country. This is not to say that the Japanese seem convinced that X-ray lithography will be an important VLSI production tool, but they are determined to find out if this will be the case and to be in a leadership position if it is.

The Japanese are also increasing the amount of basic work they are doing. Again an example is the work using synchrotron radiation from their very advanced radiation source, the Photon Factory, and photoemission or closely related techniques to study the fundamentals of surfaces and interfaces [1-4]. As might be expected, when such research builds up in a country, some of the results and their interpretations have proven to be controversial. However, the important aspect is that the fundamental work in this field is building up rapidly and growing at a much faster rate than anywhere else in the world.

Despite very rapid progress, the Japanese have a long way to go until they build the strong fundamental programs which exists in the United States and in Europe. Thus, their practical work tends to depend more on an empirical approach and on basic work done elsewhere than does work in this country despite outstanding basic contributions from individual Japanese workers. As can be seen from the organization chart of the Optoelectronics Joint Laboratory (Figure 2), approximately one-sixth of the effort of this laboratory is devoted to interface work. This has very well defined goals, contains strong applied research, and does some very fundamental work.

Let us examine one recent development - self-aligned gates for GaAs - IC's - reported by Japanese industrial workers in order to gain insight into their present approach. However, we will not attempt to compare this with parallel work going on in the United States.

Self-aligned gates for GaAs MESFETS represent an important step in reaching an optimum technology for GaAs IC's. In this technology the Schottky barrier gate is used to shield the volume under it from the ion implant; whereas, the areas adjacent to it are converted to highly conducting n+ material via the implant and subsequent anneal. By this approach it is possible to form a FET which is turned off (does not pass current) with no voltage applied. As a result energy consumption can be minimized and packing density maximized. The critical problem was to find a gate metal which could withstand the high temperature anneals necessary to activate the implants, typically 750°C to 950°C depending on time of anneal [5], which can vary from 15 minutes to 6 seconds. A group at Fujitsu Laboratory tried TiW and, finding it unstable at the annealing temperatures, turned to TiW silicide which gave promising results [6]. The choice of TiW silicide was reported [6] to be based on the success of this material for high temperature in silicon technology [5,7]. Subsequently, W_xSi was found to give quite a satisfactory gate [5,8] with optimum value of x being about 0.64. $Ti_{0.3}W_{0.7}Si$ was found to make successful self-aligned gates, but W_xSi was found more satisfactory because of the difficulty of controlling the $Ti_{0.3}W_{0.7}Si$ composition as compared to that of W_xSi. From their publications it is clear that the Fujitsu group has worked carefully and hard to find the optimum composition and annealing temperature for the gate. It is also clear that they had the device and manufacturing problems well in mind while doing this work so that products came out of the research laboratory suitable for introduction into manufacturing. It is equally apparent that this development is based on an almost purely empirical approach.

By this example one should not form the opinion that the Japanese are uniformly successful. While visiting a Japanese laboratory in 1983, I found considerable excitement about using a Si-Ge-boron alloy for a self-aligned gate. Later I was told that this work was abandoned because of lack of reproducibility. As in most research, the Japanese work on self-aligned gates seemed to be characterized by a number of groups working hard and with enthusiasm on a number of different approaches. The size of the effort is an important consideration.

VI. CONCLUDING REMARKS

Based on the material herein, I have observed that the Japanese exhibit confidence in long-range government planning and execution of these plans for "the national good". This is particularly apparent in their efforts to build up high technology and scientific resources which produce an increase in the Japanese gross national product. Note, for example, Figure 4, which illustrates the allotment of $75M of MITI funds to expand the optoelectronics industry from $0.1B in 1978 to a predicted $50B in the year 2000. Note also that this effort drew on university efforts supported and directed by MESC in the mid 1970's (Section V of this report). Thus, one can identify government-directed and funded work aimed at increasing the gross national product in a specific area spanning a period of about twenty-five years. It would be very hard to find a parallel in the United States government efforts. The parallels probably come from our defense effort and are not aimed at increasing the gross national product as are Japanese efforts. Many arguments have and undoubtedly will be made against the Japanese approach. However, the Japanese success in increasing their gross national product (particularly via exports to the United States) argue that we cannot ignore their organization to address the dual objectives of increasing their gross national product and keeping a favorable balance of payments. The Japanese have a superb job of learning from others; perhaps we need to work harder at this.

References:

1. K. L. I. Kobayoshi, H. Daiman, and Y. Murata, Phys. Rev. Lett. 50, 1701 (1983).

2. K. L. I. Kobayoshi, H. Daiman, and Y. Murata, Phys. Rev. Lett. 52, 1569 (1984).

3. K. L. I. Kobayoshi, N. Watanabe, H. Nakashima, M. Kubota, H. Daiman, and Y. Murata, Phys. Rev. Lett. 52, 160 (1984).

4. M. Taniguchi, S. Suga, M. Seki, H. Sakamoto, H. Kanzaki, Y. Akahama, S. Endo, S. Terada, and S. Narita, Solid State Commun. 49, 867 (1984).

5. T. Ohnishi, Y. Yamaguchi, T. Inada, N. Yokoyama, and H. Nishi, IEEE Electron Device Letters, EDL-5, 403 (1984).

6. N. Yokoyama, T. Ohnishi, K. Odani, H. Onodera, and M. Able, IEEE Transactions of Electron Devices, ED-29, 1541 (1982).

7. In visiting Japanese laboratories, the author has been sruck by the close proximity of GaAs and Si work and by the movement of scientists and engineers from work on one technology to another. The development of this GaAs self-aligned gate technology may have benefited from such flexibility.

8. T. Ohnishi, N. Yokoyama, H. Onodera, S. Susuki, and A. Shibatomi, Appl. Phys. Lett. 43, 600 (1983).

Properties and Applications of Multiquantum Well and Superlattice Structures

K. Hess
Coordinated Science Laboratory and
Department of Electrical and Computer Engineering
University of Illinois
Urbana, Illinois 61801

The principles and new features of electronic transport in lattice matched III-V compound semiconductor heterolayers will be reviewed. It will be demonstrated that new horizons have opened up in semiconductor device research with the possibility of varying the bandgap as well as the doping levels. In addition, light as well as electric currents can be used to transport signals in new forms of III-V compound heterolayers. A detailed discussion will be devoted to electronic transport in high electron mobility transistors. It will be shown that this device draws major advantage from significant velocity overshoot.

Further emphasis will be given to real space transfer devices and planar doped barrier structures. All of these devices profit from the possibility to design device functions from physical considerations involving Ångstrom scale dimensions.

A comparison of research involving these multiquantum-well structures in Japan and in the US will also be given.

Band-gap engineering and interface engineering: novel superlattice structures and tunable band discontinuities

Federico Capasso

AT&T Bell Laboratories
Murray Hill, New Jersey 07974

I. Bandgap Engineering, Interface Engineering and Heterojunctions for Everything.

The advent of Molecular beam epitaxy (MBE) has made possible the development of a new class of materials and heterojunctions with unique electronic and optical properties. Most notable among these are: heterojunction and doping superlattices, modulation doped superlattices, strained layer and variable gap superlattices. The investigation of the novel physical phenomena made possible by such structures has proceeded in parallel with their exploitation in novel devices. As a result a new approach or philosophy to designing heterojunction semiconductor devices, band-gap engineering, has gradually emerged.[1-3]

The starting point of band-gap engineering is the realization of the extremely large number of combinations made possible by the above mentioned superlattices and heterojunction structures. This allows one to design a large variety of new energy band diagrams. In particular through the use of band-gap grading one can obtain, starting from a basic energy band-diagram, practically arbitrary and continuous variation of this diagram. Thus the transport and optical properties of a semiconductor structure can be modified and tailored to a specific device application. One of the most powerful consequences of band-gap engineering is the ability of independently tuning the transport properties of electron and holes, using quasi-electric fields in graded gap materials and the difference between conduction and valence band discontinuities in a given heterojunction. This unique tunability and the philosophy "Heterojunctions for everything" was discussed by H. Kroemer[1] and more recently by the author.[2,3]

A new technique, capable of modifying band discontinuities at heterojunctions using doping interface dipoles has also been experimentally demonstrated.[4-5] This method allows one to modify selectively the band-diagram in the vicinity ($\lesssim 100$Å) of a heterointerface (interface engineering) and therefore has tremendous potential for basic studies and device applications.

1. Multilayer Avalanche Photodiodes and Solid State Photomultipliers

A large difference between the ionization rates for electrons (α) and holes (β) is an essential requirement for low noise APDs. We have demonstrated experimentally that in a 50 layers superlattice ($Al_{0.45}Ga_{0.55}As/GaAs$, as shown in Fig. 1), the effective impact ionization rates for electrons and holes are very different[6] (α/β=8; Fig. 2) even if they are comparable in the basic bulk materials (in GaAs $\alpha = 2\beta$). One of the physical reasons for the large α/β is the difference between the conduction and valence band edge discontinuities at the AlGaAs/GaAs interface ($\Delta E_c = 0.62\Delta E_g$ $\Delta E_v = 0.38\Delta E_g$). This effectively reduces the ionization energy of hot electrons entering the well more than the ionization energy of hot holes thus greatly increasing α/β.[7] The enhancement of α/β in superlattices has received further experimental confirmation recently.[8] It has also been proposed that the hole scattering rate is increased in quantum wells, thus also contributing to the enhancement of the α/β ratio.[7,8] Heterojunctions of interest for long wavelength (1.3-1.5μm) superlattice detector applications are $(Al_xGa_{1-x})_yIn_{1-y}As/In_{0.53}Ga_{0.47}As$ and $Al_xGa_{1-x}As_ySb_{1-y}/GaSb$, for which there is evidence of a large difference between ΔE_c and ΔE_v. Recently we have reported the first operation of an $Al_{0.48}In_{0.52}As/Ga_{0.47}In_{0.53}As$ multiquantum well avalanche photodiode.[9]

The p^+in^+ structure, grown by molecular beam epitaxy, consists of a 35 period $Al_{0.48}In_{0.52}As$ (139 Å)/$Ga_{0.47}In_{0.53}As$ (139 Å) multiquantum well *i* region sandwiched between p^+ and n^+-$Al_{0.48}In_{0.52}As$ transparent layers. DC and high-frequency multiplications of 32 and 12, respectively, have been measured; the dark current at unity gain is 70 nA. High speed of response (Fig. 3) with full width at half-maximum of 220 ps at a gain of 12 and the absence of tails are demonstrated, indicating that carrier pile-up in the wells is negligible. This result is very important, considering that the conduction band discontinuity is large (ΔE_c=0.5 eV). Hole pile-up at the valence band barrier (ΔE_v=0.2 eV) had previously been shown to be negligible.[10] Thus grading of the interface at the exit of the wells (for electrons) is not necessary to achieve high speed.

In cases where ΔE_c is equal to the electron ionization energy in the material following the conduction band step, the structure of Fig. 4 can be used to achieve the solid state analog of the photomultiplier (staircase APD).[11] Here electrons gain all the kinetic energy for the ionization at the steps since $\Delta E_c = E_i$, while holes do not ionize because the electric field is too low to cause hole-initiated multiplication. Note the low bias voltage V of this structure; for 5 stages with ΔE_c=1eV, the gain is $= 2^5 = 32$ and V=5 volts. The reason that in the staircase APD the avalanche noise is lower than in the best conventional APD (one in which only one type of carrier can ionize) can be understood as follows. In a conventional APD the avalanche is more random because carriers can ionize everywhere in the avalanche region, while in the staircase APD electrons ionize at well defined positions in space (the steps) with a high probability ($\simeq 1$) (i.e. the multiplication process is more deterministic).[11] Note that similarly in a photomultiplier tube the avalanche is essentially noise free (F=1).

A possible material system for the implementation of this device in the 1.3-1.6μm region is AlGaAsSb/GaSb as discussed in Ref. 10. In practical cases a thin ungraded layer (100-200Å) should be left after the conduction band step to account for the finite ionization mean free path ($\lambda_i \approx 50$Å).

Another structure with ultrahigh α/β ratio has also been proposed.[12] Here the enhancement of α/β is produced by the spatial separation of electrons and holes in materials of different bandgap (Fig. 5). Experiments have revealed other unique optoelectronic properties of this structure when biased at very low voltage such as the low capacitance and its step-like dependence on reverse bias.[13]

Finally, also a simple graded gap layer in the high field region of a pn junction can be used to enhance α/β, as recently demonstrated.[1] The principle of the device is illustrated in Fig. 6. It is seen that electrons experience a higher effective electric field and lower ionization energy than holes, since they are moving towards lower gap regions. The effect is a significant increase in the α/β ratio, when the grading exceeds $1\text{eV}/\mu\text{m}$.

2. Repeated Velocity Overshoot Structures, Electrical Polarization Effects in Sawtooth Superlattices and Pseudoquaternary Alloys

Other interesting applications of staircase potentials have been proposed. We shall discuss here the repeated velocity overshoot device.[14] This structure offers the potential for achieving average drift velocities well in excess of the maximum steady state velocity over distances greater than 1μ.

Figure 7(a) shows a general type of staircase potential structure. The corresponding electric field, shown in Fig. 7(b), consists of a series of high field regions of value E_1 and width d superimposed upon a background field E_0. Electrons, upon entering the high field regions, gain rapidly energy and momentum, so that the drift velocity overshoots the steady state value. The energy and momentum are then allowed to relax in the subsequent low field region via phonon conditions. With the staircase structure repeated velocity overshoot can be achieved (Fig. 7(c)). A graded gap AlGaAs structure with steps $\Delta W = 0.2\text{eV}$ could be used for this purpose.

Sawtooth superlattices have also other intriguing physical properties such as the possibility of generating a transient macroscopic electric polarization extending over many periods of the superlattice. This effect is a direct consequence of the lack of reflection symmetry.[15] The energy band diagram of a sawtooth p-type superlattice is sketched in Fig. 8(a). The layer thicknesses are typically a few hundred angstroms and a suitable material is graded gap $Al_xGa_{1-x}As$. The superlattice is sandwiched between two high doped p^+ contact regions. Let us assume that electron-hole pairs are excited by a very short pulse as shown in Fig. 8(a). Electrons experience a higher quasi-electric field due to the grading and have a much higher mobility than holes. Therefore electrons separate from holes and reach the low-gap side in a subpicosecond time ($< 10^{-13}$)sec. This sets up an electrical polarization in the sawtooth structure which results in the appearance of a voltage across the device terminals (Fig. 8(b)). This macroscopic dipole moment and its associated voltage subsequently decay in time by a combination of (a) dielectric relaxation and (b) hole drift under the action of the internal electric field produced by the separation of electrons and holes (Fig. 8(c)). This polarization phenomenon has recently been observed in AlGaAs sawtooth superlattices. The transient photovoltage showed a decay time of $\approx 150\text{ps}$.[15]

An elegant and very effective way to achieve band-gap grading, for example, in the AlGaAs system, is by growing an AlAs/GaAs superlattice with spatially varying period or duty factor, so that the average composition and band-gap varies with position. Such a technique has recently been used to grow parabolic quantum wells.[16] A similar approach was recently taken by Capasso et al., who demonstrated for the first time a pseudo-quaternary GaInAsP semiconductor consisting of a graded gap $Ga_{0.47}In_{0.53}As/InP$ superlattice.[17] The average composition and the band gap of this structure are spatially varied by gradually changing the thicknesses of the InP and $Ga_{0.47}In_{0.53}As$ layers between 5 and 55 Å while keeping constant the period of the superlattice (=60 Å). This new graded gap superlattice has been used to eliminate the interface pile-up effect of holes in a "high-low" $InP/Ga_{0.47}In_{0.53}As$ avalanche photodiode, without requiring the growth of a separately lattice-matched $Ga_{1-x}In_xAs_{1-y}P_y$ layer (Fig. 9).

Pulse response studies of conventional avalanche photodiodes with abrupt $InP/Ga_{0.47}In_{0.53}As$ heterojunctions show a long (> 10 ns) tail in the fall time of the detector due the pile-up of holes at the heterointerface (Fig. 10a). This is caused by the large valence-band discontinuity (≅0.45 eV).

Insertion of the previously described InP/GaInAs superlattice between the InP avalanche region and the $Ga_{0.47}In_{0.53}As$ absorbing layer as shown in Fig. 9(a),(b) produces an effective grading of the heterointerface, thus eliminating the hole pile-up effect and enhancing the device speed (Fig. 10b). This structure was grown by a new vapor phase technique (levitation epitaxy), which provides extremely abrupt interfaces.[18]

3. Superlattice Effective Mass Filters

We have recently reported the observation of a new extremely large photocurrent amplification phenomenon at very low voltages in a superlattice of $Al_{0.48}In_{0.52}As/Ga_{0.47}In_{0.53}As$ in the quantum coupling regime (35 Å wells, 35 Å barriers). Room temperature responsivities at $\lambda = 1.3$ μm are typically 2×10^3 and 300 Amps/Watt, at 0.3 Volt and 0.08 Volts bias respectively, while the highest measured value is 4×10^3 A/W, corresponding to a current gain of 2×10^4. This effect, which represents a new quantum type photoconductivity, is caused by the extremely large difference in the tunneling rates of electrons and holes through the superlattice layers (*effective mass filtering*; Fig. 11).[19]

The superlattice was grown by molecular beam epitaxy and consists of one hundred, undoped (n $\simeq 3\times 10^{15}/cm^3$) 70 Å periods sandwiched between two degenerately doped n^+, 4000 Å thick, $Ga_{0.47}In_{0.53}As$ layers. Short circuit photocurrent measurements were made at low light levels ($\lesssim$ 5n W). This new effective mass filtering concept can be explained as follows. In the regime of quantum coupling between the wells, the states of the carriers form minibands. When a perpendicular electric field is applied, band-like conduction via these extended Bloch states cannot be supported if the mean freepath of the carrier does not exceed appreciably the superlattice period. This occurs for example when intra- and inter-layer thickness fluctuations cause fluctuations in the subband energies of the order or greater than the miniband width ΔE. In our superlattice the electron and (heavy) hole ground-state miniband widths are smaller than the estimated in-homogeneous broadening due to size fluctuations. Hence conduction will proceed by phonon-assisted tunneling between adjacent wells, the so-called hopping conduction. Since electrons have a much smaller mass than holes, their phonon-assisted tunneling rate between adjacent wells is much larger (*effective mass filtering*). Photogenerated holes therefore remain localized in the wells (their hopping probability is negligible) while electrons propagate through the superlattice Fig. 11a. This effective mass filtering effect produces a photocurrent gain, given by the ratio of the lifetime to the electron transit time. The gain strongly decreased with increasing $Al_{0.48}In_{0.52}As$ barrier layer thickness and becomes unity when this exceeds 100 Å. This confirms effective mass filtering as the origin of the large gain, since, as the barriers are made thicker, electrons eventually become localized, thus decreasing the tunneling probability and increasing the recombination rate. The temperature dependence of the responsivity conclusively confirms hopping conduction.[19]

For superlattices made of the same two materials with wider electron minibands (achieved by using thinner barriers) the electron transport is by miniband conduction, while holes are still localized (Fig. 11b). Such superlattice effective mass filters will have a much greater gain-bandwidth product due to the much shorter electron transit time, than the other kind (Fig. 11a). We have also observed recently a large photoconductive gain (7×10^3) related to effective mass filtering of the type shown in Fig. 11b in a forward biased pn junction with a superlattice (100 periods of 23 Å thick $Al_{0.48}In_{0.52}As$ and 49 Å thick $Ga_{0.47}In_{0.53}As$) in the n layer.[20] This effect is accompanied by a blue shift in the spectral response and a reversal in the direction of the photocurrent and occurs only for applied voltages greater than the built-in voltage.[20] In summary the crucial difference of effective mass filters compared with conventional photoconductors is that the current gain and the gain bandwidth product are artificially tunable by varying the superlattice period and/or duty cycle.

4. Transistors with graded band-gap and quantum well base

Bandgap engineering can also be used to improve the performance of heterojunction bipolar transistors. The idea of introducing a quasi-electric field in the base of a transistor to reduce the transit time of minority carriers was introduced by Kroemer.[21] This effect was demonstrated by Capasso et al[22] in a wide gap emitter phototransistor with a graded base having response time $t_r < 20$ps and Levine et al. via a measurement of the velocity of electrons in a graded gap p^+ AlGaAs layer.[23] These authors reported a velocity as high as 1.8×10^7 cm/sec in a quasi-field of $= 8.8\times 10^3$ V/cm for a 0.4μm thick layer. Figure 12 represents the energy band diagram of a bipolar transistor with graded gap base at equilibrium. With the base-emitter junction forward biased electrons are injected in the base where they move by drift rather than diffusion. The short base transit time can be traded off against the base resistance by making the base layer thickness relatively large. This will reduce the access resistance to the base thus increasing the maximum oscillation frequency of the transistor.

Recently high performance AlGaAs/GaAs heterojunction bipolar transistor with graded gap base have been reported.[24] It was found that the use of an undoped setback layer of 200-500 A to offset Be diffusion in the emitter resulted in significant current gain increases. Maximum current gains of 1150 for a base width of 0.18 μm were obtained which are the highest yet reported for graded base bipolars. Zn diffusion was used to contact the base and provides a low base contact resistance. Microwave s-parameter measurements yielded $f_T = 5$ GHz and $f_{max} = 2.5$ GHz. Large signal pulse measurements resulted in rise times of $\tau_r \sim 150$ ps and pulsed collector currents of $I_c > 100$ mA which is useful for high current laser drivers.

More recently Capasso and Kiehl have proposed a new negative conductance device consisting of a heterojunction bipolar transistor with a quantum well and a symmetric double barrier or a superlattice in the base region.[25] The key difference compared to previously studied structures is that resonant tunneling is achieved by high-energy minority carrier injection into the quantum state rather than by application of an electric field. Thus this novel geometry maintains the crucial, structural symmetry of the double barrier, allowing unity transmission at all resonance peaks and higher peak-to-valley ratios and currents compared to conventional resonant tunneling structures. Both tunneling and ballistic injection in the base are considered (Fig. 13 and 14). These new functional devices have significant potential for a variety of signal processing and multiple-valued logic applications and for the study of the physics of transport in superlattices. Of particular interest is the device of Fig. 14 which represents the electronic equivalent of an optical Fabri-Perot.

5. Tunable band-discontinuities

It is clear from the previous examples of heterostructures that band offsets play a central role in the physics and in the operation of superlattice devices. Until recently band-discontinuities have been considered as basic properties of a given heterojunction. It is clear however that a technique capable of artificially varying band offsets would have a tremendous impact because it would introduce a new powerful degree of freedom in device design.

Capasso et al. recently demonstrated experimentally for the first time a technique to effectively tune barrier heights and band discontinuities at semiconductor heterojunctions using doping interface dipoles (DID).[4] The DID consists of two ultrathin ionized donors and acceptor sheets *in situ* grown within 100 Å of the heterointerface by MBE. This technique is illustrated in Fig. 15. Consider an abrupt heterojunction assumed to be undoped (ideally intrinsic). We next assume to introduce in situ by planar doping during the molecular beam epitaxial growth of a second identical heterojunction, one sheet of acceptors and one of donors of identical charge sheet concentration σ at the same distance $d/2$ ($\leq$ 100 Å) from the interface. The DID is therefore a microscopic capacitor. The electric field between the plates is σ/ϵ. There is a potential difference $\Delta\Phi = (\sigma/\epsilon)d$ between the two plates of the capacitor. Thus the DID produces abrupt potential variations across a heterojunction interface by shifting the relative positions of the valence and conduction bands in the two semiconductors outside the dipole region. This is done without changing the electric field outside the DID (Fig. 15). For example the conduction band discontinuity can be effectively increased to $\Delta E_c + e\Delta\Phi$ (Fig. 15a), if the distance between the sheets is comparable or less than electron De-Broglie wavelength. The conduction band discontinuity can be also decreased by interchanging the position of the two charge sheets. Two effects here contribute to the barrier lowering.

On the low gap side of the heterojunction a triangular quantum well is formed. Since typically the electric field in this region is $\gtrsim 10^5$ V/cm and $e\Delta\Phi \approx 0.1$-0.2 eV, the bottom of the first quantum subband E_1 lies near the top of the well. Therefore, the thermal activation barrier seen by an electron on the low gap side of the heterojunction is $\simeq \Delta E_c - e\Delta\Phi/2$.

Electrons can also tunnel through the thin triangular barrier; this further reduces the effective barrier height. In the limit of a DID a few atomic layers thick and of potential $\Delta\Phi$ the triangular barrier is totally transparent and the conduction-band discontinuity is lowered to $\Delta E_c - e\Delta\Phi$ (Fig. 16b).

Using a DID the effective conduction band-discontinuity at an $Al_{0.26}Ga_{0.74}As/GaAs$ heterojunction was lowered by $\simeq 0.15$ eV. This led to a one order of magnitude enhancement of the photocollection efficiency across the heterojunction, with respect to heterojunctions without DID, but otherwise identical. Several device applications of DID have also been discussed.[5] Recently Margaritondo and coworkers have used an alternative method to tune band offsets in CdS/Si and CdS/Ge heterojunctions, based on a metallic interlayer of aluminum which effectively changes the interface dipole.[25]

REFERENCES

[1] H. Kroemer, Proc. IEEE *70*, 13 (1982); Japan. J. Appl. Phys. *20*, Suppl. 20-1, 9 (1981).

[2] F. Capasso, *J. Vac. Sci. Technol. B1*, 457 (1983).

[3] F. Capasso, Surface Sci. *142*, 513 (1984), Physica *129B*, 92 (1985).

[4] F. Capasso, A. Y. Cho, K. Mohammed, and P. W. Foy, Appl. Phys. Lett. *46*, 664 (1985).

[5] F. Capasso, K. Mohammed, and A. Y. Cho, *J. Vac. Sci. Technol. B3*, 1245 (1985).

[6] F. Capasso, W. T. Tsang, A. L. Hutchinson, and G. F. Williams, Appl. Phys. Lett. *40*, 38 (1982).

[7] R. Chin, N. Holonyak, G. E. Stillman, J. Y. Tang, and K. Hess; Electron Lett. *38*, 467 (1980).

[8] F. Y. Juang, V. Das, Y. Nashimoto, and P. K. Battacharya, Appl. Phys. Lett. *47*, 972 (1985).

[9] K. Mohammed, F. Capasso, J. Allam, A. Y. Cho, and Al Hutchinson, Appl. Phys. Lett. *47*, 597 (1985).

[10] F. Capasso, B. Kasper, K. Alavi, A. Y. Cho, and J. Parsey, Appl. Phys. Lett. *44* 1027 (1985).

[11] F. Capasso, W. T. Tsang, and G. F. Williams, IEEE Trans. Electron Devices *ED-30*, 381 (1983).

[12] F. Capasso, Electronics Lett. *18*, 12 (1982).

[13] F. Capasso, R. A. Logan, W. T. Tsang, and J. R. Hayes, Appl. Phys. Lett. *41*, 944 (1982).

[14] J. A. Cooper, F. Capasso, and K. K. Thornber, IEEE Electron Device Lett. *EDL3*, 402 (1982).

[15] F. Capasso, S. Luryi, W. T. Tsang, C. G. Bethea, and B. F. Levine, Phys. Rev. Lett. *51*, 2318 (1983).

[16] R. C. Miller, D. A. Kleinman, and A. C. Gossard, Phys. Rev. *B29* 7085 (1984).

[17] F. Capasso, H. M. Cox, A. L. Hutchinson, N. A. Olsson, and S. G. Hummel, Appl. Phys. Lett. *45*, 1192 (1984).

[18] H. M. Cox, J. Cryst. Growth *69*, 641 (1984).

[19] F. Capasso, K. Mohammed, A. Y. Cho, R. Hull, and A. L. Hutchinson, Appl. Phys. Lett. *47*, 420 (1985).

[20] F. Capasso, K. Mohammed, A. Y. Cho, R. Hull, and A. L. Hutchinson, Phys. Rev. Lett. *55*, 1152 (1985).

[21] H. Kroemer, R. C. A. Rev. *18*, 332 (1957).

[22] F. Capasso, W. T. Tsang, C. G. Bethea, A. L. Hutchinson, and B. F. Levine, Appl. Phys. Lett. *42*, 93 (1983).

[23] B. F. Levine, C. G. Bethea, W. T. Tsang, F. Capasso, K. K. Thornber, Fulton and D. A. Kleinman, Appl. Phys. Lett. *42*, 769.

[24] R. J. Malik, F. Capasso, R. A. Stall, R. A. Kiehl, R. W. Ryan, R. Wunder, and C. G. Bethea, Appl. Phys. Lett. *46*, 600 (1985).

[25] F. Capasso and R. A. Kiehl, J. Appl. Phys. *58*, 1366 (1985).

[26] D. W. Niles, G. Margaritondo, P. Perfetti, C. Quaresima, and M. Capozzi, Appl. Phys. Lett., to be published.

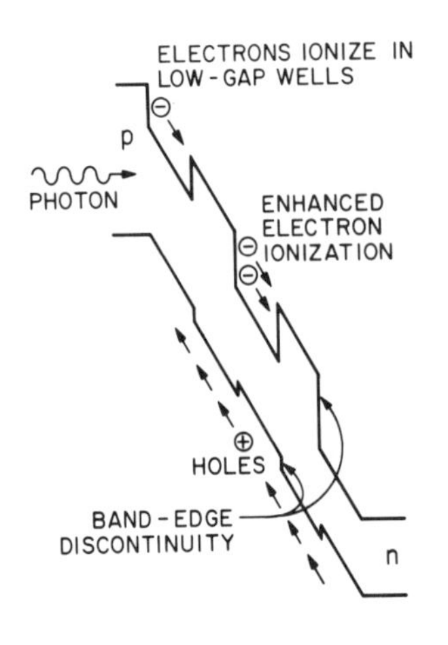

Fig. 1 Band diagram of superlattice avalanche detector.

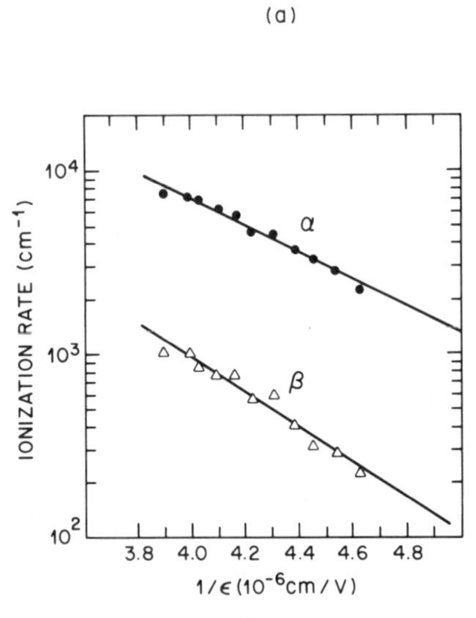

Fig. 2 Measured ionization rates for superlattice APD with 500Å $Al_{0.45}Ga_{0.55}As$ barriers and 450Å GaAs wells.

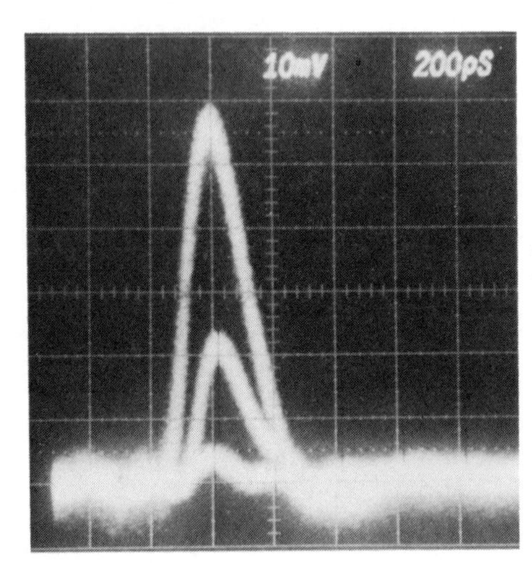

Fig. 3 Pulse response of AlInAs/GaInAs quantum well photodiode to a 100ps 1.3μm laser pulse at -20, -34, -36V bias.

(b)

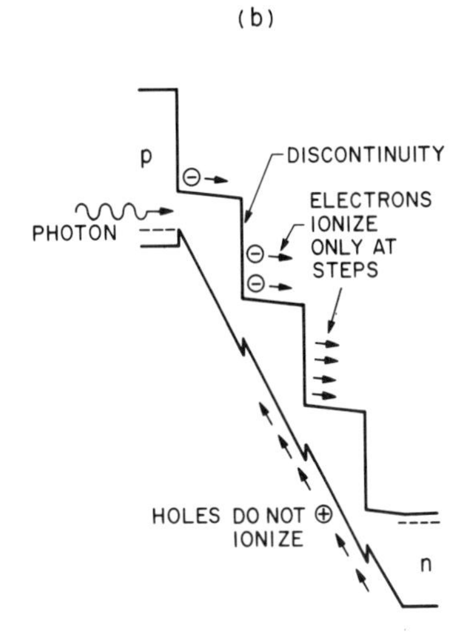

Fig. 4 Band diagram of staircase solid state photomultiplier.

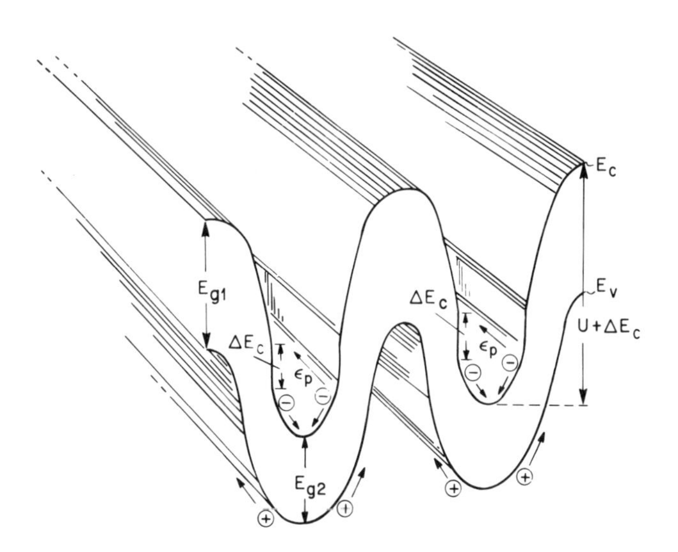

Fig. 5 Band diagram of channeling avalanche photodiode.

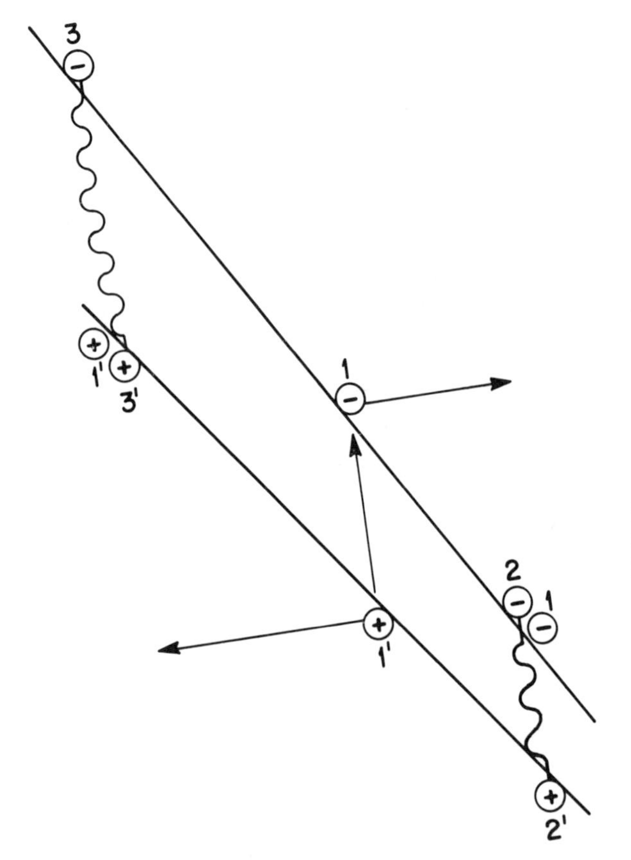

Fig. 6 Band diagram of graded gap avalanche photodiode.

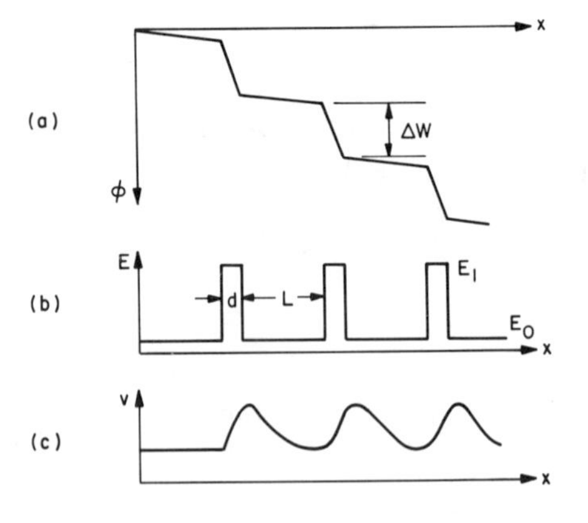

Fig. 7 Principle of repeated velocity overshoot. Staircase potential and the corresponding electric field. The ensemble velocity V as a function of position is also illustrated schematically.

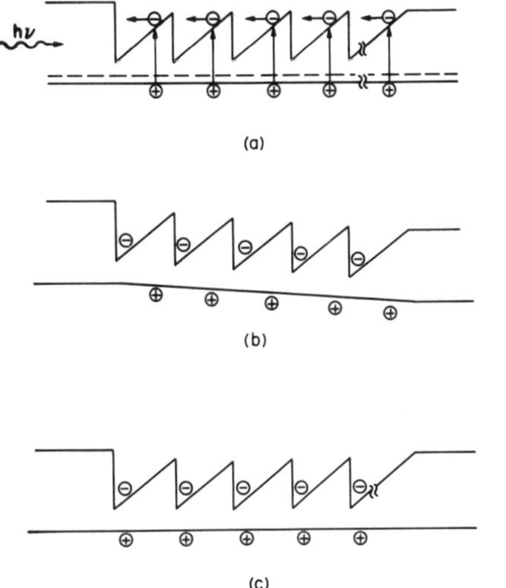

Fig. 8 Formation and decay of the macroscopic electrical polarization in a sawtooth superlattice.

Fig. 9 (a) Band diagram of pseudoquaternary graded gap semiconductor. (b) and (c) are a schematic and the field profile of a high-low APD using a pseudoquaternary layer to achieve high speed.

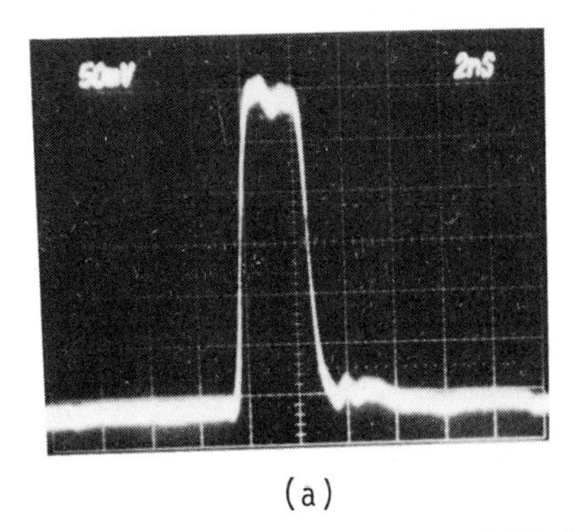

(a)

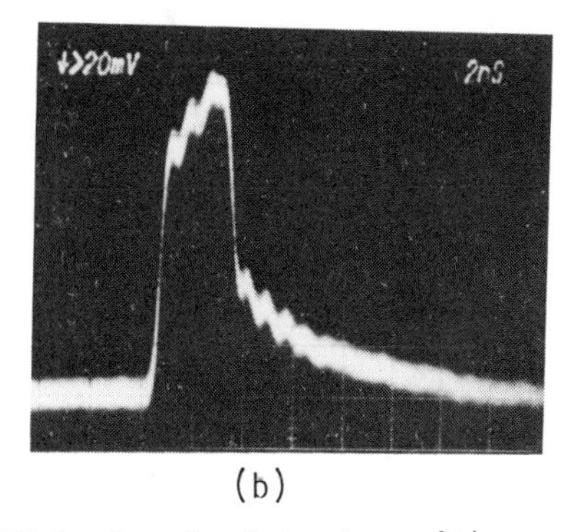

(b)

Fig. 10 Pulse response of High-Low SAM Avalanche Detector with graded gap superlattice (a) and without (b) to a ≅2ns λ=1.55μm laser pulse. The bias voltage is -65.5 V for both devices. Time scale 2ns/div.

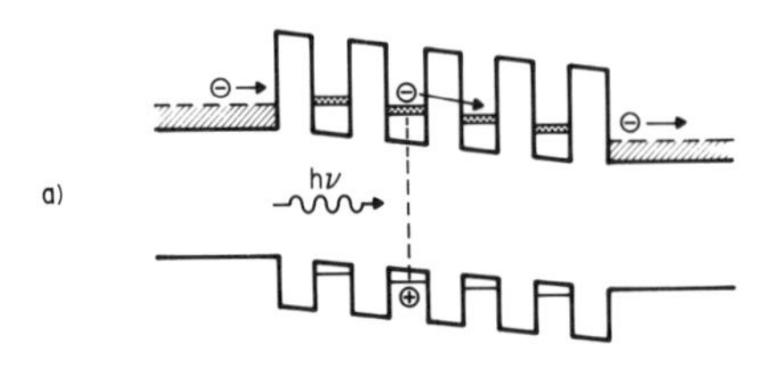

Fig. 11

Band diagram of the superlattice detector and schematic of effective mass filtering in the case of (a) phonon assisted tunneling between the wells and (b) miniband conduction.

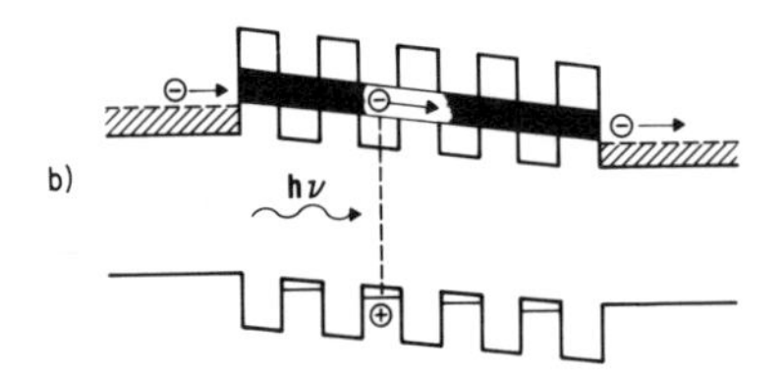

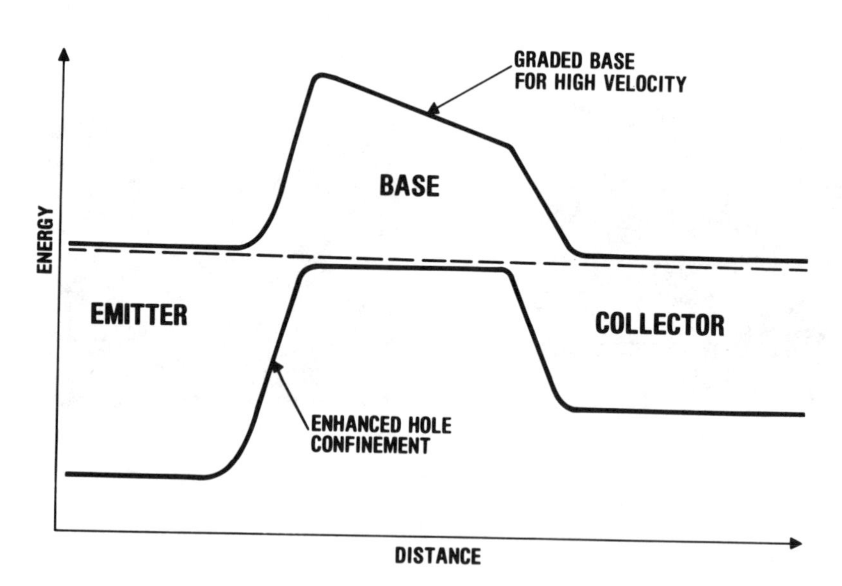

Fig. 12 Band diagram of graded gap base bipolar transistor.

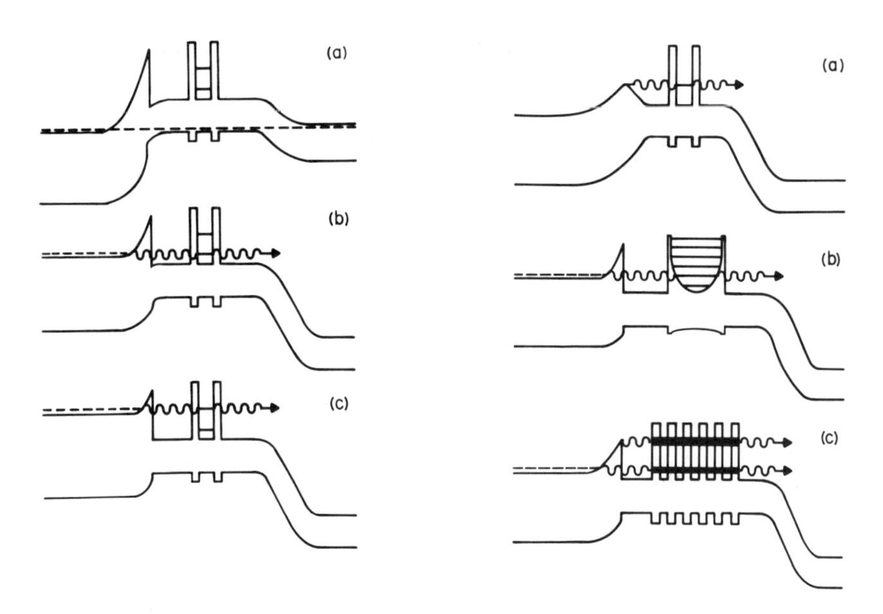

Fig. 13 Band diagram of resonant tunneling transistor with tunneling emitter.

Fig. 14 Different varieties of Resonant Tunneling Transistors: with ballistic emitter (a); parabolic quantum well (b); and superlattice in the base region (c).

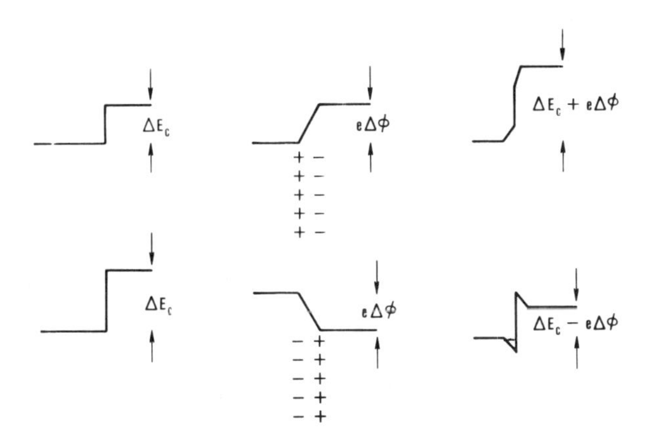

Fig. 15 Schematic doping-interface-dipole of (DID) concept. The DID can be used to either increase or decrease the band discontinuities.

A BIRD'S EYE ASSESSMENT OF THE JAPANESE OPTOELECTRONICS: SEMICONDUCTOR LASERS, LEDs, AND INTEGRATED OPTICS

W. T. Tsang

AT&T Bell Laboratories, Holmdel, NJ 07733

ABSTRACT

Within the last decade and a half, the field of optoelectronics has undergone and is still experiencing a tremendous growth in both research and product development in all industrial countries. Here, we shall examine the underlying driving-force of the Japanese optoelectronics, compare the growth of the Japanese optoelectronics with those of U.S.A., and other western countries from 1969 to present.

I. BACKGROUND

Within the last decade and a half, the field of optoelectronics has undergone and is still experiencing a tremendous growth in both research and product development. This explosive growth and vitality is fueled by the advent of various major technologies, lasers, integrated circuits and optics, lightwave communications and optical storage, just to name a few. It is impossible to survey the entire field in its completeness and detail. Thus, this study aims at providing an overall observation of the general trends of evolution. It will also try to compare between the Japanese and U.S. performance in the specific areas of semiconductor lasers, light-emitting diodes, and integrated optics.

II. THE DRIVING-FORCE OF JAPANESE OPTOELECTRONIC:

The driving-force for the Japanese optoelectronic R&D is domestic market sales with high hope for world market; and strong, co-ordinated, and systematic support from the government. The major resources of government support are the well-publicized MITI programs and the N.T.T.'s Information Network System (INS). At present, MITI's major program in optoelectronics is the Optoelectronics Joint Research Laboratory. A summary about this program is given below:

0094-243X/86/1380184-12$3.00

JAPAN OPTOELECTRONICS JOINT RESEARCH LABORATORY

- In operation October, 1981
- Monolithic OEIC technology
- communications, optical measurement and control
- High speed, high quality picture data subsystems, high speed and composite process data subsystems, Data control subsystems
- Approximately 50 researchers
- Budget $77M ($5.6M 1981, $25.7M, 1987)

The N.T.T.'s INS is expected to supply the Japanese optoelectronics industries with $8-13B for research and product development over the next 10-15 years.

The other driving force comes from domestic and world markets. The Japanese optoelectronic markets include the following major areas:

1. COMPONENTS:
   optical fibers, lasers & LEDs, photodetectors
2. EQUIPMENT:
   optical disk units, laser printers, electronic fields
3. LARGE & SMALL SYSTEMS:
   telecommunications, local area networks

Japan's performance includes a tenfold increase in sales over the past five years, mostly in component area, and components represent half of the market now.

A detailed breakdown of the Japanese optoelectronic market in 1982 and a forecast based on various predictions in journals like Photonics and others in 1992 in various areas is given below:

THE JAPANESE OPTOELECTRONIC MARKET:

	1982	1992
Optical Communication:	$85M	→ $400 - 800M
LAN:	$70M	→ $420M
Laser:	$5M	→ $18M
LED (comm.):	$10M	→ $29M
Detector:	$0.5M	→ $1.25M
Optical disk:		→ $1.25B
TOTAL OPTOELECTRONIC MARKET:		→ $4-8B

One other general observation is that the Japanese government seem to work more closely with private Japanese industrial companies than is the case in the U.S. This communication allows the private Japanese industrial companies easier access and more influence on drawing up governmental policy in terms of trade and others. This communication between government and private industries is essential to the economic welfare of the country as a whole and the Japanese appears to be doing particularly well in this respect.

The U.S. optoelectronic industry appears to be driven mostly by the market. Both the private industry and government do not appear to be in any particular effort to communicate and co-ordinate general policies. This lack of cooperation can be hurting to the future welfare of U.S. optoelectronics industry.

III. THE GROWTH OF JAPANESE OPTOELECTRONICS, 1969 - PRESENT

The Japanese industry certainly sees optoelectronics as a fast-growing technology with large potential product base. This is reflected by the R&D efforts in a large number of industrial companies and their progress through the recent years. Major Japanese optoelectronic industries include NEC, Hitachi, Toshiba, Mitsubushi Electric, Matsushita, Frijitsu, Oki, Sumitomo Electric, Furubawa, N.T.T., Sharp, Sanyo, Sony, and KDD (in arbitrary order). Some outstanding industries involved in long-wavelength (1.3 μm,

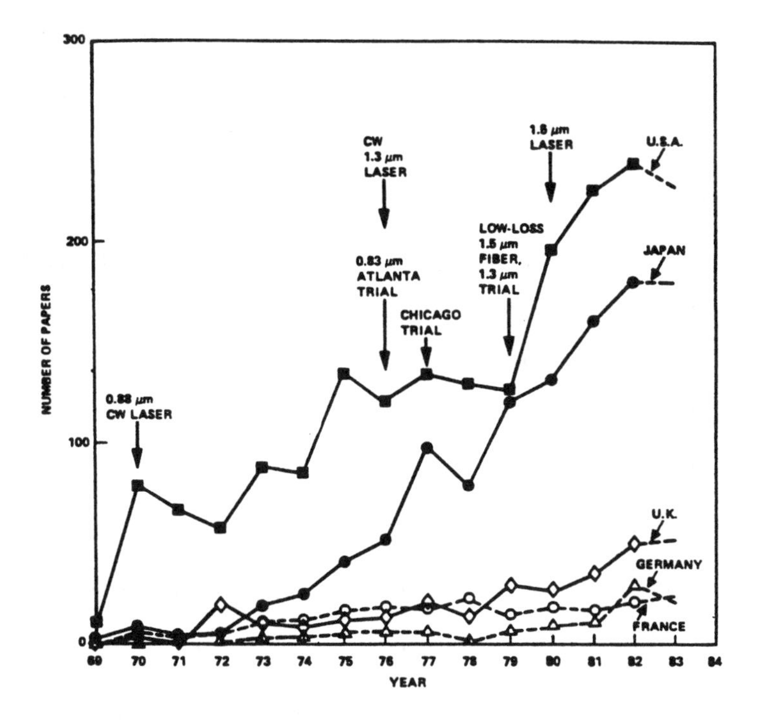

FIGURE 1 (a) PUBLISHED PAPERS ON SEMICONDUCTOR LASERS BY DIFFERENT COUNTRIES.

1.55 μm) semiconductor laser-R&D are Hitachi, NEC, Fujitsu, N.T.T., Mitsubushi Electric, Matsushita, Toshiba, and KDD (in arbitrary order), while major industries involved in developing visible ($<\sim 0.78 \mu m$) semiconductor lasers are Sharp, Sony, Sanyo, and Hitachi (in arbitrary order).

The R&D effort, achievements, and growth of a certain field can sometimes be very well monitored by plotting the total number of scientific publications as a function of year. Figure 1 shows the number of published papers (including those in Japanese journals) on semiconductor lasers by different countries from 1969 to 1982. The data-base for 1983 and 1984 are at present incomplete yet. Some obvious features can be seen:

1. The field of semiconductor lasers is fast-growing in every industrial country.

2. Before 1976, there was relatively little effort in Japan. But after 1976 growth in Japan is faster than the growth in the U.S.

3. The various ups-and-downs on the curves can be related to the various important events occurred as indicated in the figure.

A similar plot for the light-emitting diode (LED) is shown in Fig. 2. Some conclusions can be drawn as follows:

1. LED research grows to a peak in 1978 and then suddenly levels off as the technology becomes matured. The important point here is that: in high technology a fast-growing technology can suddenly be replaced by another up-coming more suitable technology without any advance warning.

2. By 1976, the Japanese R&D tracks that of U.S. very closely in synchronism.

3. The effort in LED in slowly decreasing. Figure 3 shows a similar plot.

Some general observations are:

(1) The field of Integrated Optics is rather temperamental, full of abrupt ups-and-downs. The cause of the sudden drop world-wide in 1981 is not understood by

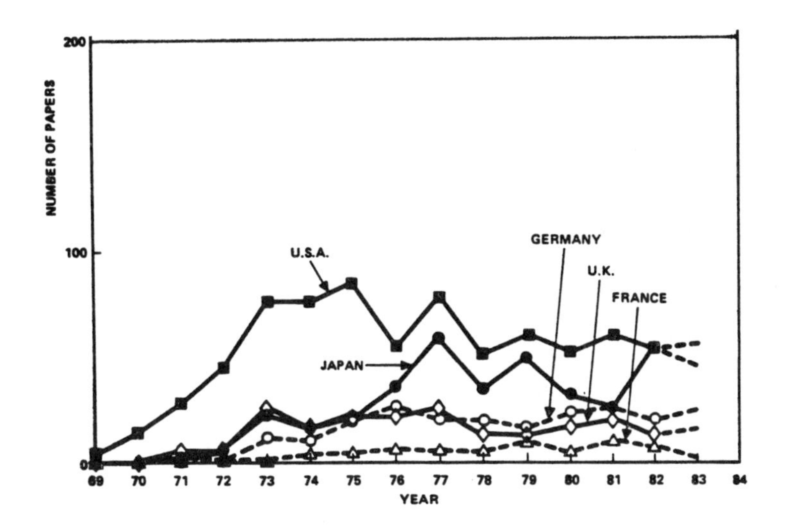

FIGURE 2 PUBLISHED PAPERS ON LIGHT-EMITTING DIODES BY DIFFERENT COUNTRIES.

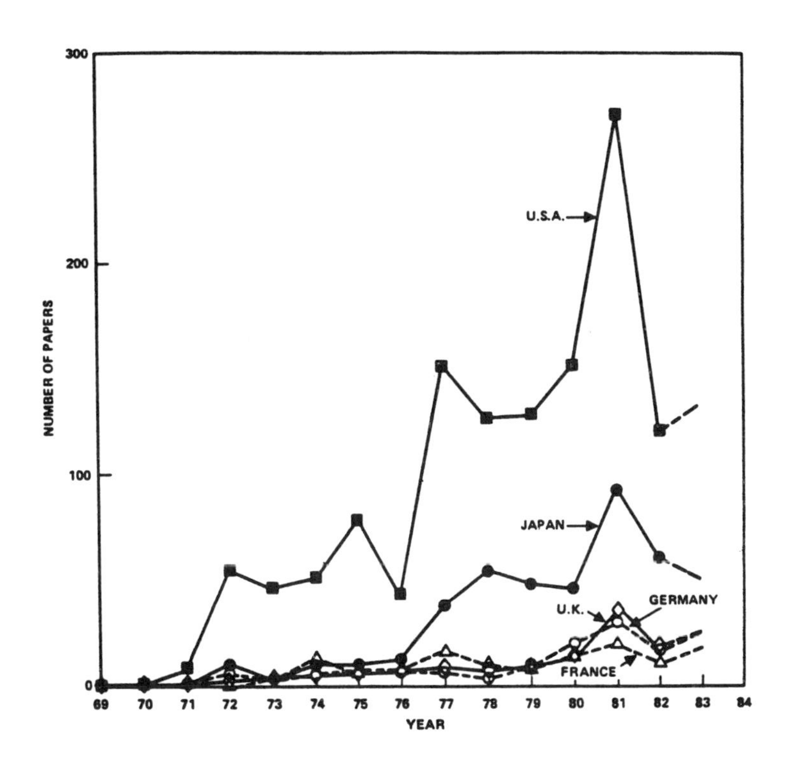

FIGURE 3 PUBLISHED PAPERS ON INTEGRATED OPTICS BY DIFFERENT COUNTRIES.

the author.

(2) The Japanese effort in Integrated Optics did not start until 1976. After 1976, the ups-and-downs occurred in exact synchronism with U.S.

Next, the performance of various industrial companies can also be studied in a similar manner. However, the data will not only reflect the R&D effort but can also be intermixed and influenced by general company policy. So some care has to be exercised in interpreting the data. Plotted in Figure 4 is the total number of published papers as a function of year for AT&T Bell Laboratories, N.T.T., Hitachi, Fujitsu, NEC, Mitsubishi, Toshiba. Some observations can be made.

(1) The up-and-downs in the curves of Figure 4 for AT&T Bell Laboratories can be related to the various events that happened in semiconductor laser research and development. The sharp spike around 1975 and the subsequent drop around 1978 are related to the initial lightwave systems based on 0.83 μm multimode technology. The rise after that is due to the new generation 1.3 μm single mode systems.

(2) On the other hand, Japanese came in late to semiconductor laser field. As a result, they were spared of the cost of the 0.83 μm technology for use in lightwave transmissions for trunk application (for LAN the story can be different).

Finally, Figures 5(a) and (b) plot the R&D performance of AT&T Bell Laboratories with respect to U.S. and N.T.T. to Japan on semiconductor lasers. The exact synchronism between AT&T Bell Laboratories and U.S., and N.T.T. and Japan show that AT&T Bell Laboratories is the dominant industrial company in semiconductor laser research in U.S. and likewise in Japan for N.T.T. Interestingly enough, both AT&T Bell Laboratories and N.T.T. account for 28 percent of total publications in their respective countries.

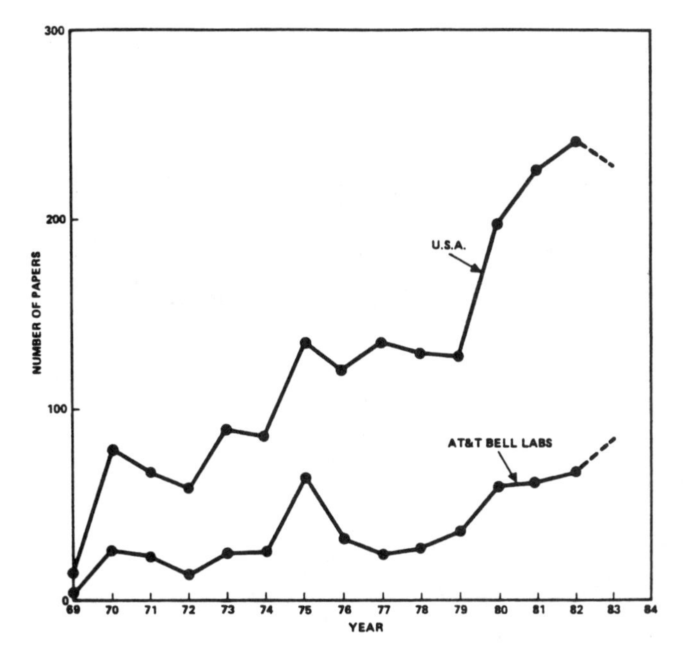

FIGURE 5 (a) PUBLISHED PAPERS ON SEMICONDUCTOR LASERS
AT&T BELL LABORATORIES ACCOUNTS FOR 28% OF TOTAL USA PUBLICATIONS IN 1982.

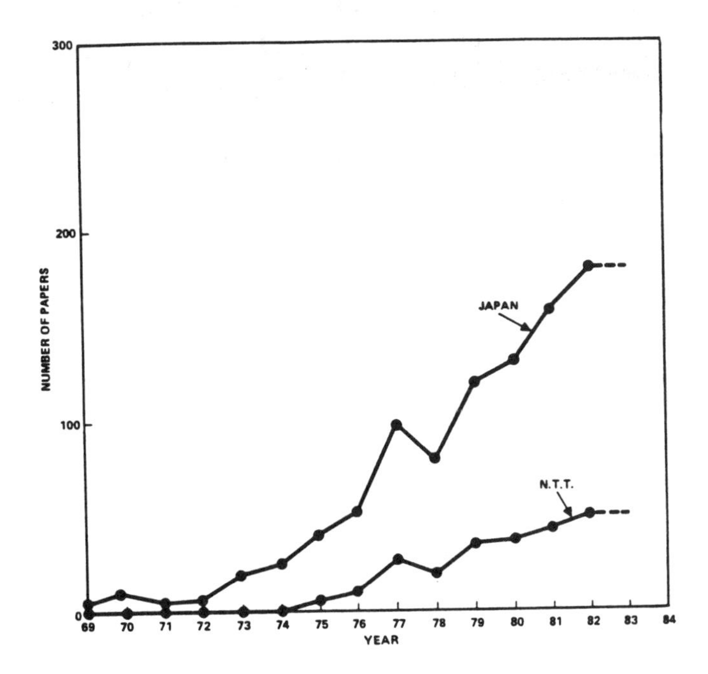

FIGURE 5 (b) PUBLISHED PAPERS ON SEMICONDUCTOR LASERS
N.T.T. ACCOUNTS FOR 28% OF TOTAL JAPANESE PUBLICATIONS IN 1982.

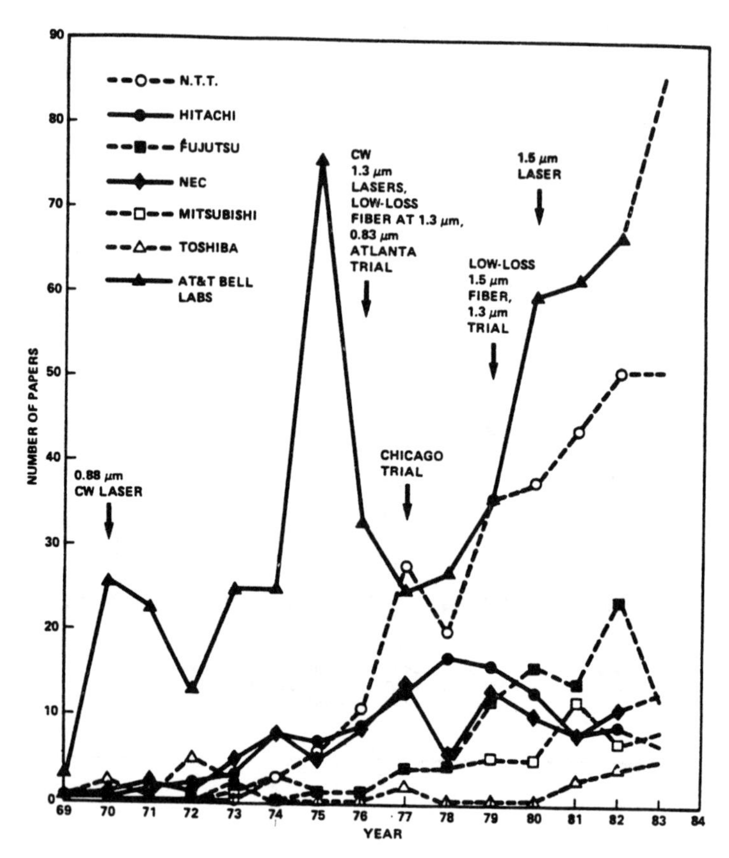

FIGURE 4 PUBLISHED PAPERS ON SEMICONDUCTOR LASERS BY DIFFERENT COUNTRIES.

In the above comparisons, one should keep in mind that Japan is about the size of California carrying on with many other industrial activities and world-dominances at the same time.

IV. THE JAPANESE VERSUS U.S.A. INVENTIVENESS IN OPTO-ELECTRONICS

Next, we investigate the Japanese versus U.S. inventiveness in optoelectronics. This is best studied by a survey of the various important first publications in the field of semiconductor laser structures as an indicative example.

The following table shows that all the major research laser heterostructures were *first reported* in the west mostly in U.S. All these papers involve more fundamental changes and of conceptual nature.

MAJOR RESEARCH LASER HETEROSTRUCTURES

Laser Type	First Reported
double-heterostructure (DH)	USSR, USA
separate confinement heterostructure (SCH)	USA
graded-index SCH (GRIN-SCH)	USA
quantum well (QW)	USA

On the other hand, if we survey the major stripe-geometry laser structures reported, we find the Japanese become more strong. Further, their reported devices usually also become commercialized products. This is shown in the table below.

MAJOR RESEARCH STRIPE-GEOMETRY LASER STRUCTURES:

Laser Type	First Reported
buried heterostructure (BH)	Japan
double-channel planar BH (DC-PBH)	Japan
channel subrate planar (CSP), V-groove	Japan, USA
transverse junctions tripe (TJS)	Japan
ridge-waveguide	Japan
strip-buried heterostructure (SBH)	USA
proton-stripe	USA
oxide-stripe	USA

One must bear in mind that this can only be considered a rather restrictive example illustrative and care must be taken not to generalize.

V. THE JAPANESE R&D EFFICIENCY

The R&D efficiency and technology-transfer are undoubtly two most important factors in describing the success of a private industrial company. R&D in general is very expensive. Thus, choosing the right area with the right timing, minimizing unnecessary waste in duplicative efforts and maximizing productivity are particular crucial. Efficient and timely transfer of technology guarantees market sales and dominance. The Japanese industry appears to do both rather well.

- Their R&D productivity remains high.
- Able to get timely results.

- Only 10-15 percent of their R&D people have Ph.D.
- Make very efficient use of expensive equipments and facilities.
- R&D results are in general of high quality.

SOLID STATE LASERS IN THE NEAR INFRARED AND VISIBLE REGIONS OF THE SPECTRUM:

A COMPARATIVE ASSESSMENT OF JAPANESE AND U.S. R&D

Robert S. Bauer and Robert D. Burnham
Xerox Palo Alto Research Center
Palo Alto, CA 94304

ABSTRACT

We present an assessment of the frontiers in III-V technology for short wavelength opto-electronics. Particular focus is given to fundamental studies by the Japanese over the last two to three years in the areas of compound semiconductor materials and devices. We discuss advances in AlGaAs lasers operating in their normal range between 900 and 780 nm. Tremendous progress in epitaxial growth techniques have allowed new flexibility in device design and improved operating characteristics. MO-CVD is emphasized because of considerable work in Japan to harness this technology for manufacturing. These improvements in materials control have allowed growth of AlGaAs quantum well structures which produce CW lasing as short as 680 nm at room temperature. For shorter wavelengths, novel III-V materials systems such as InGaAlP, InGaAsP, and AlGaN are being actively pursued with impressive results in Japan. Comparison with U.S. efforts in these areas is made.

I. INTRODUCTION

This paper presents an assessment of some of the frontiers of technology pursued by the Japanese in short wavelength opto-electronics. Focus is on advances made over the last two or three years in the areas of fundamental GaAlAs materials, devices and processing. Whereas most of the invention is done in the U.S. (where patents are held for some of the

0094-243X/86/1380196-12$3.00

fundamental breakthroughs), there is a great amount of activity in Japan following through on reducing novel phenomena to practical devices. Rather than dwelling on the 1.3/1.55 micron optical fiber window, we discuss how "standard" GaAlAs materials can be pushed into the visible range by using very sophisticated materials growth and device structures. GaAlAs quantum wells have achieved 680 nanometer emission, while at shorter wavelengths, novel III-V materials systems such as InGaAlP, InGaAsP, and AlGaN are applicable.

The visible region of the spectrum is extremely important for a broad range of applications. U.S. efforts seem to focus currently on sources for optical fiber communication because of the very large market potential. But visible lasers allow humans to see the light source in aligning and using real systems. They also have specific applications where spectral overlap with photographic and photosensitive materials is important; these of course are optimized for the visible region of the spectrum. For example, in printing applications, there are great advantages to working in the visible range. And, finally, for diffraction-limited optical beam applications, the shorter the wavelength, the more the accompanying smaller diffraction limits will result in a narrower spot size. The Japanese are attacking all segments of the market in both short and long wavelength regions of the spectrum.

II. JAPANESE MARKET TARGETING

Integrated opto-electronics, which is the integration of light-emitting devices with electronics on a single chip, was a MITI-targeted technology in the late 1970's. Over $100 M of Japanese government funding was committed to a joint industrial research cooperative under the banner of the Optoelectronics Joint Research Laboratory. This money is crucial not only for its absolute amount but also because of the funding stability that it provides. Being invested over a period of eight years, research can follow technical paths rather than being subjected to shifts in a particular marketplace. This stabilizes otherwise uncertain justifications for industrial R&D investment.

All projections of the of the opto-electronics industry estimate an enormous rate of expansion. From 1980 to the year 2000, the annual growth rate is 28%, which is about double the rate of expansion for the Si-based electronics industry over the last 10 years. Opto-electronics will grow worldwide to over 100 billion dollars in annual sales by the end of the century. The Japanese openly stated in 1981 that they intend to capture 40 to 50% of that market.

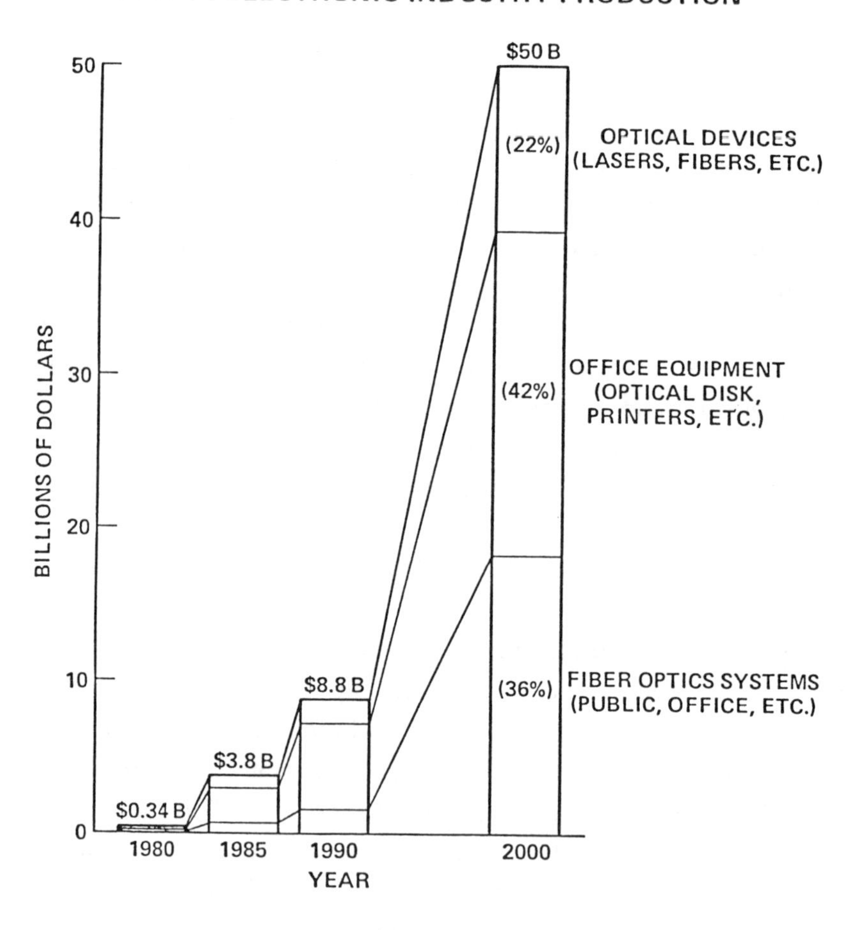

Fig.1 Japanese projection for their share of markets dependent on opto-electronics technology. This goal is 40- 50% of the estimated world-wide market place.

Individual components, such as lasers, and fiber-optic communications systems together represent about 60% of the marketplace. The largest single segment is the office equipment market. This includes such items as optical disks, printers using lasers to write on photosensitive material for computer and workstation output, and other applications in the office environment. Office automation is exploding, and opto-electronics is a key technology in that market.

The important lesson for U.S. commerce is that, when the Japanese decided to pursue opto-electronics in the middle 1970's, the markets for applications of this technology were very small, less than $100 M. Based on technological opportunity, they have created a component market in and of itself. This creates a technology privilege which can be exploited through further innovation to create a larger office systems and communications industry. This steady, long-term investment in market creation through technology push represents a different philosophy than the short-term market pull strategy employed by U.S. industry today.

III. ADVANCES IN AlGaAs LASERS (900 $> \lambda >$ 780 nm)

First we discuss "standard" AlGaAs lasers operating in their normal range, which is between 900 and 780 nanometers. The diode laser begins with a GaAs substrate. Substrate characteristics for GaAS IC's based on today's MESFET technology differ from the requirements of opto-electronics; in the integration of these two functions on the same chip this will be a key issue. The entire technology of solid state optical sources depends on growing epitaxial AlAs/GaAs layers on top of the III-V substrate. Because they are lattice-matched to the GaAs substrate, almost perfect continuation of the crystal structure is achieved. Any defects that exist in the substrate affect, or are grown into, the laser material. Through doping during growth, a p-n junction is created near the active layer. This "active" layer is that region where the electrons and holes are brought together to combine and create light emission. A striped contact is fabricated to confine electron flow into a very small region. The electrons recombine with holes as they move vertically in this structure, at high enough gain to achieve stimulated photon emission. If there is enough gain compared to the loss in the material itself, then lasing action occurs and a coherent beam of light is produced.

In this section, we discuss substrates, growth and epitaxy, and then finally some aspects of device structures to show why materials are the key to advances in opto-electronic technology requirements.

A. GaAs Substrates

The Japanese have decided that they will be the world's supplier of GaAs substrates. By 1980 they decided that they were going to build up a massive bulk GaAs manufacturing capability. In the three-year period from 1982 to 1985, there was a fourfold increase to eight tons in the production capacity of bulk GaAs by the Japanese. This would produce 10 billion lasers, if all the GaAs was used just for discrete lasers; in fact, most material will be used for the much larger die required for GaAs integrated circuits. It is reported that Sumitomo alone is going to be able to make 100 metric tons of GaAs by 1988 using Ga supplies from the Peoples Republic of China.

What is abundantly clear is that today the most reputable commercial sources of GaAs substrates are in Japan. Therefore, large American companies working on DOD programs (e.g., Rockwell International) grow their own material. But other large companies like IBM and Xerox and start-ups like Gigabit Logic do not have captive material sources; they are dependent on Japanese suppliers.

To improve manufacturability and integration, the Japanese are working very hard on growing GaAs on top of silicon substrates. M. Akiyama of OKI Electric reported MO-CVD growth of GaAs on GaAs/GaAlAs buffer layers to accommodate mismatch to a Si(100) substrate. Silicon is a much more mature technology than GaAs-based microelectronics. If high quality III-V growth on a silicon substrate is achieved, one has an enormous advantage, both in the quality of that starting material and in substrate price and availability.

Both U.S. and Japanese groups have reported GaAlAs lasers grown on a silicon substrate. [1] The ability to create light for growth of GaAlAs on top of silicon is very exciting. T. Nonaka of OKI has reported fabrication of GaAs MESFET IC's on Si substrates. The University of Illinois has achieved excellent MODFET and MESFET properties in MBE GaAs on Si substrates. [2] Even with the great emphasis on GaAs substrates, the Japanese are as active as U.S. researchers in the field of growing III-V's on standard Si substrate material.

B. GaAlAs Epitay Using MO-CVD

We emphasize metal-organic chemical vapor deposition (MO-CVD) for epitaxy. Compared to molecular beam epitaxy (MBE), we believe MO-CVD to be more appropriate for large-scale manufacturing. MO-CVD is a field that the Japanese have been late to enter, but in which they presently have a great amount of activity.

MO-CVD is based on a very simple technique. Metal atoms attached to an organic group are brought together with a source of arsine gas at an appropriate temperature. For tri-methyl compounds the reaction results in CH_4 being given off with deposition of AlGaAs in the ratios appropriate for the relative metal-alkyl gas flows. Control of both doping and composition is achieved by temperature stabilization of the sources, valve and tubing cleanliness, sophisticated flow control, and reactor design in order to heat and rotate the substrates uniformly. Compared to MBE, MO-CVD has demonstrated very similar abilities to control composition, dopant thickness, and abruptness of the interfaces. While it is difficult to control background doping, HEMT devices with high mobility grown by MO-CVD were first demonstrated by the Japanese. [3]

MO-CVD and MBE are new technologies for growing epitaxial layers that offer real advances over Liquid Phase Epitaxy (LPE) in fabricating artificial microstructures for devices. To achieve its potential, sophisticated and difficult research is required. In the literature one finds that the Japanese now seem to be leading in the material science of these growth techniques. Basic studies of pressure dependence for MO-CVD GaAs vacancy incorporation has been reported by Kobayashi. [4] At the Second International Conference on Metal-Organic Vapour Phase Epitaxy in Sheffield, England in April 1985, a number of significant Japanese papers were presented. H. Terao of NEC studied the effect of O_2 and H_2 introduction during GaAlAs MO-CVD growth; M. Akiyama of OKI reported on Cr doping to achieve semi-insulating GaAs by MO-CVD; H. Ohno and H. Hasegawa of Hokkaido University reported attempts at producing atomic layer doping of MO-CVD GaAs in order to obtain novel opto- and high-speed electronic device structures. One can find fundamental work on the effect of relative arsine concentration on GaAs and AlGaAs carrier concentration [5]and EL2 deep level concentration. [6] These are scientific, not just empirical, studies.

To make this a manufacturable technology, understanding process control variations is basic. The work of T. Nakanisi of Toshiba provides one example of the Japanese effort to achieve this. One grows MO-CVD material tipically with only 10% arsine in hydrogen. Nakanisi started with a tank of 100% arsine, and divided it into 10 tanks with high-purity hydrogen to study effects of variations in the arsine source. A series of growths was conducted where nothing but the small arsine sources were changed. When the bottle was changed after six runs, the luminescent output of the epitaxial material dropped by three to four orders of magnitude. Such layers would not be suitable for devices. This unpredictable change in arsine characteristics is a well-known phenomenon which limits MO-CVD device yield. No one has characterized what is causing the arsine to be "bad." Moisture monitors are used on all MO-CVD reactors and contaminants are checked. What is done is to throw the arsine bottle away and start again. In Nakanisi's work, a very tedious program to characterize the fundamentals of the material sources is being pursued as a necessary condition to harness MO-CVD technology for manufacturing. One does not see much work of this type going on in the U.S. today.

C. Quantum Well Device Structures

Precise control of III-V epitaxial layers is required to achieve uniform, low threshold, efficient opto-electronic devices. By sandwiching a thin GaAlAs region of lower aluminum concentration between higher percent Al alloy regions, an electron confining region with bound quantum well levels is created by the energy band structure variations of the resulting materials structure. Electrons which are injected will be captured in this well where they can subsequently recombine with holes to produce photons. The basis for making efficient devices is to produce very thin wells in which the electrons are captured; the population of carriers can be very high, producing enhanced gain in the active region for better lasing characteristics. There are reports of wells that have intermediate band gap barriers within the wells, these barriers being tens-of-Angstroms in thickness. Such structures can be used for separately optimizing optical confinement (by index of refraction changes) and carrier confinement. The ratio of the various compositions and their spacings is very critical in determining how a device behaves.

High power lasers are one application where not only must the material in a local region be controlled, but also the uniformity across the wafer must be maintained. A structure that was invented at Xerox [7] has achieved 2.6 watts of optical emission from a single chip. Its basis of operation is to take a large area (250 x 400micron) of material composed of multiple quantum well layers and fabricate 40 adjacent lasers. The lasing regions are pumped in parallel and are close enough to couple the optical fields of neighboring lasing regions. These emitters all phase-lock into a single lasing mode if the material is uniform and devices are processed properly.

There is considerable research into the proper device geometry for such array structures. Material uniformity over 100,000 square microns is required to avoid filimentary hot spots. Surveying the literature, we find that the Japanese employ both MO-CVD and MBE for such devices. At the biannual international laser conference in Brazil in July 1984, Hitachi announced their use of multiple quantum wells for high-power devices. Where the U.S. invented novel structures just a couple of years ago, the Japanese are very quick to apply these advanced technologies for practical components.

IV. GaAlAs FOR VISIBLE LASERS (780 >λ > 680 nm)

Most of the lasers that are used in the world are helium-neon gas lasers operating at 632.8 nm; they are used in such applications as supermarket scanners and the original optical videodiscs. These will be replaced with solid-state lasers. There already is evidence of this in digital-audio compact discs and video disc players.

In order to move into the visible range with GaAlAs, one needs to work with quantum wells grown by controlled epitaxial materials technologies. The quantum size effect in thin layers provides the ability to achieve visible laser operation. [8] Using the quantum-size effect, electron states inside a well of tens-of-angstroms thickness occur at energies which are dependent on the well width and barrier height (i.e. alloy composition). This effect was first discussed in detail by U.S. researchers. [9] The narrower the well, the higher the energy state of the electron level and consequently the shorter the wavelength of the photon emission.

One of the nicest demonstrations of this effect was reported by a Japanese group. [10] A series of GaAs quantum wells was grown 30 **Å**, 40 **Å**, 70 **Å** and 100 **Å** in thickness between 500 **Å** $Al_{0.54}Ga_{0.46}As$ barriers. The corresponding luminescence peaked at 710nm, 730nm, 780nm, and 800 nm, respectively. For GaAs active layers thinner than 50 **Å**, lasing occurs at wavelengths shorter than achievable in standard double heterostructures. Lasing action as short as 680 nm at room temperature has been demonstrated by applying this principle. Japanese companies intensely pursuing the quantum size effect for such near IR lasers include both at Sony [11] and Sharp. [12]

V. OTHER III-V VISIBLE LASERS (680 nm >λ)

The Japanese establish the state-of-the-art in III-V materials other than GaAlAs for solid state light emitters at wavelengths shorter than 680nm. To achieve these shorter wavelengths, one has to use materials that have larger bandgaps than AlAs. Phosphide-based III-V's offer such a possibility but the lattice constant for GaP or AlP is very different from the GaAs which might provide the substrate for epitaxial layer growth. GaAlAsP heterojunctions cannot be grown on GaAs without producing a strain-induced effect which keeps the material from working satisfactorily as an opto-electronic device. Complex quaternary materials of Ga, P, and In with Al or As have larger lattice constants capable of matching with the GaAs crystal structure.

The Japanese have a clear lead and are the most active in applying substantial resources to such III-V quaternary alloys [13-19]. For the phosphorous-based III-V quaternary [16-19], lasing in the yellow (579 nm) has been achieved at room temperature.

By using reactive MBE, alloys of AlGaN can be grown for materials with band gaps which cover the entire visible spectrum. The Japanese have reported cathodo-luminescence from AlGaN into the UV up to 6 eV [20]. The intensity of Japanese publication activity compared to U.S. efforts suggests a major thrust by them in this area of visible light emitters.

VI. SUMMARY AND CONCLUSIONS

The Japanese are striving to perfect semiconductor lasers. They are not driven by immediate economic return, but they are investing in the future. Since their motivation is based on technology push, Japanese R&D programs are not as sensitive to particular cycles in the marketplace. Further, their government (through MITI) provides stability with long-term funding and promotes industry-wide cooperation as well as the pooling of resources.

The R & D programs that flowed from this government commitment in the 1970's and early 80's clearly indicate that the Japanese expect to seize a major portion of this market; they have openly stated this as their goal. Not only will they continue their dominance of individual opto-electronic components, but also across this industry they are positioning to capture 50% of this growing market by the end of the century.

Japanese GaAs laser activities are very broad-based. Major companies are working in such diverse areas as III-V epitaxial growth techniques, novel processing, new materials alternatives, and device physics. They are not doing this empirically, but have sophisticated techniques which allow high-level scientific work.

The Japanese will pursue U.S. innovations as well as their own. They do not seem to have difficulties with the "not-invented-here" syndrome that slows technology advances into the marketplace in the U.S. There are numerous examples of U.S. invention reduced to practice first by the Japanese. They will continue to bring products from the development stage to the marketplace rapidly, and they will not hesitate to capitalize on U.S. developments.

There is an explosion in the information that is coming out of Japan on opto-electronics. They report their latest research results (much of it in English) as soon as possible. There is no more secrecy among Japanese industrial R&D than is present in the U.S.

The Japanese level of investment in the GaAs field suggests strong activity by them in the future. The gap that is opening up in III-V opto- and micro-electronics threatens to become a permanent disadvantage for this country.

REFERENCES

1. T. H. Windhorn, G. M. Metze, B-Y. Tsaur, J. C. C. Fan, Appl. Phys. Lett. **45**, 309 (1984); T. H. Windhorn, G. M. Metze, Appl. Phys. Lett. **47**, 1031 (1985)
2. R. Fischer, T. Henderson, J. Klem, W. Kopp, C. K. Peng, H. Morkoc, J. Detry, S. C. Blackstone, Appl. Phys. Lett. **47**, 983 (1985);S. Sakai, T. Soga, M. Takeyasu, M. Umeno, Jpn. J. Appl. Phys. 24, L666 (1985)
3. N. Kobayashi, T. Fukui, Elec. Lett. 20, 887 (1984), Y. Takanashi, N. Kobayashi, IEEE Elec. Device Lett. Vol. **EDL-6**, 154, (1985)
4. N. Kobayashi, T. Fukui, Y. Horikoshi, Jpn. J. Appl. Phys. Pt.2, **21**, L705 (1982)
5. Y. Mori, H. Sato, M. Ikeda, O., Matsuda, K. Kaneko, N. Watanabe, Appl. Phys. Lett. **40**, 293 (1982)
6. M. Watanabe, et al., Jpn. J. Appl. Phys. 22, 923 (1982)
7. D. R. Scifres, C. Lindstrom, R. D. Burnham, W. Streifer, Elet. Lett. **19** 169 (1983).
8. N. Holonyak, Jr., R. M. Kolbas, R. D. Dupuis, PO. D. Dapkus, IEEE J. Quantum Elec. **QE-16**, 170 (1980)
9. R. D. Burnham, W. Streifer, T. L. Paoli, J. Crystal Growth**68**, 370 (1984)
10. H. Kawai, et al., J. Appl. Phys. **56**, 463 (1984)
11. Y. Mori and N. Watanabe, J. Appl.. Phys. **52**, 2792 (1981)
12. S. Yamamoto, Appl. Phys. Lett **41**, 796 (1982)
13. A. Usui, T. Matsumoto, M. Inai, I. Mito, K. Kobayashi, H. Watanabe, Jpn. J. Appl. Phys. **24**, L263, (1985)
14. S.Mukai, H. Yajima and J. Shimada, Jpn. J. Appl. Phys. **20**, L729 (1981).
15. T. Suzuki, I. Hino, A. Gomyo, K,. Nishida, Jpn. J. Appl. Phys. **21**, L731 (1982)
16. M. Ikeda, Y. Mori, H. Sato, K,. Kaneko, N. Watanabe, Appl. Phys. Lett. **47**, 1027 (1985)
17. M. Ikeda, M. Honda, Y. Mori, K. Kaneko, and N. Watanabe, Appl. Phys.Lett. **45**, 964 (1984)
18. K. Kobayashi, S. Kawata, A. Gomyo, I. Hino, T. Suzuki, Elec. Lett. **21**, 932 (1985)
19. Y. Kawamua, H. Asahi, N. Nagai, and T. Ikegami, Elec. Lett. **19**, 163 (1983)
20. S. Yoshida, S. Misawa, and S. Gonda, J. Appl. Phys. **53**, 6844 (1982)

SYNTHESIS OF III-V COMPOUND SEMICONDUCTOR MATERIALS: AN ASSESSMENT OF RESEARCH AND DEVELOPMENT ACTIVITIES IN THE UNITED STATES AND JAPAN

Douglas M. Collins
Hewlett-Packard Laboratories, Palo Alto, CA 94304

ABSTRACT

A comparative assessment of U.S. and Japanese research and development activities in the field of III-V compound semiconductor materials is presented. Work on both epitaxial materials (i.e., molecular beam epitaxy (MBE) and organometallic vapor phase epitaxy (OMVPE) of materials in the (Al,Ga)As and (In,Ga)(As,P) systems) and bulk materials (i.e., liquid-encapsulated Czochralski (LEC) growth of GaAs) is described with an emphasis on high-speed and optoelectronic device and integrated circuit applications. The assessment is based primarily on widely available information including that which is disseminated in both English-language and Japanese-language journals and conferences. Key points which are addressed include (1) a comparison of U.S. and Japanese technical achievements, (2) the equity of technical exchanges between the U.S. and Japan, and (3) the relative effectiveness with which these technologies are commercialized in the U.S. and Japan.

INTRODUCTION

The importance of compound semiconductors in microelectronics and optoelectronics is primarily based on (1) the particularly suitable electrical or optical properties of these materials (e.g., band-gap, electron mobility, etc.), (2) the ability to tailor these properties to those desired by forming ternary or quaternary compound semiconductor alloys, and/or (3) the ability to form heterojunction structures of two or more different compounds. In the least complex applications (an example of which is the GaAs MESFET), nothing more than high-quality bulk substrate material may be required. For more complex device structures (e.g., heterojunction high-speed devices such as the MODFET or light emitting devices such as the multiquantum-well laser), it is necessary to use sophisticated epitaxial growth techniques such as molecular beam epitaxy (MBE) or organometallic vapor phase epitaxy (OMVPE) to produce complex, well-controlled, highly uniform heterojunction material structures.

The Japanese and U.S. sources of technical information that have been surveyed for this report include those available through early 1985. The scope of the survey has been limited to compound semiconductor materials that are used in high-speed and optoelectronic device and integrated circuit applications. Thus, the emphasis will be on epitaxial materials grown by MBE and OMVPE as well as on the growth of the bulk, liquid-encapsulated

0094-243X/86/1380208-15$3.00 Copyright 1986 American Institute of Physics

Czochralski (LEC) GaAs substrate material which is required for many of these applications. There will be a sub-section devoted to each of these technologies guided by the following objectives: (1) to provide an objective assessment of the technical information disseminated by the Japanese, in both English-language and Japanese-language publications, in comparison to the open technical literature published as a result of research efforts in the United States and (2) to indicate areas in which (a) U.S. efforts lead Japanese efforts, (b) there is equity between U.S. and Japanese efforts, and (c) Japanese efforts lead U.S. efforts.

An additional major objective of this report is to identify the sources of information which are the most valuable for keeping abreast of recent Japanese contributions in these three areas of research and development. This topic is addressed in the final sub-section of this report.

MOLECULAR BEAM EPITAXY

The technique of molecular beam epitaxy (MBE) was pioneered at Bell Laboratories in the U.S.[1] It was also at Bell Laboratories that the two most promising materials structures were invented. These are the multiquantum-well heterostructure[2] which has led to important applications in the area of optoelectronic devices (e.g., the quantum-well laser[3-5]) and the modulation-doped superlattice[6] which has led to important high-speed device and integrated circuit applications (e.g., the MODFET[7-9]). Currently, MBE is being widely developed in both the U.S. and Japan for application to optoelectronic and high-speed devices and integrated circuits.

In optoelectronics, two major areas of research and development will be discussed here: (1) MBE materials for short-wavelength (650 - 870 nm) emitters and detectors and (2) MBE materials for long-wavelength (1,300 - 1,600 nm) emitters and detectors. The short-wavelength work was pioneered in the U.S. in the (AlGa)As/GaAs material system.[10,11] For a time, the U.S. held a significant lead in this technology. However, most of the U.S. MBE activity in this area has subsided in favor of OMVPE (see the next sub-section). In contrast, the Japanese have been pursuing this technology quite enthusiastically. In the (AlGa)As/GaAs materials system, Fujitsu,[12-13] NTT,[14,15] Mitsubishi,[16] and the Optoelectronics Joint Research Laboratory (OJRL[17]) have pursued multiquantum-well laser R&D while the Electrotechnical Laboratory has reported a surface emitting laser.[18] In addition, NTT has reported lasers with (AlGaIn)P active layers operating between 610 and 660 nm.[19] There is also R&D, primarily in Japan, on the MBE growth of II-VI materials for short-wavelength emitters. One major application of these materials is thin-film electroluminescent displays.

There are a number of laboratories which are pursuing the MBE growth of InP, (InGa)As, (InGa)(AsP), or related materials for long-wavelength optoelectronic applications. These include Bell Laboratories in the U.S.,[20,21] British Telecom in the U.K.,[22] and NTT,[23] Tokyo Institute of Technology,[24] and Sumitomo Electric[25] in

Japan. The work at Bell Laboratories is the most advanced in the sense that workers there have reported MBE growth of (InGa)(AsP).[21] This was accomplished through the use of gas sources (arsine and phosphine) which are cracked by passing them through a high temperature zone as they are admitted into the MBE system. Because this technique results in As and P atoms in the incident group V flux it has thus far been the only method to offer the control over the simultaneous incorporation of As and P required for MBE growth of this alloy. Conventional MBE As and P sources result in sublimation of the As and P tetramer molecules and, when followed by a "cracking" furnace, As and P dimer molecules result. Because in both of these latter cases As molecules displace the P molecules at the growth interface, workers have been unsuccessful in controlling the relative amounts of As and P in the MBE film with enough precision to routinely achieve suitable lattice-match to the InP substrate. Another way of getting around this problem is through the use of quaternary alloys which contain three group III elements and one group V element. Workers at Tokyo Institute of Technology have reported the growth of (InGaAl)As with band-gaps corresponding to the 1,000 - 1,650 nm wavelength range.[24] There is also R&D, primarily in the U.S. and in Europe, on the MBE growth of II-VI materials for long-wavelength devices. One major application of these materials is very long-wavelength detectors.

There are two primary heterojunction materials systems which are being developed by MBE for high-speed device and integrated circuit applications: (1) (AlGa)As/GaAs (on GaAs substrates) and (2) (InAl)As/(InGa)As (on InP substrates). In both cases, the principal applications are MODFET's and MODFET integrated circuits. In the (AlGa)As/GaAs material system, Japanese and U.S. laboratories are on a roughly equal footing. Advanced MBE technology which is being largely dedicated to this application exists at Bell Laboratories,[26-28] Rockwell,[29,30] TRW,[31] Hewlett-Packard,[32] Honeywell,[33,34] IBM,[35] the University of Illinois,[36,37] and Cornell University[38,39] in the U.S. and at Fujitsu,[40-42] NEC,[43] and Hitachi[44] in Japan. Overall, the U.S. laboratories hold a slight lead in speed (ring oscillators with 8.5 ps gate delay at 77 K (Honeywell[34]), frequency dividers operating at 10.1 GHz at 77 K (Bell Laboratories[27]), and discrete MODFET devices with f_t = 80 GHz at 300 K for a 0.2 μm gate length (Cornell[39]) and f_t = 33 GHz at 300 K for a 0.8 μm gate length (Hewlett-Packard[32])). The Japanese hold the lead in the complexity of MODFET integration (4 K SRAM with a 2 ns access time at 77 K (Fujitsu[42])) as well as in MESFET integration (16 K SRAM with a 4.1 ns access time at 300 K (NTT[45])). A potentially important advance reported recently by NEC[43] is the substitution of an AlAs/GaAs superlattice structure with the Si doping in the GaAs layers (to overcome the deep-donor or deep-level effects in Si-doped AlAs and (AlGa)As) for the normal Si-doped (AlGa)As layer in the MODFET structure. NEC's promising results have already been repeated at Bell Laboratories.[28] However, key problems continue to face laboratories in both countries: the elimination of oval defects (the lowest densities are ~200 cm^{-2} in both the U.S. and Japan)[46] and availability of improved GaAs substrate materials.

There is much less activity in the MBE growth of (InAl)As/(InGa)As. While the initial work on these materials was carried out in the U.S., primarily at Bell Laboratories[47,48] and Cornell University,[49] NEC has recently reported mobilities of 11,000 cm^2/V-s at 300 K and 56,000 cm^2/V-s at 77 K for modulation-doped (InAl)As/(InGa)As heterojunctions grown by MBE.[50] These results are nearly identical to those reported from Cornell.[49] Other related work in Japan includes the chloride vapor phase epitaxial growth of high mobility (μ = 9,400 cm^2/V-s at 300 K and μ = 71,200 cm^2/V-s at 77 K) modulation-doped structures at Fujitsu.[51]

There are a number of novel techniques related to MBE which are being pursued in both the U.S. and Japan. However, most of the more ambitious, longer range work such as reactive MBE (Electrotechnical Laboratory[52]), cluster-ion beam epitaxy (Kyoto University[53]), radical beam epitaxy (Hiroshima University[54]), and maskless ion-beam doping (OJRL[55]) were all pioneered and are currently most advanced in Japan. A major area which was pioneered in the U.S. and which has been described above is the use of gas sources for the growth of (InGa)(AsP) (e.g., cracked arsine and phosphine [21,56]). In addition, very recently a number of laboratories have begun using organometallics such as trimethylgallium, trimethylindium, trimethylarsenic, and trimethylphosphorous as gas sources of Ga, In, As, and P in MBE systems. This technique, which is being called MO-MBE (metal-organic MBE) or CBE (chemical beam epitaxy), is being studied at Tokyo Institute of Technology,[57,58] Bell Laboratories,[59] and Aachen Technical University.[60]

Other items which bear mentioning are (1) the development in the U.S. of an indium-free mount for substrates during MBE growth,[61-63] (2) work in the U.S.[64-66] and Japan[67] to develop passivating overlayers to facilitate MBE regrowth following processing, and (3) efforts in the U.S.[68-70] and Japan[71,72] to understand the causes and effects of oval defects in MBE films.

Finally, it is important to note that most of the advanced MBE R&D being performed in both the U.S. and Japan uses commercially manufactured MBE systems. The most common systems are manufactured in the U.S. (Varian Associates (~8 systems in Japan) and Perkin-Elmer (~1 system in Japan)) and in Europe (Riber (~30 systems in Japan) and Vacuum Generators (~12 systems in Japan)).[73] The MBE systems manufactured in Japan by Anelva have not yet been widely accepted by Japanese researchers. However, Anelva's most recently introduced system (January, 1984) has all of the important features of MBE systems manufactured in the U.S. and Europe. It will be important to monitor the acceptance of this system in the Japanese market place.

ORGANOMETALLIC VAPOR PHASE EPITAXY

Organometallic vapor phase epitaxy (OMVPE), alternatively referred to as metal-organic chemical vapor deposition (MOCVD), is being developed primarily for the growth of the III-V compound materials GaAs, (AlGa)As, InP, (InGa)As, and (InGa)(AsP) for optoelectronic device applications. There is also some R&D activity on (AlGa)As/GaAs and (InGa)As/InP heterojunction structures for high-speed device and integrated circuit applications as well as on other compounds such as the II-VI's.

As with MBE, OMVPE technology was pioneered in western countries. The most advanced results for visible quantum-well lasers have been achieved at the Xerox Palo Alto Research Center;[74,75] whereas for long wavelength emitters using the InP based III-V alloys, the work at Thomson-CSF in France[76] is currently the most advanced. The use of OMVPE for the growth of modulation-doped (AlGa)As/GaAs heterojunctions has been most actively pursued at LEP[77] and Thomson-CSF[78] in France and at Hewlett-Packard[79] and Cornell University[80] in the U.S. These four laboratories have all succeeded in producing two-dimensional electron gas mobilities close to 100,000 cm^2/V-s at 77 K, thus being suitable for MODFET device applications.

While western countries may currently hold the lead in some aspects of OMVPE technology, the gap is narrowing. For example, Sony has reported the OMVPE growth of superlattice structures with very abrupt interfaces,[81] multiquantum-well lasers operating at 782 nm,[82] and MODFET's with a noise figure of 1.47 dB with 9 dB associated gain at 12 GHz.[83] Hitachi has reported a rather unique, undoped, self-aligned enhancement-mode MODFET device.[44] The OJRL has reported the growth of high purity OMVPE GaAs ($\mu(77\ K)$ = 100,000 cm^2/V-s).[84] NEC has reported the low pressure (~70 Torr) growth of very high uniformity GaAs (±2.6% in thickness and ±3.5% in carrier concentration over 3-inch diameter GaAs substrates).[85] And, NTT has reported OMVPE modulation-doped (AlGa)As/GaAs heterojunctions with mobilities of 450,000 cm^2/V-s at 2 K as well as MODFET's with transconductances as high as 330 mS/mm at 300 K.[86] In addition, there are areas in which the Japanese have made novel contributions to OMVPE technology. These include the growth of alloys such as (AlGaIn)P for application to visible light emitters (Sony has reported 300 K pulsed operation of 626.2 nm lasers[87]) and extraordinary results in the growth of GaAs on Si substrates at OKI.[88,89]

This latter accomplishment warrants further discussion. OKI has reported several "firsts" in their GaAs/Si work. They have reported the growth of GaAs on Si substrates by using (1) an interfacial layer of Ge deposited by cluster-ion beam epitaxy and (2) an interfacial layer consisting of an (AlGa)As/GaAs superlattice. In addition, they have reported vanadium doping of OMVPE GaAs to yield semi-insulating (> $10^8\ \Omega$-cm) buffer layers. Finally, they have fabricated E/D MESFET ring oscillators on the GaAs/Si films with gate delays of 51 ps for 0.5 μm gate lengths.

GaAs/Si R&D is also being pursued in the U.S. The most successful work to date has been reported by MIT Lincoln Laboratory in which double-heterojunction lasers have been successfully fabricated on MBE grown (AlGa)As/GaAs layers grown on Ge coated (~0.15 μm of Ge) Si substrates[90] and at the University of Illinois where both MESFET's and MODFET's have been fabricated on GaAs or (AlGa)As/GaAs MBE films grown on Si substrates.[91]

Finally, it should be noted that essentially all OMVPE work reported to date has been carried out in reactors which were custom designed and built in the laboratories in which they are used. Although commercially manufactured reactors have been available for several years (vendors include Cambridge Instruments in England and Crystal Specialties and Spire in the U.S.) these have not been widely accepted by established OMVPE researchers. Recently SPC Electronic Corporation in Japan has announced their entry into the OMVPE reactor market. The initial specifications for this reactor look impressive, making it a product to watch.[92] In addition, one should not fail to consider that the Japanese (i.e., Sumitomo Chemicals) supply the highest purity alkyl source materials for use in OMVPE. Thus OMVPE activities in Japan surely warrant continued close observation.

LIQUID-ENCAPSULATED CZOCHRALSKI GaAs

Ultimately, the promise of using III-V materials for high-speed LSI or VLSI circuits as well as for complex integrated optoelectronic circuits will depend on the ability to produce high-quality bulk substrate material. This is because acceptably high yields in both direct ion-implanted and epitaxially-based integrated circuits will require highly uniform, low defect density substrates. This is well recognized by the Japanese as is evident by the prominence of the bulk substrate growth program within the MITI sponsored Optoelectronic Joint Research Laboratory. The importance of substrates has not been overlooked by researchers in the U.S. either. In fact, numerous activities are underway throughout the U.S. and Japan.[93-102] Much of this work is being carried out in commercial LEC crystal pullers. The most common commercial LEC pullers are manufactured by Cambridge Instruments in England. These are high-pressure pullers and are in use at Rockwell[93,94] and Westinghouse[95,96] in the U.S., Cominco[97] in Canada, and the OJRL[98-101] and Sumitomo Electric[102,103] in Japan. In addition, a number of laboratories are using commercial or internally developed low-pressure LEC pullers. These include Texas Instruments[104] and Hewlett-Packard[105] in the U.S.

There are five main approaches being reported for the improvement of bulk LEC GaAs. These are (1) improved thermal design to reduce thermal gradients, (2) indium doping (~1-3% indium in the melt), (3) growth in a magnetic field, (4) active arsenic injection during growth to control stoichiometry, and (5) computer controlled growth (especially automatic diameter control). Most of the laboratories in both the U.S. and Japan are investigating two or more of these approaches in the same puller.

The indium doping of LEC GaAs, which increases the critical resolved shear stress for plastic deformation in the crystal, was pioneered by Westinghouse.[96] Other laboratories have followed this lead, in particular Sumitomo Electric[103] which is one of the world's leading suppliers of GaAs substrate material. Sumitomo Electric was the first company to market (deliveries began in early 1985) semi-insulating In-doped GaAs substrates with very low dislocation densities. Sumitomo has used an improved thermal geometry in addition to the indium doping. While Sumitomo was the first company to market low-dislocation density In-doped GaAs substrates, other Japanese companies soon followed. These include Sumitomo Metals and Mining, Shin-Etsu Handotai, and Furukawa Electric. U.S. commercial substrate suppliers have been much slower in offering In-doped GaAs substrates for sale.

The application of a magnetic field to the LEC (i.e., MLEC) growth of GaAs was pioneered by the OJRL in Japan.[99-101] They have reported significant improvements in the homogeneity of the crystals grown under fields greater than 1300 G. NTT has also reported the growth of MLEC GaAs[106] and has reported the use of indium-doped, MLEC, GaAs to achieve 99.8% bit yields in 16K MESFET SRAM's.[45] To date, no other laboratories have reported significant magnetic LEC activities.

The bulk GaAs program at the OJRL is a good example of the effectiveness of long-range R&D planning in Japan. The OJRL was established in 1979 with funding committed through 1986. This is typical of R&D programs in Japan which often have funding committed for as long as ten years. The accompanying expectation is that the technologies that are developed will have commercial impact in a five to twenty year time frame. The June 1984 progress report of the OJRL[101] is testimony to the success of such long-range planning: two-inch diameter undoped semi-insulating GaAs with dislocation densities less than 1,000 cm^{-2} is reported in addition to indium-doped semi-insulating material which is dislocation-free.

Overall, the sophistication of the R&D activity in the U.S. and in Japan is comparable. However, Japanese laboratories are pioneering novel approaches to LEC GaAs growth which are proving technologically important. In addition, Japanese companies (especially Sumitomo Electric) are in a good position to profitably commercialize this technology and they remain the only companies with their sights clearly set for world-wide domination of this market.

RESOURCE ASSESSMENT

In the course of this survey it has become clear that there is a vast resource of literature published by the Japanese. This is neither surprising nor are the specific journals which are most useful surprising. The journals which are most popular for the U.S. MBE, OMVPE, and bulk GaAs researchers (e.g., the *Journal of Applied*

Physics and *Applied Physics Letters*) are also popular among Japanese researchers. Also, Japan's English-language journals, primarily the *Japanese Journal of Applied Physics* and the *Japanese Journal of Applied Physics Letters* are very valuable sources in these areas of research.

English-language conferences in the U.S. and abroad provide additional, and more timely, information about Japanese technological developments. Recent examples of such conferences include the 2nd International Conference on Molecular Beam Epitaxy (Tokyo, Japan, August 27-30, 1982), the 5th U.S. Molecular Beam Epitaxy Workshop (Atlanta, Georgia, October 6-7, 1983), the 3rd International Conference on Molecular Beam Epitaxy (San Francisco, California, July 31 - August 3, 1984), the 2nd International Conference on Metal-Organic Vapor Phase Epitaxy (Sheffield, England, April 10-12, 1984), the 3rd International Conference on Semi-Insulating III-V Materials (Warm Springs, Oregon, April 24-26, 1984), the annual Device Research and Electronic Materials Conferences, the International Electron Devices Meeting (IEDM), the International GaAs Integrated Circuit Symposium, and the International Conference on GaAs and Related Compounds.

In addition, other extremely valuable and timely semi-open sources of information about Japanese technological advances may be available to those motivated enough to seek them out. These include translations of abstracts, figure captions, summary reports, etc., of conferences held under the auspices of the Japanese Applied Physics Society and of program reviews or meetings held by Japanese government agencies. A key difference between Japanese conferences and U.S. conferences is that many Japanese researchers have a good command of the English language and, thus, can profit by attendance at English-language conferences; whereas most U.S. researchers have little knowledge of the Japanese language and, thus, can not profit from attending the Japanese-language conferences. However, this may not be as large a problem as it first appears. First, translations of conference or meeting abstracts could be generated and widely distributed. Second, U.S. companies and government laboratories could identify individuals within their organizations who do understand Japanese and sponsor their attendance at these conferences in order to bring back information of importance. Information-gathering is simplified at Japanese conferences because it is common practice for attendees to photograph visual material presented by the speakers. Very complete reports can result.

Personal visits by active Japanese and U.S. researchers to each other's laboratories can also provide opportunities to learn of Japanese technological achievements. Naturally, the amount of valuable technical information gleaned from these visits varies widely from visit to visit. The use of good judgment by the visitor and the host can insure that technical exchanges during these visits remain on a level which does not compromise proprietary or classified information while facilitating a mutually beneficial technical exchange.

SUMMARY AND INTERPRETATION

There is extensive active research and development in the U.S., Japan, and Europe in the areas of MBE, OMVPE, and LEC growth of GaAs and related III-V compound semiconductors. In the case of the epitaxial growth techniques MBE and OMVPE, the pioneering work was done in the U.S. As with any emerging technology, the open technical literature and open technical conferences played an important role in disseminating advances in these areas which stimulated additional R&D in the U.S. and new efforts in Japan and Europe.

In the case of LEC growth of bulk GaAs, the demand for higher quality substrates for use in MBE and OMVPE, as well as other applications, became clear to researchers throughout the world and stimulated original work concurrently in the U.S. and Japan. Again, reports of technical advances in the open technical literature and at open technical conferences stimulated additional work and cross-fertilized ongoing work, resulting in rapid progress in both the U.S. and Japan.

There is no question that the Japanese benefited from the availability of MBE and OMVPE research results published by pioneering researchers in the U.S. In addition, other U.S. companies have benefited from the availability of these early publications, as well as publication of later work in Japan and Europe. In fact, because the Japanese now lead in many aspects of these technologies, the U.S. now stands to benefit more than ever from Japanese contributions, which are being widely disseminated in the open technical literature and at technical conferences. Thus, the mechanisms that exist for the dissemination of technical information which are outlined in the previous sub-section seem to provide an equitable technical exchange between the U.S. and Japan as well as other countries active in R&D in these technologies. This type of exchange has played a major role in the rapid growth of all high-technology industries.

In contrast to the seemingly equitable exchange of technical information between the U.S. and Japan, key differences between the U.S. and Japanese governments' approach to R&D funding of these technologies became apparent during the course of this survey. These differences are the extent of government and industrial committment (1) to the commercialization of these technologies and (2) to the establishment of long-range goals and plans needed to meet this committment.

One need only take a cursory look at the differences between the U.S. and Japanese R&D activities surveyed in this report to see one of the most important impediments to their commercialization in the U.S. as compared to Japan. Most Japanese companies which accept Japanese government R&D funding are in an excellent position to apply the developed technologies to their expanding commercial markets. For example, Fujitsu receives Japanese government funds for semiconductor laser R&D (potential commercial application: high-speed optical data links for their computer systems), NTT receives Japanese government funds for short and long-wavelength laser R&D (potential commercial application: optical communication systems), and Sony receives Japanese government funds for visible laser R&D (potential commercial application: optical video and audio disc players). In contrast, most U.S. government R&D funds support work at companies that have no objectives for the commercial application of these technologies. It would appear that this is not the most efficient way of converting U.S. government R&D investments into products which result in commercial market dominance and thus corporate profits for U.S. industry.

However, it is clear that it is not just the technology that the Japanese have gleaned from the U.S. literature nor the product oriented use of government R&D expenditures which have led to the Japanese domination of markets based on that technology. Additionally, the Japanese establishment of and committment to specific long-range goals (on the order of ten years) is a very important factor. The establishment of the OJRL for the development of optoelectronic integrated circuits typifies this long-range planning. U.S. industrial managers and U.S. government policy-makers might well look to the Japanese example in the establishment of long-term goals and the long-range plans required to meet these goals as an effective route to commercialization of technologies pioneered in the U.S.

In sum, when compared with Japan, the U.S. continues to lead in the development of many aspects of MBE, OMVPE, and LEC technologies, but has lost its lead in numerous other aspects of these technologies. Overall, the U.S. is losing ground to the Japanese in the development and commercialization of these advanced crystal growth techniques. This trend is primarily a result of U.S. industry's failure to set long-range goals for the commercialization of these technologies. Therefore, U.S. government and industry officials should (1) consider, in addition to research excellence, the potential for commercial products in appropriating U.S. government R&D funds and (2) commit to long-range goals and establish long-range plans for implementing these advanced technologies in the commercial market-place.

AKNOWLEDGMENTS

I am grateful to J. Miller, D. Houng, M. Scott, and D. Ilic for critically reading the manuscript.

REFERENCES

1. A. Y. Cho and J. R. Arthur, Prog. Solid State Chem. 10, 157 (1975).
2. R. Dingle, W. Wiegmann, and C. H. Henry, Phys. Rev. Lett. 33, 827 (1974).
3. W. T. Tsang, C. Weisbuch, R. C. Miller, and R. Dingle, Appl. Phys. Lett. 35, 673 (1979).
4. R. D. Dupuis, P. D. Dapkus, R. Chin, N. Holonyak, and S. W.

Kirchoefer, Appl. Phys. Lett. 34, 265 (1979).
5. N. Holonyak, R. M. Kolbas, R. D. Dupuis, and P. D. Dapkus, IEEE J. Quantum Electronics QE-16, 170 (1980).
6. R. Dingle, H. L. Störmer, A. C. Gossard, and W. Wiegmann, Appl. Phys. Lett. 33, 665 (1978).
7. T. Mimura, S. Hiyamizu, T. Fujii, and K. Nanbu, Jap. J. Appl. Phys. 19, L225 (1980).
8. D. Delagebeaudeuf, P. Delecluse, P. Etienne, M. Laviron, J. Chaplart, and N. T. Linh, Electron. Lett. 16, 667 (1980).
9. S. Judaprawira, W. I. Wang, P. C. Chao, C. E. C. Wood, D. W. Woodard, and L. F. Eastman, IEEE Electron Device Lett. EDL-2, 14 (1981).
10. A. Y. Cho and H. C. Casey, Appl. Phys. Lett. 25, 288 (1974).
11. W. Tsang, Appl. Phys. Lett. 34, (1979).
12. T. Fujii, S. Hiyamizu, O. Wada, T. Sugahara, S. Yamakoshi, T. Sakurai, and H. Hashimoto, J. Crystal Growth 61, 393 (1983).
13. T. Fujii, S. Hiyamizu, S. Yamakoshi, and T. Ishikawa, J. Vac. Sci. Technol. B3, 776 (1985).
14. H. Iwamura, T. Saku, H. Kobayashi, and Y. Horikoshi, J. Appl. Phys. 54, 2692 (1983).
15. H. Iwamura, T. Saku, T. Ishibashi, M. Naganuma, and H. Okamoto, Collected Papers of MBE-CST-2, p.47 (Tokyo, 1982).
16. K. Mitsunaga, K. Kanamoto, M. Nunoshita, and T. Nakayama, Abstracts of the 3rd Int. Conf. on MBE, p.70 (San Francisco, 1984).
17. M. Mannoh, T. Yuasa, K. Asakawa, S. Naritsuka, K. Shinozaki, and M. Ishii, Abstracts of the 43rd Annual Device Research Conference (Boulder, 1985).
18. M. Ogura and T. Yao, Abstracts of the 3rd Int. Conf. on MBE, p.139 (San Francisco, 1984).
19. H. Asahi, Y. Kawamura, K. Wakita, and H. Nagai, Abstracts of the 3rd Int. Conf. on MBE, p.25 (San Francisco, 1984).
20. W. T. Tsang, Appl. Phys. Lett. 44, 288 (1984).
21. M. B. Panish and H. T. Temkin, Abstracts of the 3rd Int. Conf. on MBE, p.45 (San Francisco, 1984).
22. E. G. Scott, D. Wake, A. W. Livingstone, D. A. Andrews, G. J. Davies, Abstracts of the 3rd Int. Conf. on MBE, p.49 (San Francisco, 1984).
23. Y. Kawamura, H. Asahi, and H. Nagai, J. Appl. Phys. 54, 841 (1983).
24. K. Masu, T. Mishima, S. Hiroi, M. Konagai, and K. Takahashi, J. Appl. Phys. 53, 7558 (1983).
25. Y. Matsui, H. Hayashi, K. Kikuchi, K. Yoshida, Abstracts of the 3rd Int. Conf. on MBE, p.15 (San Francisco, 1984).
26. R. H. Hendel, S. S. Pei, R. A. Kiehl, C. W. Tu, M. D. Feuer, and R. Dingle, IEEE Electron Device Lett. EDL-5, 406 (1984).
27. S. S. Pei, N. J. Shah, R. H. Hendel, C. W. Tu, and R. Dingle, IEEE GaAs IC Symp. Tech. Dig., p.129 (Boston, 1984).
28. C. W. Tu, J. Chevallier, R. H. Hendel, and R. Dingle, Abstracts of the 3rd Int. Conf. on MBE, P.156 (San Francisco, 1984).

29. S. J. Lee, C. P. Lee, D. L. Hou, R. J. Anderson, and D. L. Miller, IEEE Electron Device Lett. EDL-5, 115 (1984).
30. C. P. Lee, D. Hou, S. J. Lee, D. L. Miller, and R. J. Anderson, IEEE GaAs IC Symp. Tech. Dig., p.162 (Phoenix, 1983).
31. J. Berenz, K. Nakano, and K. Weller, IEEE Microwave and Millimeter Wave Circuits Symposium, p.83 (San Francisco, 1984).
32. M. Hueschen, N. Moll, E. Gowen, and J. Miller, IEDM Tech. Dig., p.348 (San Francisco, 1984).
33. N. C. Cirillo, J. K. Abrokwah, and M. S. Shur, IEEE Electron Device Lett. EDL-5, 129 (1984).
34. N. C. Cirillo and J. K Abrokwah, Abstracts of the 43rd Annual Device Research Conference (Boulder, 1985).
35. P. M. Solomon, C. M. Knoedler, and S. L. Wright, IEEE Electron Device Lett. EDL-5, 379 (1984).
36. T. J. Drummond, H. Morkoç, and A. Y. Cho, J. Crystal Growth 56, 449 (1982).
37. T. J. Drummond, S. L. Su, W. Kopp, R. Fischer, R. E. Thorne, H. Morkoç, K. Lee, and M. S. Shur, IEDM Tech. Dig., p.586 (San Francisco, 1982).
38. W. I. Wang, C. E. C. Wood, and L. F. Eastman, Electron. Lett. 17, 36 (1981).
39. L. Camnitz and L. F. Eastman, 11th Int. Symp. on GaAs and Related Compounds (Biaritz, France, 1984).
40. S. Hiyamizu and T. Mimura, J. Crystal Growth 56, 455 (1982).
41. S. Hiyamizu, J. Saito, K. Kondo, T. Yamamoto, T. Ishikawa, and S. Sasa, Abstracts of the 3rd Int. Conf. on MBE, p.52 (San Francisco, 1984).
42. S. Kuroda, T. Mimura, M. Suzuki, N. Kobayashi, K. Nishiuchi, A. Shibatomi, and M. Abe, IEEE GaAs IC Symp. Tech. Dig., p.125 (Boston, 1984).
43. T. Baba, T. Mizutani, and M. Ogawa, Jap. J. Appl. Phys. 22, L627 (1983).
44. Y. Katayama, M. Morioka, Y. Sawada, K. Ueyanagi, T. Mishima, Y. Ono, T. Usagawa, and Y. Shiraki, Jap. J. Appl. Phys. 23, L150 (1983).
45. Y. Ishii, M. Ino, M. Idda, M. Hirayama, and M. Ohmori, IEEE GaAs IC Symp. Tech. Dig., p.121 (Boston, 1984).
46. Private communications.
47. T. P. Pearsall, R. Hendel, P. O'Conner, K. Alavi, and A. Y. Cho, IEEE Electron Device Lett. EDL-4, 5 (1983).
48. K. Alavi, A. Y. Cho, and W. R. Wagner, Abstracts of the 3rd Int. Conf. on MBE, p.34 (San Francisco, 1984).
49. T. Griem, M. Nathan, G. W. Wicks, J. Huang, P. M. Capani, and L. F. Eastman, Abstracts of the 3rd Int. Conf. on MBE, p.86 (San Francisco, 1984).
50. T. Mizutani and K. Hirose, Abstracts of the 27th Annual Electronic Materials Conference, p.15 (Boulder, 1985).
51. M. Takikawa, J. Komeno, and M. Ozeki, Appl. Phys. Lett. 43, 280 (1983).
52. S. Yoshida, CRC Critical Reviews in Solid State and Materials Sciences 11, 287 (1984).

53. T. Takagi, I. Yamada, K. Matsubara, and H. Takaoka, J. Crystal Growth 45, 318 (1978).
54. S. Miyazaki, H. Hirata, and M. Hirose, Extended Abstracts of the 16th Int. Conf. on Solid State Devices and Materials, p.447 (Kobe, 1984).
55. Y. Bamba, E. Miyauchi, K. Kuramoto, A. Takamori, and T. Furuya, Jap. J. Appl. Phys. 22, L331 (1983).
56. A. R. Calawa, Appl. Phys. Lett. 38, 701 (1981).
57. E. Tokumitsu, Y. Kudou, M. Konagai, and K. Takahashi, J. Appl. Phys. 55, 3163 (1984).
58. E. Tokumitsu, Y. Kudou, M. Konagai, and K. Takahashi, Abstracts of the 3rd Int. Conf. on MBE, p.89 (San Francisco, 1984).
59. W. T. Tsang, Abstracts of the 3rd Int. Conf. on MBE, p.90 (San Francisco, 1984).
60. N. Putz, E. Veuhoff, H. Heinecke, M. Heyen, H. Luth, and P. Balk, Abstracts of the 3rd Int. Conf. on MBE, p.91 (San Francisco, 1984).
61. L. P. Erickson, G. L. Carpenter, D. D. Seibel, P. W. Palmberg, P. Pearah, W. Kopp, and H. Morkoç, Abstracts of the 3rd Int. Conf. on MBE, p.20 (San Francisco, 1984).
62. B. Caffee, T. Hierl, G. Ross, G. Muraoka, Varian Associates, Inc., private communications (June - September, 1984).
63. S. C. Palmateer, B. R. Lee, J. C. M. Hwang, J. Electrochem. Soc. 131, 3028 (1984).
64. S. P. Kowalczyk, D. L. Miller, J. R. Waldrop, P. G. Newman, and R. W. Grant, J. Vac. Sci. Technol. 19, 255 (1981).
65. D. L. Miller, R. T. Chen, K. Elliott, and S. P. Kowalczyk, Abstracts of the 3rd Int. Conf. on MBE, p.30 (San Francisco, 1984).
66. Y.-J. Chang and H. Kroemer, Abstracts of the 3rd Int. Conf. on MBE, p.8 (San Francisco, 1984).
67. N. J. Kawai, T. Nakagawa, T. Kojima, K. Ohta, and M. Kawashima, Electron. Lett. 20, 47 (1984).
68. Y. G. Chai and R. Chow, Appl. Phys. Lett. 38, 796 (1981).
69. G. D. Pettit, J. M. Woodall, S. L. Wright, P. D. Kirchner, and J. L. Freeouf, J. Vac. Sci. and Technol. B2, 241 (1984).
70. S.-L. Weng, C. Webb, Y. G. Chai, and S. G. Bandy, Abstracts of the 27th Annual Electronic Materials Conference, p.140 (Boulder, 1985).
71. Y. Suzuki, M. Seki, Y. Horikoshi, and H. Okamoto, Jap. J. Appl. Phys. 23, 164 (1984).
72. M. Shinohara, T. Ito, K. Wada, and Y. Imamura, Jap. J. Appl. Phys. 23, L371 (1984).
73. Private communications.
74. R. D. Burnham, C. Lindstrom, T. L. Paoli, D. R. Scifres, W. Streifer, and N. Holonyak, Appl. Phys. Lett. 42, 937 (1983).
75. C. Lindstrom, T. L. Paoli, R. D. Burnham, D. R. Scifres, and W. Streifer, Appl. Phys. Lett. 43, 278 (1983).
76. M. Razeghi, R. Blondeau, K. Kazmierski, M. Krakowski, B.

de Cremoux, and J. P. Duchemin, Appl. Phys. Lett. 45, 784 (1984).
77. J. P. André, A. Brière, M. Rocchi, and M. Riet, J. Crystal Growth 68, 445 (1984).
78. S. D. Hersee, J. P. Hirtz, M. Baldy, J. P. Duchemin, Electron. Lett. 18, 1076 (1982).
79. Y.-M. Houng and A. F. Sowers, Abstracts of the 26th Annual Electronic Materials Conference, p.112 (Santa Barbara, 1984).
80. L. F. Eastman, Cornell University, private communication (1984).
81. K. Kajiwara, H. Kawai, K. Kaneko, and N. Watanabe, Jap. J. Appl. Phys. 24, L85 (1985).
82. H. Kawai, O. Matsuda, and K. Kaneko, Jap. J. Appl. Phys. 22, L727 (1983).
83. H. Takakuwa, Y. Kato, S. Watanabe, and Y. Mori, Electron. Lett. 21, 125 (1985).
84. S. Takagisha and H. Mori, Jap. J. Appl. Phys. 22, L795 (1983).
85. Japan Science and Technology News 3, 20 (1984).
86. N. Kobayashi, Y. Kadota, T. Fukui, and Y. Takanashi, Abstracts of the 27th Annual Electronic Materials Conference, p.31 (Boulder, 1985).
87. K. Kobayashi, I. Hino, and T. Suzuki, Appl. Phys. Lett. 46, 7 (1985).
88. M. Akiyama, Y. Kawarada, and K. Kaminishi, J. Crystal Growth 68, 21 (1984).
89. T. Nonaka, M. Akiyama, Y. Kawarada, and K. Kaminishi, Abstracts of the 43rd Annual Device Research Conference (Boulder, 1985).
90. T. H. Windhorn, G. M. Metze, B.-Y. Tsaur, J. C. C. Fan, Appl. Phys. Lett. 45, 309 (1984).
91. R. Fischer, W. Kopp, T. Henderson, C. K. Peng, W. T. Maselink, and H. Morkoç, Abstracts of the 43rd Annual Device Research Conference (Boulder, 1985).
92. Japan Science and Technology News 3, 54 (1984).
93. R. T. Chen and D. E. Holmes, J. Crystal Growth 61, 111 (1983).
94. D. E. Holmes and R. T. Chen, J. Appl. Phys. 55, 3588 (1984).
95. R. N. Thomas, H. M. Hobgood, G. W. Eldridge, D. L. Barrett, and T. T. Braggins, Solid-State Electron. 24, 387 (1981).
96. H. M. Hobgood, D. L. Barrett, L. B. Ta, G. W. Eldridge, and R. W. Thomas, Abstracts of the 25th Annual Electronic Materials Conference, p.8 (Burlington, 1983).
97. Private communications.
98. T. Fukuda, K. Terashima, and H. Nakajima, Inst. Phys. Conf. Ser. No. 65, 23 (1982).
99. K. Terashima and T. Fukuda, J. Crystal Growth 63, 423 (1983).
100. T. Shimada, K. Terashima, H. Nakajima, and T. Fukuda, Jap. J. Appl. Phys. 23, L23 (1984).
101. R&D Progress Report of the Optoelectronics Joint Research Association (June, 1984).
102. M. Sekinobu and K. Matsumoto, J. Electronic Engineering, p.32 (September, 1983).

103. K. Tada, S. Murai, S. Akai, and T. Suzuki, IEEE GaAs IC Symp. Tech. Dig., p.49 (Boston, 1984).
104. R. L. Lane, Semiconductor International, p.68, (October, 1984).
105. A. G. Elliot, C.-L. Wei, R. Farraro, G. Woolhouse, M. Scott, and R. Hiskes, J. Crystal Growth 70, 169 (1984).
106. J. Osaka and K. Hoshikawa, Abstracts of the 3rd Int. Conf. on Semi-Insulating III-V Materials, p.21 (Warm Springs, 1984).

METAL CONTACTS TO III-V SEMICONDUCTORS

J.M. Woodall
T. J. Watson Research Laboratory, Yorktown Heights, NY 10598

ABSTRACT

Over the past decade III-V materials have been successfully commercialized for optoelectronic applications requiring LED's lasers and photodetectors. The success of these materials for these applications is based primarily on the use of heterojunction structures formed by epitaxial techniques. More recently, III-V materials, notably GaAs, have been studied in the R&D environment as possible materials for use in high speed devices and circuits including VLSI. However, the use of epitaxially grown structures alone will not assure success. One of the several important challenges facing the commercialization of these materials for this application is the unique problem of metal contacts. Stated simply, the problem is that at nearly all GaAs/metal interfaces the Fermi level is pinned near mid-gap in the GaAS. This causes two major problems in high speed devices: high ohmic contact resistance and poorly controlled Schottky barriers. This talk will discuss the contact problem, examine how U.S. and Japanese device technologists are coping with it. and present some of the future trends in the R&D of contacts to GaAs.

SCIENCE AND TECHNOLOGY DEVELOPMENT: HISTORICAL PERSPECTIVE

Compared with Japan, the U.S. is leading in GaAs science and device physics and lagging in GaAs technology development. I think the major reason is that the Japanese electronics industry has a stronger belief that GaAs will play an important role in future high speed data processing and communications systems. In addition there are some historical reasons discussed below. The issue of current leadership in product development, manufacturing and sales is unclear.

The science and technology development of GaAs materials and devices is prototypical of the contrast that has occurred in the last two decades between Japan and U.S. industry in the development of electronics. Namely, the U.S. has the "get rich quick and get out" mentality, whereas Japan seems to plan and operate for the long haul. For GaAs technology, the Japanese have taken the best of the U.S. (and European) discoveries, developed economically viable technologies, and successfully made and sold their products.

Some notable examples of U.S./European breakthroughs/early reports include: horizontal Bridgman growth of oxygen and Cr doped semi insulating GaAs; LPE; VPE; MOCVD; MBE; in the area of materials fabrication, GaAs, GaP, and GaAsP LEDs; GaAs, GaAlAs, GaInAs injection lasers; Gunn devices; FETs; heterojunction bipolar transistors in the area of new devices, and ion implantation, diffusion, lithography and etching, and Au-Ge-Ni ohmic contacts, in the area of device processing. The U.S./European effort in each of these accomplishments (save perhaps the GaAsP LEDs and MBE growth) can be characterized by a period of frenetic research and rapid publication. This was generally followed by a period of almost no activity and then a period of resurgence precipitated by a concern of lagging behind Japanese technology. The consequence of this scenario was that in addition to the current status of generally lagging the Japanese in nearly all areas of GaAs technology development, the U.S. lost their experts, especially industrial experts, to other areas of R&D. Thus, our current R&D effort is generally dependent on a work force trained within the last decade by a few select universities.

Much of the early and valuable practical knowledge and experience of GaAs technology gained by such leading companies as TI, RCA, GE, IBM, Bell and Howell, Westinghouse has been lost. Fortunately, Stanford, UC Berkeley, Illinois, and Cornell Universities had professors who were versed in early GaAs R&D and trained the people who now manage or influence GaAs R&D. Another factor contributing to the current U.S. lag in GaAs technology development is that ironically most of the companies involved in the early R&D have been profitably marketing products which either do not or only peripherally utilize GaAs devices. Much of the GaAs R&D in the past decade has been spawned by U.S. military needs and hence much of the current R&D in the U.S. is being done by companies such as Rockwell, Hughes, TRW and Lockheed which have been manufacturers of military equipment but not the centers of expertise of early GaAs science and technology. The Japanese seem to have committed to long term development plans and have a stable and growing work force devoted to GaAs R&D. Unless the U.S. makes similar commitments it will end up as an importer of GaAs products. This could lead to the position that the U.S. will have to import future generation data processing and communication systems.

METAL CONTACTS: SCIENCE AND TECHNOLOGY

It is my opinion that the U.S. and Europe have a current lead in the science of the metal/GaAs interface. The U.S. and Europe are about equal to Japan in the art and technology of metallizing GaAs circuits, except for a possible edge to Japan in high temperature stable Schottky barrier gate metallurgy. The W-Si, and Ti-W-Si gate metallurgy developed in Japan is used with state-of-art performance everywhere. Likewise, the Au-Ge-(Ni) ohmic contact was invented in the U.S., but state-of-the-art performance is successfully practiced everywhere. However, most workers agree that the Au-Ge-Ni technology is dead-ended with respect to LSI applications due to a low eutectic temperature ($<400^{\circ}C$) among other things. The U.S. is currently leading in the area of developing possibly better and more thermally stable ohmic contacts. Furthermore surface science in the U.S. and Europe is starting to impact technological innovation in contacts, especially new concepts in Schottky barrier contacts. The U.S. is in a good position to take advantage of its knowledge in this area. This can be accomplished by encouraging U.S. companies to pursue strong patent protection for contact metallurgy. Also U.S. government contracting agencies should be encouraged to fund R&D in both fundamental surface science and innovation in contacts. Finally there should greater incentives to U.S. companies who are not active in GaAs R&D but which have considerable expertise. This is important because, other than GaAs substrate quality and ion implantation technology, the successful commercialization of of GaAs high speed device technology will depend on a useful and high yield contact technology. To my knowledge even the Japanese do not yet have a complete contact technology suitable for the high yield manufacture of LSI circuits.

HIGHLIGHTS OF THE CURRENT STATE-OF-THE-ART

Lessons From Surface Science

The U.S. leads in innovative concepts in contacts to III-V semiconductors. This is largely due to an awareness by the technologists of the comprehensive surface science work on III-V surfaces and interfaces of the past decade which has been reported in Physical Review, Journal of Vacuum Science and Technology and the annual PCSI meetings. The findings can be summarized as follows. The barrier at most but not all metal/III-V semiconductor interfaces is caused by Fermi level pinning. For GaAs this pinning position of about 0.8 eV below the conduction band causes great difficulty in forming both low resistance ohmic contacts and rectifying contacts with precisely controlled barriers. Most device technologists agree that in order to realize the speed promised by GaAs based devices in either a monolithic

IC or LSI format, the contact resistance at the source contact must be less than 2 microohm - cm^2, and the standard deviation in barrier height of the gate electrode of not more than 10 mV. For an 0.8 eV barrier height, a doping level or a space charge density of about 10^{20} cm^{-3} is needed to produce a tunneling ohmic contact that meets the source contact resistance criteria. This doping level is not easily achieved by most doping methods. A few special reports of high doping and high space charge densities are mentioned below. It is somewhat ironical that the property of the workhorse ohmic contact, i.e. Au-Ge-Ni, which leads to low resistance ohmic behavior is not understood at this time even though it was invented almost twenty years ago! The W-Ti-Si gate contact even with its high temperature stability can produce an unacceptable spread in barrier heights due to both processing and substrate variables.

Even though there is not yet universal agreement on the origin of Fermi level pinning, the research to track its origin has resulted in a recent increase in innovative concepts to form ohmic and Schottky barrier like contacts. Since nearly all of these ideas (discussed below) are the product of US researchers, it appears that Japan has not been aware of the implications of US and European surface science research on contact technology.

Au-Ge-Ni

The Au-Ge-Ni contact is currently the most widely used n-type ohmic contact for GaAs devices and circuits. Even though its ohmic properties are not understood at this time, it is capable of meeting the source resistance criteria. The contact was invented in the mid 60s at IBM for use in Gunn diode studies. It is characterized as an alloyed contact in that it strongly reacts with GaAs at temperatures above 400°C to form a variety of phases nonuniformly dispersed at the GaAs interface. The resulting interface is nonplanar. This has led to the theory that the observed proportionality of specific contact resistance upon the reciprocal of the doping level is due to the spreading resistance at protrusions in the non planar interface. Both an advantage and disadvantage of this metallurgy is that it is invasive up to several hundred nm from the original interface. This feature is critical to the performance of current HEMT type devices, but is thought to be undesirable for other applications e.g. lasers and bipolar devices. Also, further device processing at temperatures greater than 400-500°C cause rapid degradation in contact morphology and resistance. However the biggest drawback is its lateral dimensional instability upon alloying. It is thought that as gate lengths and source-to-gate spacing become less that 1 micron this instability will cause both shorting and significant variation in device performance.

There is little evidence that Japan has made significant contributions to the Au-Ge-Ni contact technology. Their main contribution is to successfully employ it in high performance monolithic IC and LSI applications. There have been two minor contributions: a study showing the aging characteristics of Au-Ge-Ni, and a study showing that Ge dopes GaAs n-type by LPE from Au solutions rather than the normally observed p-type behavior of Ge in LPE from Ga melts. My view of Japanese noteworthy accomplishments are listed in Table I.

TABLE I

NOTEWORTHY JAPANESE ACCOMPLISHMENTS
METAL CONTACTS TO GaAs

- IMPLEMENTATION OF Au-Ge-Ni OHMIC CONTACT TECHNOLOGY FOR LSI (16K SRAM) - (NTT)
- AGING STUDIES OF Au-Ge-Ni CONTACTS (NEC)
- HIGH TEMPERATURE STABLE REFRACTORY (W-Ti-Si) SCHOTTKY CONTACTS - (Fujitsu)

RECENT U.S. INNOVATIONS:

The following section is a partial list of U.S. innovation in contacts to GaAs, especially ohmic contacts. This is an important part of the assessment of Japanese GaAs technology since it represents at this time an area almost totally dominated by U.S. R&D. It is my view that those who are responsible for setting U.S. R&D policy ought to bear in mind that the work listed below represents a sizable U.S. advantage and that we should take steps to maintain it.

Ge-GaAs Heterojunction Contact

An important lesson from interface science and technology is that lattice matched isoelectronic heterojunctions, e.g. GaAlAs/GaAs, do not exhibit Fermi level pinning at the interface when properly made. Indeed, this property has been the reason for the success of many recent high speed and optoelectronic devices. The Cornell group has successful applied the lattice matching concept by developing the Ge/GaAs heterojunction interface to make low resistance ohmic contacts. The reason for success is not entirely understood but is due in part to a low conduction band discontinuity at the Ge/GaAs interface (reported to range from 0.050 to 0.3 eV.), high As doping of the Ge and a metal/Ge barrier of only 0.5 eV (see Fig. 1). It appears capable of meeting required contact resistance criteria. Based on Ge/GaAs phase diagrams this contact may not be stable above 725°C.

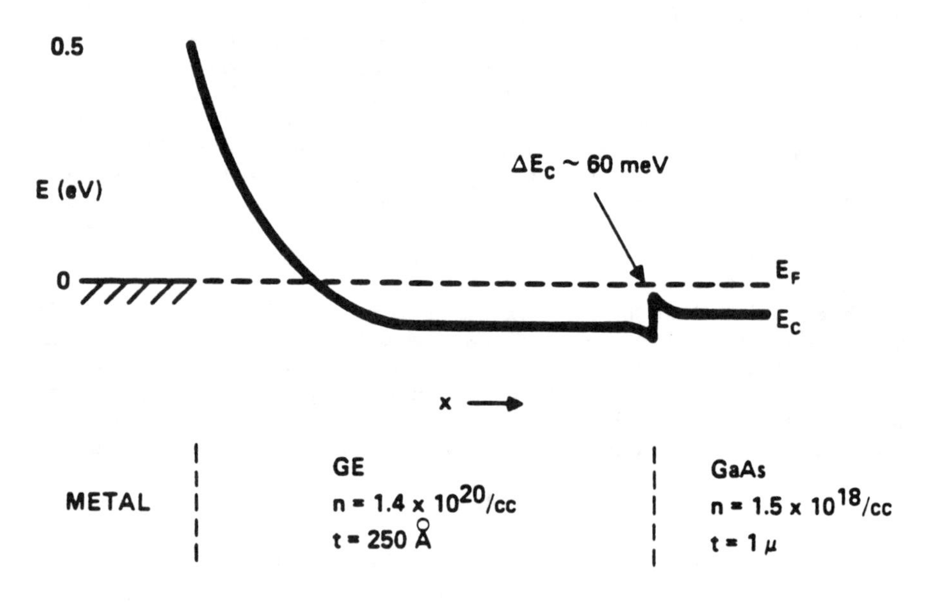

FIGURE 1. AFTER STALL, WOOD, BOARD AND EASTMAN ELECTRON LETT. 15 (24) 801-2 (1979).

Graded Gap GaInAs/GaAs Contact

It is known that Fermi level pinning occurs in the conduction band at InAs surfaces. The consequence of this is that there is an electron accumulation layer at metal/n-InAs interfaces. This leads to an ideal ohmic contact since there is no barrier to electron flow (see Fig. 2a). Since Fermi level pinning produces A 0.8 eV Schottky barrier at metal/n-GaAs interfaces the question is "can InAs be used to eliminate this barrier?" The answer is "not directly" There are at least two reasons for this. First, it is thought that the band gap difference between InAs and GaAs is about 60-65% in the conduction band. This may produce a large band offset. (see Fig. 2b) and thus would not have any advantage over a metal/GaAs interface. Second, there is a 7% lattice constant difference between InAs and GaAs. A heterojunction between the two would contain a large density of defects which are known to cause Fermi level pinning. However, this problem has been solved by a continuous grading in composition from GaAs to InAs. This produces a structure with no barrier to electron flow (see Fig. 2c) and a measured contact resistance of about 1 microohm cm^2.

Sn Doped GaAs by MBE

A barrier height of 0.8 eV requires a doping level of 10^{20} cm^{-3} to produce a tunneling contact resistance of about 1 microohm cm^2. These values have been approached in layers grown by MBE and doped with Sn. A doping level of 6×10^{19} cm^{-3} produced a non alloyed contact resistance of about 2 microohm cm^2. A disadvantage of this method is that for this doping level the Sn clusters on the surface during growth; this produces a rough morphology.

Si doped GaAs by MBE

It has been recently found that non alloyed contacts to MBE grown GaAs doped with 10^{20} Si atoms cm^{-3}. have a resistance of 1.1 microohm cm^2. This is a surprising and important result since the measured electron concentration is only about 5×10^{18} cm^{-3}. Previously, it was observed that the electron concentration saturated at doping levels of about 10^{19} cm^{-3} for nearly all n-type dopants and crystal growth methods. This had lead many workers to doubt that a low non alloyed contact resistance could be achieved by high doping. These new results should have a positive effect on continued research on high doping.

New Approaches to Rectifying Contacts

It is important to note the innovative work being done in the US on Schottky barrier modification and multilayer majority carrier rectifiers. There is work on barrier height modification through near surface layer doping being done at Stanford University and the University of Illinois. Cornell has pioneered a diode structure known as planar doped barriers in which the rectification properties are control by doping and layer thickness using MBE. AT&T Bell Labs has studied a similar structure using compositionally graded layers of GaAlAs. The Japanese strategy seems to be a combination of trial and error, variations of current schemes and adaptive engineering in this area.

SUMMARY

The U.S. generally lags the Japanese in GaAs technology in both optoelectronic and high speed device applications. The reasons are complicated but include: 1) a lack of a perceived imminent need for a GaAs technology by major U.S. computer companies. This relegates GaAs R&D to companies which rarely have large scale silicon system implementation experience; usually these are smaller or medium sized companies concerned with the fabrication of optoelectronic devices or devices for military applications, 2) a loss of expertise after an early U.S. R&D period. With respect to contact technology the U.S. and Japan are both using state-of-the-art in the use of Au-Ge-Ni ohmic and W-Ti-Si rectifying contacts. The US has a considerable advantage in innovative concepts at this time. Action should be taken to maintain

Figure 2a.

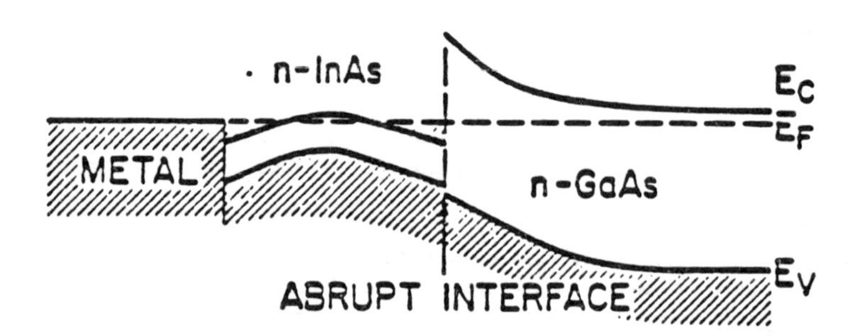

Figure 2b.

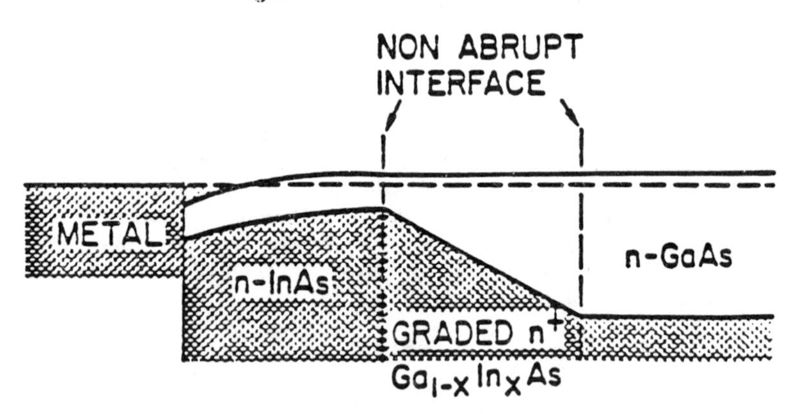

Figure 2c.

Graded Gap Contact.

this advantage and assure early integration of these ideas into leading US companies whose future depends on GaAs LSI and VLSI.

REFERENCES

Ge-GaAs Heterojunction

1. Stall, Wood, Board and Eastman; Electron. Lett. 15(24) 800-1 (1979).
2. Stall, Wood, Board, Dandekar, Eastman and Devlin; J. Appl. Phys. 52(6) 4062-9 (1981).
3. Ballingal, Stall, Wood and Eastman; J. Appl. Phys. 52(6) 4098-103 (1981).
4. Metze, Stall, Wood and Eastman; Appl. Phys. Lett. 37(2) 165-7 (1980).
5. Katnani Chiaradia, Sang and Bauer; J. Vac. Sci. Technol. B2(3) 471-5 (1984).

InAs-GaAs Graded Heterojunction

6. Woodall, Jackson, Pettit, Freeouf and Kirchner; J. Vac. Sci. Technol. 19(3) 626-7 (1981).

Sn Doping

7. Barnes and Cho, Appl. Phys. Lett. 33(7) 651-3 (1978).
8. DiLorenzo, Niehaus, and Cho; J. Appl. Phys. 50(2) 951-4 (1979).

Si Doping

9. Kirchner, Jackson and Woodall; to be published.
10. Kirchner; Electronic Materials Conference, Santa Barbara, CA (1984).
11. Miller, Zehr and Harris; J. Appl. Phys., 53(2) 744-8 (1982).
12. Casey, Panish and Wolfstirn; J. Phys. Chem. Solids 32, 571-80 (1971).

Ohmic Contact Resistivity

13. Chang, Fang and Sze; Sol. St. Electron. 15, 541-50 (1971).
14. Schroder and Meier; IEEE Trns. Electron. Dev. Ed-31(5) 637-47 (1984).
15. Braslan, J. Vac. Sci. 19, 803 (1981)

III-V COMPOUND SEMICONDUCTOR INSULATED GATE FIELD EFFECT TRANSISTORS

H. H. Wieder
Electrical Engineering and Computer Sciences Department
C-014
University of California at San Diego
La Jolla, CA 92093

ABSTRACT

A perspective of research and development in Japan on insulated gate field effect transistors employing III-V compound binary and ternary alloys and related technological issues.

INTRODUCTION

The projected applications of intermetallic semiconducting III-V compounds for high speed digital and analog field-effect transistors (FET's) require that a judicious choice be made among the available binary, ternary and quaternary alloys of these compounds. The synthesis and material technology of such alloys are more complicated than those of the elemental semiconductors. Preservation of their stoichiometry, reduction of residual impurities, control of their homogeneity, impurity diffusion and ion implanting require basic information and data which is available only to a limited extent. The bandstructure parameters, the physical and chemical properties of the surfaces and the interfaces between them and dielectric or metal overlayers impose severe constraints on their technological utilization. They also provide hitherto unavailable options for the implementation of high frequency and microwave devices which may not be matched by the ubiquitous silicon-based FET and integrated circuit technology.

The surface and interface properties of III-V compound semiconductors are among the most important material parameters which define and constrain the available technological options for making FET. Among these are the surface barrier ϕ_B of Schottky barrier gate metal-semiconductor-field-effect transistors (MESFETs), the surface Fermi level E_F^* and the surface or interface state density N_{SS} of dielectrically-insulated gate metal-insulator semiconductor field-effect transistors (MISFETs). Of particular interest is the extent of displacement of the surface potential Ψ_s produced by V_g, the voltage applied to the gate of such transistors and the feasibility of driving the surfaces of such compounds into inversion, depletion, and through flatband, into accumulation.

The potential advantages of a III-V compound semiconductor insulated gate field effect transistor and integrated circuit technology intended to emulate the ubiquitous metal-oxide-silicon (MOS) technology have been recognized for some time. However, serious attempts to make such devices began only in the 1970's. The impetus for these investigations is the expected improvement in the gain-bandwidth product of metal-insulator-semiconductor field effect transistors (MISFET) available in III-V semiconductors whose low field and peak electron velocities are substantially greater than those of Si. The evolution of such a MISFET technology requires a thorough understanding and technological control of the interfaces between these semiconductors and their native oxides which evolve upon contact with the ambient environment and of syntehtic dielectric layers deposited on their surfaces. In many instances the ~ 20 Å thick native oxide cannot be removed and such structures have an interfacial native oxide layer between the semiconductor and a synthetic dielectric layer. A comprehensive model of such interfaces is not yet at hand; however, sufficient information has been accumulated during the past few years on their physical and chemical properties than similarities and differences between them and SiO_2-Si interfaces are evident. First order correlations can also be established between the electrical properties of such interfaces and the corresponding properties of MISFET structures.

An arbitrary classification of insulating dielectric layers used for III-V compound MISFET can be made in terms of their preparation and their macroscopic properties. Homomorphic dielectric layers are those produced by subjecting various III-V compound semiconductors to electrochemical anodization, thermal or plasma oxidation or ion bombardment processes. Such layers are often compositionally inhomogeneous, they contain variable quantities of the atomic constituents of the respective III-V compounds, oxygen and impurities such as C introduced during their growth. Their crystalline phase, order and morphology change as a function of preparation. Native oxides are neither stoichiometric nor spatially homogeneous and are much too conductive to qualify as gate dielectric layers adequate for MISFET.

Heteromorphic or synthetic dielectric insulating layers are usually deposited by means of chemical vapor phase transport reactions, pyrolisis or sputtering, are usually amorphous, are more nearly homogeneous and their dielectric-semiconductor interfaces are more nearly abrupt than those of homomorphic layers.

An evaluation of the electronic properites of metal-insulator-semiconductor (MIS) two-terminal capacitors employing homomorphic and heteromorphic layers can be made to first order within the context of the theoretical framework developed for the interpretation of experimental measurements made on the silicon-silicon oxide system.

The large electron to hole mobility ratio in III-V compounds precludes the use of complementary n-channel and p-channel structures such as those employed in the Si technology. It is only n-doped semiconductors that are suitable for such applications. However, both depletion mode and enhancement mode MISFET are feasible. A depletion mode MISFET (D-MISFET) is "normally on" i.e. a large source-drain current appears for zero gate voltage and this current can be reduced by the depletion of electrons from the conducting channel by an applied negative gate voltage, V_g. An enhancement MISFET can be made by forcing the channel to be, with $V_g = 0$, thin enough so as to be totally depleted. In that case a positive V_g is required to decrease the channel depletion and thus to produce a source-drain current. Alternatively, with a MISFET made of a p-type semiconductor with n-type source and drain contacts, no current flows between them because they are reverse biased, unless a large enough positive V_g is applied to the gate to cause surface inversion of a thin conducting n-type region between them.

In any case, it is clear that a MISFET, unlike a Schottky barrier gate field effect transistor (MESFET) will not draw gate current even though the applied gate voltage is such as to force the surface into, or above, its flatband value.

MISFET allow the application of large gate voltages of either polarity; these are limited primarily by the breakdown strength of the gate dielectric layer. Large V_g values contribute to the large dynamic range of MISFET and simplify the task of circuit designers. An insulated gate also provides a margin of safety; switching transients which might affect, adversely, the gate current of MESFET, have a small or negligible effect on MISFET. The insulator prevents such current flow. Furthermore, metal-semiconductor interdiffusion and degradation of the electrical properties of the interface inherent in MESFET are also prevented by the gate insulating layers. It is evident that the considerations enumerated above apply not only for discrete devices but also for integrated circuits (IC's). A considerable effort has been expended in Japan as well as in the U.S. on developing such a technology based primarily on GaAs and InP because substrates of these binary semiconducting materials are available in the form of semi-insulating (SI) wafers.

GaAs-INSULATOR INTERFACES AND MOS STRUCTURES

Homomorphic and heteromorphic dielectric layers grown on or deposited on the surfaces of n and p-type GaAs have been the most extensively investigated III-V compound MIS structures with the expectation that a suitable dielectric interface will be found to make a GaAs MISFET technology feasible. Homomorphic layers have been made by anodic oxidation, microwave or RF plasma oxidation or by means of magnetically confined plasma-beam oxidation. A large number of investigations were made on anodically oxidized dielectric layers using an aqueous solution of glycol as well as other electrolytes. To first order, the C-V curves measured on all homomorphic MIS capacitors exhibit a strong frequency dispersion whose characteristics resemble those of Si MOS structures only superficially. Methods of analysis, usually applied to Si MOS structures using 1 MHz sinusoidal potentials to obtain C-V curves in order to derive their surface state densities must be used with caution; in GaAs MIS structures fast surface states may respond to frequencies in excess of 10 MHz.

Hetermorphic MIS structures on GaAs have been made by a variety of methods which include the pyrolisis of silane, chemical vapor deposition of silicon nitride and of silicon oxynitride layers, and growth of aluminum oxide by the pyrolisis of aluminum isopropylate. In an attempt to reproduce the interfacial properties of the silicon-silicon oxide system a wide variety of methods have been tried; none have yielded thus far the hoped-for characteristics. A review and assessment of the properties of GaAs MIS structures and of the corresponding MISFET technology has been published in Thin Solid Films[1,2].

In Japan, research on the oxidation of GaAs surfaces and on the properties of the resultant oxide layers was intended to provide at least two specific functions: the passivation of the surfaces of discrete devices and ICs and suitable gate insulators for MISFET. Such work and various facets pertaining to it, including the surface physics and chemistry of homomorphic and heteromorphic insulators and fabrication of prototype devices, was carried out in University Research Laboratories as well as in Industrial Research Centers.

Sugano and Mori[3] described apparatus and techniques used to oxidize GaAs and $GaAs_xP_{1-x}$ in a high frequency oxygen plasma at a pressure of 0.1 to 1 Torr. They found the oxidation rate to be strongly dependent on the plasma characteristics. The oxides thus produced were chemically stable although they had relatively low resistivities of 10^8 to 10^{10} ohm-cm. This work, performed at the University of Tokyo, was supported by a Grant-in-Aid from the Ministry of Education.

The anodization of GaAs by liquid procedures was carried forward in Japan, initially by Hasegawa[4] and his associates at the University of Hokkaido following his return from the University of Newcastle-on-Tyne in the U.K. where, working with Hartnagel and his collaborators they found good electrical and dielectric properties of oxide layers grown in a solution of glycol and water, sufficient to demonstrate elementary aspects of field effect transistors and speculated on improving the GaAs-oxide interface in order to develop an appropriate MISFET technology. However, Sawada and Hasegawa[5] found subsequently that the capacitance-voltage (C-V) characteristics of GaAs-oxide MOS structures were anomalous. They attributed this to a high density of interface states present in n-type GaAs within 0.8 eV below the conduction band edge. This limited the feasible excursion of the surface potential (from its equilibrium position for $V_g = 0$) towards the conduction band edge. They presumed that no such impediments are present on p-type GaAs MOS structures. This work was supported by the Ministry of Education and was also performed at Hokkaido University, Sapporo. Further work, performed by Hasegawa and Suzuki[6], provided details of the electrochemical anodization under dark and illuminated conditions. Other investigations of the process of anodic oxidation of GaAs and GaP and the properties of their corresponding MOS devices were made by Ikoma et al[7] at the University of Tokyo. They found that using various electrolytes, the oxide on p-GaAs grows in a constant field while on n-type GaAS the oxide grows under a constant resistivity condition. Oxide films produced in this manner were found to have a high breakdown strength of ~ 7×10^6 V/cm and a resistivity of the order of 10^{14} ohm-cm.

In a paper presented at the 7th International Vacuum Congress in Vienna in 1977, Hasegawa and Sawada[8] correctly identified the displacement of the surface potential in anodized n and p-type GaAs MOS structures to be limited to the lower half of the fundamental bandgap i.e. between the valence band edge and midgap. However, they also stated that n-GaAs exhibits normal surface inversion while p-type GaAs exhibits normal accumulation. These last two statements are erroneous; subsequently obtained evidence, primarily in the U.S., demonstrates that neither accumulation nor inversion can be obtained in GaAs MOS or MIS structures.

Searching for an improvement in the electrical properties of wet chemically anodized GaAs, Tokuda et al[9] proposed that such oxide layers be sugjected to annealing at about 600°C which lowers their effective resistivity from about 5×10^{14} ohm-cm to ~ 2×10^{8} ohm-cm. By a reanodization at a higher potential they managed to increase the resistivity of these annealed oxide layers to ~ 6×10^{12} ohm-cm. This work was performed at the University of Tokyo with partial support of the Ministry of Education.

Investigations on the potential of anodically oxidized surface layers on GaAs for use as an encapsulator intended for post-ion implantation annealing at temperatures up to 700°C was investigated by Yokomizo and Ikoma[10] at the University of Tokyo. The results appeared promising but not necessarily better than synthetic dielectric layers deposited on GaAs for the same purpose. The potential application of the glycol-water wet anodization process to gallium phosphide was investigated by Hasegawa and Sakai[11]. They found that such GaP MOS structures have a very long charge retention and that the oxide may serve as a mediator for ion diffusion. This work was performed at Hokkaido University as part of a continuing grant from the Ministry of Education.

Investigations on the properties of homomorphic insulating layers grown on GaAs were also performed in Japanese Industrial Laboratories. At the Nippon Telegraph and Telephone Corporation Shinoda and Yamaguchi[12] investigated the preparation and properties of oxide layers grown by first bombarding GaAs surfaces with positive ions producing a Ga-rich oxide layer. Thereafter these layers were anodized by a 13.6 MHz RF plasma anodization process. They found that such oxide layers have a low pin-hole density and high breakdown strength of ~ 3×10^{6} V/cm and exhibited, as well, less hysteresis in their C-V characteristics.

Chemical vapor phase transport procedures were also used by Shinoda and Kobayashi[13] of the Nippon Telephone and Telegraph, Musashino Electrical Communication Laboratory to grow Al_2O_3 layers on GaAs using aluminum tri-isopropoxide as a source material. The deposited layers were presumably non-uniform following ion beam bombardment; however, no electrical data was presented.

A comparison of oxide layers grown on GaAs by thermal, anodic and plasma oxidation was made by Watanabe et al[14] of Hokkaido University using Auger electron spectroscopy and secondary ion mass spectroscopy. They found that oxides grown by thermal oxidation consisted only of Ga_2O_3. By contrast, oxides formed by means of anodic and plasma oxidation contained both Ga_2O_3 and As_2O_3. The former had a large pile-up of arsenic at the substrate-oxide interface while the latter had a much smaller concentration of As at the interface.

A different technique for the synthesis of hetermorphic Al_2O_3 layers on GaAs was used by Yokoyama et al[15] of Hiroshima University. They employed a photochemical reaction between aluminum and oxygen molecular beams excited by ultraviolet light to produce such insulating layers. However, the capacitance-voltage data obtained on such films indicated the presence of the same anomalous frequency and voltage-dependent dispersion of the capacitance found on other heteromorphic or homomorphic layers.

Although by 1977 it had appeared to many of us in the U.S. that there are serious impediments to improving the dielectric-GaAs interface so as to make a GaAs MISFET technology feasible, in Japan the work continued towards this end and continued, furthermore, well after the series of papers published in the U.S. suggested that the problem is related to the pinning of the surface Fermi level E_F^* of GaAs surfaces. The results obtained on both n-type and p-type MIS structures by Meiners[16] on both homomorphic and heteromorphic dielectric layers, which have confirmed the earlier measurements of Hasegawa and Sawada[8], indicate that the position of the surface Fermi level, E_F^* for $V_g = 0$ calculated from the frequency dispersion of C-V and G-V measurements[17] is between 0.8 and 0.9 eV below the conduction band minimum (CBM) of n-type and between 0.6 and 0.7 eV above the valence band maximum (VBM) of p-type GaAs, in good agreement with the XPS measurements of Fermi level pinning obtained on cleaved (110)-oriented GaAs surfaces[18,19]. The surface density N_{ss} dependence on energy is approximately U shaped. The surface potential excursion is limited to ~ 0.45 eV in the lower half of the band gap with a minimum, $N_{ss} \approx 2 \times 10^{12}/cm^2$-eV, reaching values in excess of $N_{ss} = 10^{13}/cm^2$-eV at the boundaries of this zone. Neither flatband nor steady-state surface inversion was achieved with either n-type or p-type MOS capacitors with electric fields of the order of 10^6V/cm. The surface states appear to be quite slow, and thus tend to affect the low frequency response of MIS structures far more than their high frequency response.

GaAs MISFET

At the Fujitsu Laboratories, Mimura et al[20] investigated the potential application of plasma-grown homomorphic gate dielectric layers for MISFET. By controlling the etching depth of the channel between the n^+ source and drain electrodes they were able to produce either depletion mode MISFET or by reducing the channel depth below its full depletion value in a MISFET they were able to obtain enhancement-mode operation. The microwave response characteristics of such enhancement-mode devices were reported to yield 3 dB unilateral gain at 10 GHz. At the University of Tokyo Tokuda et al[21] using 1.5 μm long gates deposited on $n = 1.8 \times 10^{17}/cm^3$ GaAs with an anodically grown gate insulating layer found a maximum oscillation frequency, $f_{max} \approx 22$ GHz and a noise figure of 6.1 dB. They measured a higher transconductance, g_m, at 50 Hz as well as in the μ-wave region than at intermediate frequencies and attributed this to a high density of fast surface states.

While the increase in g_m in the microwave region can be attributed to the inability of these states to charge and discharge at a high rate, the higher g_m in the low frequency region is an artifact. In a subsequent paper describing work performed at the University of Tokyo (Institute of Industrial Science) Ikoma et al[22] inferred from C-V measurements and DLTS measurements that the surface Fermi level of GaAs MIS structures changes only within a narrow range of the bandgap at room temperature. By assuming that in MISFET the difference in g_m between its DC value and that obtained at 1 GHz is due to interface states they estimated these to be between 3×10^{12} and 10^{13} cm^{-2} eV^{-1}.

A paper by Mimura et al[23] of the Fujitsu Laboratories published in 1979 described an evaluation of the native oxide-GaAs interface covered by a plasma-grown oxide layer and subsequently in a paper presented at the 37th Annual Device Research Conference at the University of Colorado, Boulder (1979) Mimura et al[24] described the use of enhancement type MISFETs made of plasma anodized GaAs using conventional photolithographic technology to fabricate devices with 1.5 μm gate length in n 13 stage ring oscillator with a speed power product of 26 fJ per gate and with a propagation delay of 380ps. This work was also supported by MITI.

In 1978 Mimura and his associates[25] published a review paper on the microwave properties of GaAs MISFETs made by means of the above cited techniques, including an analysis of the equivalent circuit of these devices, performance criteria, and a comparison with GaAs MESFETs. They found that depletion mode devices with gate lengths of the order of 2 μm used as a class A amplifier produced 0.4 W output power at 6.5 GHz. These MISFETs had a maximum frequency of oscillation of 22 GHz which is 10% larger than the best analogous GaAs MESFET and an intrinsic current gain cutoff frequency of 4.5 GHz which is some 22% higher than that of analogous MESFET. Enhancement mode MISFET demonstrated useful unilateral power gain in the 2-8 GHz frequency range and a maximum frequency of oscillation of 22 GHz.

The evolution of a planar GaAs MISFET integrated logic circuit technology based, primarily, on the Fujitsu developed and promoted gate insulators produced by plasma oxidation was described by Yokoyama et al[26]. Selective multiple ion implantation into semi-insulating GaAs was used to produce deep depletion enhancement and depletion mode MISFETS with 1.2 μm gate lengths. A propagation delay of 72ps and a power-delay product of 139 fJ was achieved with a 27 stage ring oscillator. This paper is interesting not only on account of its subject matter and contents but also because of relevant material not included in it. It must have been clear to the authors that no DC or bias control can be obtained with such GaAs MISFETs and that, in consequence, these devices and integrated circuits would not be useful for conventional logic circuit applications. Although dynamic logic circuits have been demonstrated in Si they require more complex circuitry and higher power consumption than conventional logic circuits and therefore have not been adopted. One is left with the assumption that in Japan there must have been still some hope of solving the lack of DC response and adequate stability of such MISFET. This, however, was not borne out by subsequent events.

The present day situation with respect to GaAs MISFET and integrated circuits, worldwide, is not a favorable one. Unless and until Fermi level pinning at dielectric-GaAs interfaces can be lifted there is little hope of a viable GaAs MIS technology. Research on obtaining a better understanding of such interfaces is underway in Japan as well as elsewhere. However, the driving impulse for technological utilization has attenuated considerably and most of the principals involved in industrial research laboratories in Japan have turned to other more promising approaches. As we shall see subsequently, research on homomorphic and heteromorphic gate dielectric layers using other III-V compound semiconductors continues. In Japan the principal focus appears to reside in Universities, in particular, at the University of Tokyo.

DIELECTRIC-InP INTERFACES

The properties of thermally grown oxides in dry oxygen on InP were investigated in the U.S. by Wager and Wilmsen[27]. They found that the oxides, composed of 70% to 75% In_2O_3 and 25% to 30% P_2O_5, grow very slowly at temperatures below 340°C and rapidly above this temperature and that the composition of such oxides differs from that of the thin (~ 60 Å) native oxides. Wilmsen[28] was the first to investigate the properties of MIS capacitors made with electrochemically anodized insulating layers and found that the surface potential can be changed by an applied gate voltage in contrast with MIS capacitors made with sputter-deposited SiO_2 which had their Fermi level pinned in accumulation probably by a high density of fixed positive charge trapped in the oxide as well as by the generation of a high density of surface donor centers on the InP substrate. Although the initial impetus for concentrating on the favorable properties of InP came from the U.S., demonstrating its applications for MISFET, research in Japan, after a slow start, continued to emphasize both the fundamental aspects of InP-oxide and synthetic dielectric interfaces as well as that of various relevant MISFET structures and properties.

In the U.S., Meiners[29] found that MIS structures made of low temperature chemical vapor phase-deposited (CVD)SiO_2 on n-InP can be accumulated, that 1 MHz is a sufficiently high frequency to which no surface states can respond, that the Fermi level is pinned to within 0.2 eV of the CBM and that the interface state density is an order of magnitude smaller near midgap than that of GaAs. Essentially the same results were obtained with electrochemically anodized or plasma oxidized surfaces. MIS structures made in a similar manner with p-type InP have a strongly depleted surface; $E_F^* = 1.12$ ev in good agreement with that determined by Spicer[30] on (110)-oriented surfaces and by Waldrop et al[31] on (100)-oriented surfaces, $0.9 < E_F^* < 1.2$ eV.

In Japan, Kawakami and Okamura[32] of Nippon Telegraph and Telephone Public Corporation used chemical vapor deposited Al_2O_3 as gate insulators to investigate the properties of inversion layers made on p-type InP. Yamamoto and Uemura[33] of the same organization investigated the interfacial properties of anodic oxide-InP grown from a tartaric acid, ammonium hydroxide-glycol solution. They found an interface state density, near midgap of the order of 10^{11} cm^{-2} eV^{-1} rising rapidly near the conduction band edge to more than 10^{13} cm^{-2} eV^{-1}. Yamaguchi[34], also affiliated with NTT, of their Ibaraki Electrical Communication Laboratory, produced thin insulating layers on InP by the direct thermal reaction of InP with ammonia gas at temperatures between 530 and 560°C. The dielectric layers were found to be primarily phosphorus nitride and indium nitride and had a rather low resistivity of 10^{11} to 10^{12} ohm-cm. Thermal oxidation of InP and the resultant properties of the oxide layers were investigated by Yamaguchi and Ando[35] of NTT. They found that the oxide layer is composed of polycrystalline $InPO_4$ and that above 620°C they form lower oxides such as In_4O_2 and In_2O due to the evaporation of phosphorus. The activation rate for oxidation is 2.06 eV attributed to the oxygen diffusion through the oxide and the resistivity of such layers is low, typically of the order of 10^8 to 10^9 ohm-cm. In order to improve the characteristics of the oxide Yamaguchi[36] proposed a modified thermal oxidation process using phosphorus pentoxide vapor at temperatures between 400 and 500°C. This improved oxide layer resistivities to 10^{11} or 10^{12} ohm-cm. However the C-V data appeared to be anomalous and the advantages, if any, of this process are highly questionable.

The extensive investigations on plasma anodization of GaAs begun at the University of Tokyo by Sugano and his associates were continued subsequently with specific emphasis on developing a suitable MISFET gate dielectric layer. These will be discussed in the following section. A specific emphasis on dielectric-InP interfaces was presented by Hirayama et al[37] in their paper concerned with plasma anodization of Al-InP, i.e. anodization in an oxygen plasma of a vacuum deposited Al layer on InP. The characteristics of the Al_2O_3 layers produced in this manner were found to be dependent on the end-point of the anodization process. C-V measurements made at both room temperature and 77°K suggested that the minimum of the interface state density was ~ 4×10^{11} cm^{-2} eV^{-1} while the average value for nearly the entire bandgap range is ~10^{12} cm^{-2} eV^{-1}. Electron transport through the Al_2O_3 layer was interpreted in terms of the Poole-Frenkel effect; its dynamic dielectric constant was estimated to be ~4.8 and its resistivity to be between 10^{10} and 10^{12} ohm-cm.

InP MISFET

The favorable dielectric-InP interface properties, compared to those of GaAs make depletion-mode, accumulation-mode and inversion mode MISFET feasible. Depletion and accumulation mode MISFET can be made by direct, selected area n^+ ion implantation into SI InP of the Ohmic source and drain contacts and of the conducting channel in the depletion mode devices which require an additional shallow implantation of donors. Alternatively, epitaxial growth techniques can be used to process devices. D-MISFET were found to have microwave power gain, noise figure and maximum power added efficiency comparable to those of GaAs D-MESFET. Unlike GaAs the transconductance and gain of such MISFET is essentially frequency independent and their operating points can be controlled by means of dc bias. InP enhancement-type MISFET (E-MISFET) can be made either in the form of accumulation-mode transistors (A-MISFET) or inversion-mode transistors (I-MISFET). The latter require p-type substrates and blocking source and drain electrodes. For integrated circuit applications they appear to offer, at this time, fewer advantages in comparison with A-MISFET which can be incorporated into planar monolithic integrated circuits by means of relatively simple procedures. Most of the early pioneering work on such MISFET was performed in the U.S. However, a substantial amount of effort has been and is being spent in Japan to develop a viable InP mISFET technology and Japanese researchers had made important contributions to this end.

Enhancement-type InP MISFETs were made by Hirayama et al [43] using plasma anodized aluminum oxide layers deposited on p-type InP or on semi-insulating (100)-oriented InP. The former were, clearly, I-MISFET while the latter are accumulation mode MISFET. Si ion implantation was used to form the source and drain electrodes and photolithographic lift-off techniques were used to make the discrete devices. An effective electron mobility in the channel of 8×10^2 cm^2/V-sec was obtained with a threshold voltage, $V_T = 0.95$ V. C-V measurements suggested that the Fermi level is not pinned within the bandgap.

Further investigations on MISFETs mode by the same procedure were performed by Hirayama et al[44] at the University of Tokyo. Devices made by means of the techniques he described earlier were investigated, at room temperature and at 80°K. They found an increase in the effective electron mobility and a decrease in the threshold voltage V_t of the MISFETs at 80°K compared to their room temperature values attributed to a decrease in the density of interface states. At 300°K they found that the DC drain current drifts for a few minutes and thereafter becomes stable (for a few hours). We shall return subsequently to this drift/instability problem because it constitutes one of the hurdles that must be overcome if a digital MISFET integrated circuit technology is to become viable. Details of the research on InP surfaces and on MISFETs performed under the direction of professor Sugano at the University of Tokyo are contained in their annual research reviews (in English). At Hokkaido University, Professor Hasegawa and his associates have continued their investigations on anodically grown oxide MIS structures and on InP MISFETs[45]. Sawada and Hasegawa[46] obtained field effect mobilities of 1.5×10^3 to 3×10^3 cm^2/v-sec on enhancement-type MISFET mode with a double layer gate insulator: an initial anodically grown native oxide layer followed by the anodization of a vacuum-deposited Al layer. They claimed that this type of device configuration increased the effective channel mobility and decreased the instability and drift at DC of the drain current. Furthermore, they suggested that in contrast with the suggestions made by Okamura and Kobayashi[47], that the presence of the native oxide at the synthetic dielectric InP interface is the cause of the instability, the specific incorporation of the native anodized oxide layer does not affect adversely the electronic properties of the MISFET and may, in fact, improve their performance.

The properties of enhancement-type MISFET made with Al_2O_3 gate insulating layers deposited by chemical vapor phase transport procedures were investigated at the Nippon Telegraph and Telephone Corporation in Tokyo by Kawakami and Okamura[48]. These were made by the use of sulfur-diffused n-type source and drain contacts on Si InP. They found that these devices had a source-drain capacitance two orders of magnitude smaller than those of inversion mode devices made on p-type InP of the same dimensions. It is, however, noteworthy that only curve-tracer data was presented as evidence for improvement in the MISFET characteristics. The advantages of such a reduction of the capacitance have not been explored in terms of the high frequency properties or gain bandwidth product of MISFET.

The contact resistance of the source and drain contacts of InP MISFET is an important parameter which determines the transconductance and gain-bandwidth product of such transistors. At the Tokyo Laboratories of NTT, Yamaguchi et al[49] investigated the nature of contacts made to Si-implanted n^+ junction on Si InP. They found that Au-Ni/Au-Ge-Ni overlayers yielded a specific contact resistance $\rho_c = 10^{-5}$ ohm-cm using a transmission line method of analysis for a Si implantation dose of 2 x $10^{14}/cm^2$ at 10^2 or 2 x 10^2 keV and an alloying cycle of 400°C for 7 sec.

At the Basic Research Laboratories of the Nippon Electric Co. in Kawasaki, Ohata et al[50] made accumulation mode MISFET using chemical vapor phase transport procedures to deposit SiO_2 gate insulating layers on SiInP. They also made a variety of measurements on two-terminal MIS capacitors with this type of dielectric layers on n-type InP. Their results are rather peculiar in that they find a minimum interface state density near flatband increasing towards midgap, in contrast with evidence obtained by others, worldwide. They reported a field-effect mobility of 1.2 x 10^3 cm^2/V-sec for 1.5 μm long and 28 μm wide gates with ~ 0.09 μm thick dielectric layers, a transconductance of 64 mS/mm and a 7 dB maximum power gain at 4 GHz with a cutoff frequency of 9 GHz of such MISFET.

Itoh and Ohata[51] of the NEC Corporation's Kawasaki Laboratories found that self-aligned InP A-MISFET devices with high transconductances and microwave response are feasible. For this purpose, chemical vapor deposition was used to provide a SiO_2 self-aligned mask for the vacuum deposition of Al gates onto ~ 650 Å thick SiO_2 gate insulating layers. These led to a channel length of 0.8 μm, gate length of 1 μm and width of 280 μm. A typical gm = 17 mS/mm, a minimum noise figure of 1.87 dB with an associated gain of 10 dB was obtained at 4 GHz with power outputs of 1/7 W/mm and 1.0 W/mm at 6.5 and 11.5 GHz, respectively. The maximum power added efficiency at 6.5 GHz was 43.5%. However, these devices were also found to have long term drift in their DC characteristics. A comparison with two-dimensional electron gas field effect transistors (TEGFET) at room temperature might favor the A-MISFET.

Matsui et al[52] at the University of Tokyo investigated the properties of A-MISFET using using a double layer gate insulator: an oxide layer grown by plasma anodization followed by a plasma anodized Al_2O_3 layer. Such A-MISFET were found to have electron field effect mobilities typically 2.1 x 10^3 to 2.6 x 10^3 cm^2/V-sec and threshold voltages of 0.3 to 0.9 V. Their most significant finding is that such a structure has a sharp reduction of the DC drain current instability from 5 μs to ~ 5 x 10^4 s of ±4% compared to that of conventional A-MISFET. Further investigations of such two-layer plasma-oxidized structures were made by Fuyuki et al[53] at the University of Kyoto. The structure and formation of these layers were investigated by Auger spectroscopic methods. In the first stage a native oxide of InP grows to thicknes of ~ 0.05 μm within about 20 min. in a magnetically confined plasma. Then Al sputtered from inner discharge electrodes deposits on this native oxide taking the form of aluminum oxide. They obtained in this manner a typical resistivity of 10^{11} ohm-cm^2 and breakdown strength of 1.5 x 10^6 V/cm. The interface state density, measured in the vicinity of the Fermi level is between 2.5 and 5 x 10^{12} cm^{-2} eV^{-1}.

The uniformity of MISFET characteristics over a wafer and between wafers as well as their reproducibility under ostensibly similar conditions has not been investigated, as yet, insufficient detail. Non-uniformity and lack of reproducibility may be attributed to uncontrolled variations in the deposition of the gate insulating layers, the pre-deposition surface preparation of the semiconductor and spatial fluctuation in the properties of the semi-insulating substrates. Steady progress is being made in overcoming these difficulties and there is little doubt that these issues can be reduced to acceptable levels. More serious, however, is the experimentally observed long term drift in MISFET characteristics, in particular, the drift in drain current with applied steady state drain voltage and gate voltage. It is assumed that the experimentally observed hysteresis in C-V curves of MIS structures of the same materials is closely related to the same phenomena which gives rise to the drift and instabilities of MISFET. Fritzsche[54] interpreted the change in channel conductivity with time of I-MISFET in terms of electron tunneling into traps distributed through the native oxide. Okamura and Kobayashi[55] have analyzed these phenomena in terms of a model which contains one trapping level in the synthetic dielectric Al_2O_3 layer and a second trapping level located in the native oxide. Tunneling into the former is considered the primary transport mechanism with a long time constant and this trap, which is located below the Fermi level, is presumed to be insensitive to temperature. The trap in the oxide is above the Fermi level; it is presumed to be temperature dependent and elimination of the native oxide should, therefore, reduce the contribution of these traps. These assumptions appeared to be confirmed by their experimental data; using HCl to etch and remove the native oxide prior to the deposition of the synthetic dielectric reduced the instability. However, their subsequently published data[56] obtained by first oxidizing the InP surface and then depositing the Al_2O_3 gate dielectric layer also led to a reduction in drift. It may be assumed that a procedure that reduces the surface and interfacial disorder also reduces MISFET instability.

Lile et al.[57,58] described their investigations made on A-MISFET using a variety of surface preparation techniques, including HCl etching and oxidation; they concluded that these procedures do not improve the drift. Furthermore, they found a considerably greater drift in MISFET using Al_2O_3 dielectric layers than those with SiO_2 layers. The drain current, after an initial steady-state period varies exponentially with reciprocal temperature and logarithmically with time. They concluded that these phenomena can be represented by a model which consists of a thermally activated tunneling process, however, the physical origin of the states involved remain undetermined.

Matsui et al.[52] have reduced, as stated earlier, the drift of A-MISFET to less than ± 4% for periods from 5s to 5 x 10^4s by growing an ~ 0.01 μm thick native oxide by plasma anodization, thick enough to prevent tunneling into the superposed Al_2O_3 layer which was also grown by plasma anodization. 1This suggests that at least some options are becoming available for controlling drift. In fact some A-MISFET with SiO_2 gate insulators selected from a batch exhibit no drift whatsoever and it is not yet possible to correlate this with specific fabrication procedures. The solution of these problems might be considered the most important impediment to progress towards an integrated circuit technology employing complementary D-MISFET and A-MISFET.

REFERENCES

1. "Special issue on Semiconducting III-V compound MIS Structures" Thin Solid Films, 56, No. 1/2 (1979), H. H. Wieder and C. W. Wilmsen, ed.
2. "Second Special issue on Semiconducting III-V Compound MIS Structures" Thin Solid Films, 103, No. 1/2 (1983), H. H. Wieder and C. W. Wilmsen, ed.
3. T. Sugano and Y. Mori, J. Electrochem. Soc. 121, 113 (1974).
4. H. Hasegawa et al, Appl. Phys. Lett. 26, 567 (1975).
5. T. Sawada and H. Hasegawa, Electron. Lett. 12, 471 (1976).
6. H. Hasegawa and T. Suzuki, Japan J. Appl. Phys. 15, 2489 (1976).
7. T. Ikoma et al, Proc. 8th Conf. Solid State Dev. Tokyo (1976) and Japan J. Appl. Phys. Suppl. 16-1, 475 (1977).
8. H. Hasegawa and T. Sawada, Proc. 7th Internat. Vac. Congress 1977 and 3rd Conf. Solid Surfaces, Vienna, P.O. Box 300/A-1082 Vienna, Austria, pp. 549-552.
9. Tokuda et al, Electron. Lett. 14, 163 (1978).
10. H. Yokomizo and T. Ikoma, Japan J. Appl. Phys. 17, 1685 (1978).
11. H. Hasegawa and T. Sakai, J. Appl. Phys. 49, 4459 (1978).
12. Y. Shinoda and M. Yamaguchi, Appl. Phys. Lett. 34, 485 (1979).
13. Y. Shinoda and T. Kobayashi, Japan. J. Appl. Phys. 19, L299 (1980).
14. K. Watanabe et al, Thin Solid Films, 56, 63 (1979).
15. S. Yokoyama et al, Thin Solid Films 56, 81 (1979).
16. L. G. Meiners, J. Vac. Sci. Technol. 15, 1402 (1978).
17. L. G. Meiners, Colorado State Univ. Rept. No. SF19 (1979).

18. I. Lindau et al, J. Vac. Sci. Technol. 15, 1337 (1978).
19. W. E. Spicer et al, Phys. Rev. Lett. 44, 420 (1980).
20. T. Mimura et al, Proc. 9th Conf. Solid State Dev. Tokyo 1977; Japan. J. Appl. Phys. 17-1, 153 (1978).
21. H. Tokuda et al, Electron. Lett. 13, 761 (1977).
22. T. Ikoma et al, Proc. 10th Conf. Solid State Dev. Tokyo 1978) Japan J. Appl. Phys. 18-1, 131 (1979).
23. T. Mimura et al, Appl. Phys. Lett. 34, 642 (1979).
24. T. Mimura et al, paper MP-4, 37th Ann. Dev. Research Conf. Univ. of Colorado, Boulder (1979).
25. T. Mimura et al, IEEE Trans. Electron. Dev. ED-25, 573 (1978).
26. N. Yokoyama et al, IEEE Trans. Electron. Dev. ED-27, 1124 (1980).
27. J. F. Wager and C. W. Wilmsen, J. Appl. Phys. 51, 812 (1980).
28. C. W. Wilmsen, Crit. Rev. Solid St. Sci. 5, 313 (1975).
29. L. G. Meiners, Thin Solid Films, 56, 201 (1979)
30. W. E. Spicer et al, J. Vac. Sci. Technol. 16, 1422 (1979).
31. J. R. Waldrop et al, Appl. Phys. Lett. 42, 454 (1983).
32. T. Kawakami and M. Okamura, Electron. Lett. 15, 502 (1979).
33. A. Yamamoto and C. Uemura, Electron. Lett. 18, 64 (1982).
34. M. Yamaguchi, Japan J. Appl. Phys. 19, L401 (1980).
35. M. Yamaguchi, and K. Ando, J. Appl. Phys. 51, 5007 (1980).
36. M. Yamaguchi, J. Appl. Phys. 52, 4885 (1981).
37. Y. Hirayama et al, J. Electron. Mater. 11, 1011 (1982).
38. L. Messick, D. L. Lile, and A. R. Clawson, Appl. Phys. Lett. 32, 494 (1978).
39. L. Messick, Solid-State Electron. 22, 71 (1979).
40. D. L. Lile and D. A. Collins, Thin Solid Films 56, 225 (1979).
41. L. Messick, Solid-State Electron. 23, 51 (1980).
42. L. G. Meiners, D. L. Lile, and D. A. Collins, Electron. Lett. 15, 578 (1979).
43. Y. Hirayama et al, Inst. Phys. Conf. Ser. No. 63, Internat. Conf. on GaAs and Related Compounds Oiso, Japan (1981), p. 431.
44. Y. Hirayama et al, Appl. Phys. Lett. 40, 712 (1982).
45. T. Sawada et al, Proc. 13th Conf. Solid State Dev. 1981 Tokyo, Japan J. Appl. Phys. Suppl. 21-1, 397 (1982).
46. T. Sawada and H. Hasegawa, Electron. Lett. 18, 742 (1982).
47. M. Okamura and T. Kobayashi, Japan. J. Appl. Phys. 19, 2143 (1980).
48. T. Kawakami and M. Okamura, Electron. Lett. 15, 743 (1979).
49. E. Yamaguchi et al, Solid State Electron. 24, 263 (1981).
50. K. Ohata et al, Inst. Phys. Conf. Ser. No. 63, Internat. Conf. on GaAs and Related Compounds, Oiso, Japan, 1981, p.353.
51. T. Itoh and K. Ohata, IEEE Trans. Electron. Dev. ED-30, 811 (1983).
52. M. Matsui et al, IEEE Electron dev. Lett. EDL-4, 308 (1983).
53. T. Fuyuki et al, Japan. J. Appl. Phys. 22, 1574 (1983).
54. D. Fritzsche, Inst. Phys. Conf. Ser. No. 50, 258 (1980).
55. M. Okamura and T. Kobayashi, Japan, J. Appl. Phys. 19, 2151 (1980).
56. T. Kobayashi, M. Okamura, E. Yamaguchi, Y. Shinoda and Y. Hirota, J. Appl. Phys. 52, 6434 (1981).
57. D. Lile, M. Taylor and L. Meiners, Japan J. Appl. Phys. 22, Supl. 22-1, 389 (1983).
58. D. L. Lile and M. J. Taylor, J. Appl. Phys. 54, 260 (1983).

EQUIPMENT AND MANUFACTURING TECHNIQUES FOR OPTOELECTRONICS, Robert I. Scace, National Bureau of Standards, Center for Electronics and Electrical Engineering, Gaithersburg, MD 20899. Specialized equipment for manufacturing advanced optoelectronic devices is briefly discussed, with emphasis on the state of its development in Japan. Illustrative examples of development of other Japanese equipment for silicon microcircuit manufacturing are described, to establish the time scale from conception to wide commercial use of such equipment. The effects of that timing on the ability of industry, either Japanese or elsewhere, to deliver products based on today's laboratory research are discussed. The principal conclusion, based on the author's previous experience in industry and the information that is available, is that at least two years will elapse between laboratory demonstration of a process or device which requires new equipment and the time when that process or device can be produced in quantity. This incubation period is often longer. In view of the meager literature on this topic and the speculative nature of the conclusions, the author has elected not to publish a full-length manuscript in the topical conference proceedings.

CHAPTER V

MICROELECTRONICS PACKAGING

MULTILEVEL VLSI INTERCONNECTION - AN OPTIMUM APPROACH?

K.V. Srikrishnan and P.A. Totta
IBM General Technology Division
Hopewell Jn., New York 12533

ABSTRACT

The wirability of circuit elements is a key ingredient in the success of the very large scale integration technology. Multilevel wiring eliminates the need to use extensive areas of the silicon surface simply for wiring channels. Increasing the number of wiring planes significantly improves the possibility of achieving the goals of the VLSI, i.e. the interconnection of the maximum number of devices in the smallest possible area. Extensive modelling has shown the need to optimize the wiring pitch, number of wiring planes and electrical properties of the materials used (e.g-low resistivity for conductors and low dielectric constant for insulators). The choice of the interconnection technology is also influenced by other factors. Some of these are: cost and reliability objectives; in house expertise and practice; new process/equipment availability and a desire to maintain process commonality. The selected strategy is sometimes an optimum approach for an individual situation which is not universally optimum. In IBM, for example, two different but successful multilevel wiring technologies are being used extensively. The first is used for bipolar circuits; it is a three-level metallization design, with sputtered SiO_2 as the insulator. The second ,for FET devices, has two-levels of metal and polyimide as the insulator. Both technologies use area array input/output terminal connections and lift off line definition. The process/material set of each is reviewed to emphasize the mechanics of reaching an "optimum" solution for the individual applications.

INTRODUCTION

Multilevel interconnection technology is the coordinated design of materials, processes and structures to efficiently wire the device components of a silicon chip for the required logic and memory functions. Several studies on large scale integration show that interconnection or personalization effectiveness is key to realizing the advances in the vertical and horizontal shrinking of the silicon devices[1-5]. How this interconnecting is done affects the speed of the circuits and the density of wired circuits. To accomplish improved wirability and circuit performance, the emphasis is on multilevel wiring, optimized wiring configurations and the selection of suitable materials. The interconnection design (the choice of materials and processes in some sequence) is, however, influenced by the cost, manufacturability and reliability objectives. Reliability here includes wear out failures as well as those due to process created defects. In reality, the new needs have to be carefully matched with the existing constraints such as in-house expertise and practice, desire for commonality of tools and processes etc. This leads to a selection of process/material elements which are optimum for the individual cases but not optimum universally. These optimum approaches are considered successful so long as they satisfy the minimum needs of the program's cost, yield and reliability objectives.

This paper will review very briefly the process & material options available and used for interconnection and the two different designs practiced in IBM. Extensive discussions on these designs are reported elsewhere[6,7], and should be referred to for details. The focus of this review is to highlight the divergent needs of the two technologies and how they resulted in different choices of the elements for the interconnection. The modifications that were introduced to meet the yield, reliability and functional requirements will also be discussed.

PROCESS & MATERIAL OPTIONS

The material set used for multilevel interconnection consists of conductors, insulators and contact materials. The processes are those of deposition and delineation of these materials and include those steps that enable the build up of multilevel structures, such as sputter cleaning, sintering, planarization etc.

The contact metallurgy is in direct contact with silicon surfaces and provides ohmic, low barrier and high barrier contacts. Diffusion barriers are usually employed between Al conductors and the contacts to prevent Al-Si mixing (penetration) during the thermal excursions of normal processing. The contact metallization process is critical in semiconductor manufacturing. The conductors (Aluminum based metallurgy most commonly used for global wiring) provide power to the devices and also serve as signal carriers. The choice of the conductor is determined by the electrical resistivity and current carrying capability of the thin film structure (alloy film or sandwich). Refractory conductors (silicides, W, Mo and silicon) are typically used at the lower levels of wiring for only short lengths because of their higher electrical resistivity. Other considerations for the selection include adhesion, minimum tendency for extrusion or hillocks formation. Insulators provide electrical isolation between lines, reduced capacitance (low dielectric constant) and hermetic sealing. The interlayer insulation allows more than one level of wiring. The more popular materials used for semiconductor applications are listed below.

Table 1
Some commonly used materials in Interconnection

Conductors	Insulators	Contact Materials
Al-Si-Cu	Sputtered SiO_2	PtSi, $PdSi_2$
Al-Cu, Al-Si	CVD and PECVD	TiW, W, Mo^2
Silicides	SiO_x, SiN_x, BPSG	TiN, CVD W
poly Si	Polyimides	Cr-CrO_x

The common deposition processes for metallization are evaporation, sputtering and chemical vapor deposition. The definition of the conductors and contact metallurgy is done by subtractive etching(wet or reactive ion etch), additive (lift off), self aligned or selective processes. The processes of delineation and deposition and the choice of the materials are not independent variables. The selection of one element may restrict the choice of another. One example is that of lift-off. Since it is an additive process, the etchability of the conductor is not a concern, so that a wide variety of composite conductors can be considered. However, the deposition technique is restricted to directional evaporation; neither CVD or sputter deposition is compatible with lift off. Another example of the choice of material limiting the choice of processing options, is the selection of TiW as a barrier; only sputter deposition followed by subtractive etching is possible.

Inorganic insulators are commonly deposited by CVD, PECVD processes and also by RF sputtering (SiO_2). Organic insulators, notably polyimides, are spin coated and cured. They provide lower cost and better planarity. Insulators are conveniently wet or dry etched depending on the dimensions. This is a brief review of the options available and more details can be found elsewhere[8-11].

THREE METAL LEVEL BIPOLAR DESIGN

Logic and array chips made with this design are used in the thermal conduction modules (TCM) of the IBM 308Xs and and also in other systems such as IBM 4300 and System/38. The densest of these has over 1500 logic circuits and experimental chips with 5000 circuits have been built using the same technology[12]. A sectional drawing is shown in Fig.1. The important objective of this interconnection design is to provide a capability to process multiple part number custom logic chips with a short manufacturing cycle time. In addition, the design should provide high reliability under the bipolar use conditions of high current and power (watts), and must be usable under TCM conditions. The key elements of the design are

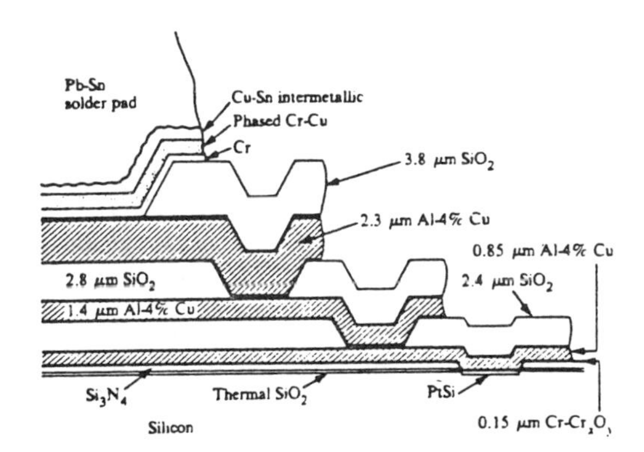

Figure 1. Sectional Drawing of the Bipolar Design with 3 levels of wiring (reference 6)

evaporated AlCu metallurgy, RF sputter deposited SiO_2 insulator and the C4 (controlled collapse chip connection) bumps for area array terminals.

a. Contact and Thin Film Metallization:

PtSi contact metallization is used on all n+ or p+ ohmic contacts and n- Schottky barrier contacts. The PtSi contacts were formed by a self-aligned process by reacting the Pt metal only at the contacts with the Si and stripping the unreacted metal elsewhere. The change from AlCuSi to the PtSi contacts was driven by the need for stable high barrier Schottky contacts as well as by the need to obtain consistently low resistance ohmic contacts. The Si from AlCuSi formed epitaxial p doped Si precipitates. These gave variable contact resistance in the smaller contacts and also an unwanted rectifying layer at the n+ contacts. The aluminum to PtSi reaction forming Al_2Pt and Si dissolution in the Al degraded the electrical behavior of the contacts. Therefore a diffusion barrier was needed.

A 1500 A thick film of Cr-CrO_x cermet was developed[13] as a diffusion barrier. The doping of the Cr grain boundaries with oxygen made the structure an effective barrier. This and the AlCu film (no silicon was needed any longer) formed a sandwich first level metal structure. The presence of the cermet film serendipitously provided an eightfold increase in electromigration life time over that of Al-4%Cu. The upper level connections are thicker and wider to take into account the cumulative topography and the current carrying needs of those levels.

The delineation of metal lines was done by lift-off using a multilayer structure. A stencil consisting of an underlay, a barrier and an imaging resist is formed. The patterns on the top resist are made by e beam or optical lithography and are transferred to layers below by reactive ion etching such that a stencil with an upper ledge layer is formed. A blanket metal film is evaporated and the stencil with the metal is lifted off, leaving the desired film behind. The e beam direct write capability is used to accomodate the need for a large number of custom logic parts. The lift off process gives good resolution without the concern for substrate etching selectivity and conductor corrosion sometimes seen in RIE. The third level metallization, which has relatively larger dimensions, is defined by wet etch.

b. Insulators and Via Technology:

RF sputtered SiO_2 thin film is used as the interlayer insulating film. A parallel plate 13.56 MHz sputtering system is used. Impressed biasing of the wafer during sputter deposition causes sputter removal of the deposited oxide and redistribution. This eliminates cusping of the oxide at the edge of the metal pattern and also partially planarizes the insulator over narrow lines. This allowed the etching of "zero overlap" design of the via holes as the overetch needs are minimized by this partial planarization. Figure 2 illustrates the "zero overlap" design. The avoidance of the metal border of a "covered" via design allows closer line spacing and improvs the wiring pitch. Along with reactive ion etching of the via, the first interlayer insulator was modified to have a dual insulator structure with a layer of PECVD deposited SiN film over the planar silicon oxide layer[14]. This significantly reduced the losses in yield and reliability due to interlayer defects.

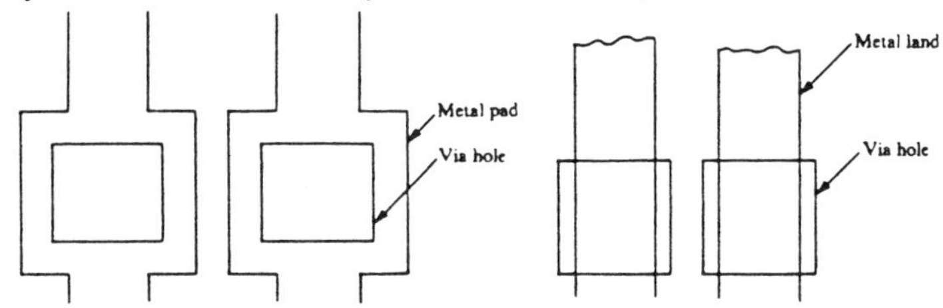

Fig.2 Conventional and "zero overlap" Via Design

A low resistance of the vias is obtained by the use of in-situ RF sputter cleaning in the evaporator just before deposition. This provides better control than the previous approach of using buffered HF etch and a 400 deg. C sinter to break through the oxide.

TWO LEVEL METAL FET DESIGN

A second design meant to produce a family of memory chips uses a metal gate technology and is built with the Silicon and Aluminum Metal Oxide Semiconductor (SAMOS) process. It was intended to serve as a low cost, large volume technology through the combination of one-device cells. Some of the key process concepts are backside ion implant, multi dielectric gate insulator (oxide-nitride), and p type polysilicon field shield. The polysilicon shield is at substrate potential and minimizes the effect of particulate caused shorting and surface leakage. The personalization is essentially two levels of Al-Cu metal with a C4 termination. Figure 3 shows a sketch of the cross section.

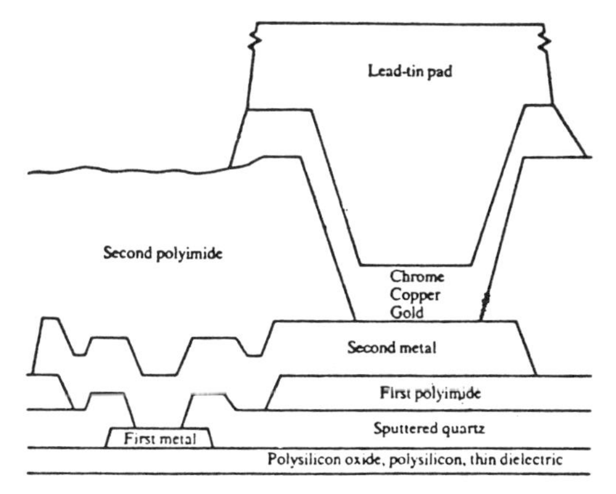

Figure 3. Sketch of the Cross section of a SAMOS structure (reference 7)

a. Metallization:

The lift off stencil used to pattern the conductors is wholly organic and is formed by a lithographic process. The first metal is evaporated Al-Cu-Al sandwich layer. The Al layer adjacent to Si is kept thin so that a temporary barrier layer, $CuAl_2$, is formed close to the gate interface and retards the Si migration. This is preferred over Al-Cu-Si metallurgy, where the occurrence of the Si precipitates caused variable contact resistance. The second level of metal is co-evaporated AlCu and is defined by subtractive etch process. The second metal also provides the redundancy fuses in both bit-line and word-line directions. After the test, the failing addresses are stored and the fuses are blown electrically on chips that could be "fixed". The C4 termination is formed by depositing Cr-Cu-Au and Pb-Sn through a metal mask.

b. Insulators & Vias:

The first interlayer insulator is a composite of sputtered SiO_2 that is overlayed by spin-coated and cured polyimide. The polyimide is first masked and the via is opened by a wet etch process. After the resist strip, a second mask with a smaller via is used to etch holes in the SiO_2. This two step process results in lower sensitivity to defects and thereby improves the yield. The vias also have a "terrace" structure that improves the step coverage of the second metal. The polyimide gave partial planarization and the SiO_2 layer keeps the ionic species away from the silicon interface. The insulator over the second metal is entirely polyimide. It is sufficiently thick to cover all the metal exposed from fuse blowing and is also the final chip seal. Terminal vias are etched into polyimide to the 2nd metal and the Cr-Cu-Au metallurgy totally seals the via and covers all the aluminum.

DISCUSSION

It was proposed earlier that the interconnect strategy was influenced by a variety of factors: product needs and objectives, in-house expertise, desire for commonality of process and equipment (in manufacturing) and reliability objectives. The product needs of the bipolar and SAMOS programs are distinctly different. SAMOS was designed to serve as the process vehicle to produce large volumes of a limited number of memory parts. Cost, which is determined by yield and by the process complexity, was key to the success of the program. Reliability issues were few and defects could be eliminated by burn-in. On the other hand, the primary aim of the bipolar design was to provide a variety of custom designed parts (of varying volume) to meet the needs of the large and intermediate processors. In addition, a fast turn around in manufacturing was essential to accomodate the engineering changes that were needed. The reliability issues were equally important. The higher current densities and temperatures associated with the bipolar circuits together with the inability to burn-in the logic circuits, necessitated a design that should be robust even if somewhat costly. Some of the key areas in which IBM had developed a strong expertise were, the lift-off technology to form fine lines, the addition of Cu in small amounts to Al to improve electomigration and reduce hillock formation and sputtering of SiO_2 from a planar silica target. IBM had also developed area array interconnection (C4) technology and had recognized some of the benefits, such as increased number of I/O s, lower cost of joining (self aligning characteristic), and the excellent reliability of the joints. Both designs incorporated most of these elements. The lift off technology allowed the use of zero overlap via design which improved wirability. For SAMOS, the lower cost coupled with less demanding reliability of the use conditions, meant it was possible to use a lower cost insulator- polyimide. The possibility of ionic mobility in polyimide was recognized as unacceptable to the silicon devices and a layer of SiO_2 was used as insurance to separate the polyimide from the substrate and the first metal. The first interlayer insulation was therefore a dual dielectric. Other enhancements such as polysilicon shield, fusible redundant design and dual etch step were all incorporated to significantly enhance the final yield of the devices.

The bipolar design also incorporated the key elements of in-house expertise such as Al-Cu metallurgy, lift off process for line definition and sputtered SiO_2 for insulation. But the design needs of Schottky contacts and a tight distribution of contact resistance, meant use of self aligned PtSi metallization. The SiO_2 insulator was preferred to be used at all levels in spite of its relatively higher cost of deposition compared to organic insulator. It was chosen due to its improved thermal conductivity and resistance to water and ionic permeability. As additional needs were recognized, the design was changed. An interesting example is the penetration barrier. The use of the lift-off technology made it difficult to use a sputter deposited barrier. The application of a Cr-CrO_x barrier (reactively evaporated) and its satisfactory barrier properties, allowed the continued use of the lift off process. The need for a large number of parts was supported by the development of e beam lithographic technology (tools, resists and processes). The need and benefit of a dual dielectric structure was recognized and the use of a PECVD SiN_x film in conjunction with the planar sputtered SiO_2 was implemented.

CONCLUSION

The major elements of two different interconnection strategies used in IBM have been reviewed. The detailed examination of the elements in the strategy show that the primary rationale for selection was the ability to meet the essential needs of the program and to do it with specially developed in-house technologies. It was also shown that new elements such as polyimide, Cr-CrO_x barrier etc., were introduced to meet the product objectives. Changes were also incorporated to enhance manufacturability and to respond to unexpected problems. Changes were avoided if solutions can be found within the framework of the main strategy (for example the Cr-CrO_x cermet and lift off process). In our discussion, the focus was mainly on the major elements. The materials and equipment to accomplish the major elements continually change with the availability of better tools and materials in the market place (resists and automated tools). The selection of the major process and material factors for a multilevel interconnect strategy is thus influenced by several considerations and this leads to strategies that are optimum for individual cases and may not be optimum universally. However, the strategies are to be considered successful so long they meet the product objectives. This is seen clearly with the diversity of approaches existing in the industry today. The future efforts in multilevel interconnection technology will lead to many approaches and will undoubtedly result in more than one successful approach.

ACKNOWLEDGEMENTS

The authors gratefully acknowledge the significant technical contributions of many individuals within IBM that enabled the development and successful practice of the above two technologies. Thanks are due also to G. Schwartz, R.M. Geffken and D.A. Grose for their review and comments. The authors also appreciate the assistance of Ed Carroll of the technical communication department.

REFERENCES

1. A.K. Sinha, J.A. Cooper Jr., and H.J. Levinstein, IEEE Electron Device Letters, 90 (1982).
2. K.C. Saraswat and F. Mohammadi, IEEE Trans. on Electron Devices vol ED-29, 645 (1982)
3. P.E. Cottrell, E.M. Buturla and D.R. Thomas, IEEE-IEDM, 548 (1982).
4. N. Sasaki and A. Anzai, IEEE-IEDM, 545 (1983)
5. W.R. Heller, G.C. Hsi and W.F. Michaill, IEEE Des. Test Comput. v1, 43 (1984).
6. L.J. Fried, J. Havas, J.S. Lechaton, J.S. Logan, G. Paal and P.A. Totta, IBM J. of Research and Dev., v26, 362 (1982).
7. R.A. Larsen, IBM J. Research and Dev., v24, 268 (1981)
8. S.T. Mastroianni, Solid State Technology, 155 (1984)
9. A.K. Sinha, Electrochem Soc.-Proceedings of Symposium on VLSI Science and Technology, vol 82-7, 173 (1982).
10. A.N. Saxena, IEEE V MIC Conf. Proceedings, 1 (1984)
11. P.S. Ho, Semiconductor International, 128, August (1985).
12 A.H. Dansky, IBM J Research and Dev., v25, 116 (1981).
13. J.S. Jaspal and H.M. Dalal, 19th Annual Proceedings of Reliability Physics, 238 (1981).
14. G. Gati, in 3rd International Conf. on Solid Film Surfaces, Sydney, Australia(1984), published in Special Issue on Applications of Surface Science, elsevier Science Publishers.

Materials and Design for Advanced MLC Packages

E.H. Kraft
Kyocera International, Inc. San Diego, CA 92123

INTRODUCTION

Increasingly severe requirements are being placed on ceramic packages by the escalation of level of integration, speed and power of microelectronic chips and circuits. At the same time, additional constraints are being placed on the manufacture of these packages by the need for cost control, improved assembly yield and device reliability. We will first review the background of requirements, materials and processing involved with multilayer ceramic packaging and then discuss the current capability and future trends in package architecture.

PACKAGE REQUIREMENTS

Interconnection

The most immediate effect of increased level of integration is on the number of input/output (I/O) connections between chip and package, and package to board. Whereas medium scale IC's require only from 12 to 55 I/O's, and large scale requires from 55 to 250, VLSI chips require on the order of 250 to over 1000 chip and board bonding points. Equation (1) is one of the expressions normally used to predict the lead count of logic circuits:

$$N_T = \alpha N_c^{\beta} \qquad (1)$$

where N_T = I/O count
N_c = gate count

α and β are constants

If $\alpha = 2.5$, $\beta = 2.3$, as for a large computer system, then a single chip package would require 200 to 400 I/O's. To accomodate these interconnects on a package without greatly increasing die cavity size or package footprint size requires much finer pitch for bond pads and, in some cases, two tiers of bond pads.

0094-243X/86/1380255-12$3.00

Electrical Considerations

VLSI circuits also require minimum signal propagation delay and minimum distortion. In such circuits however, the impedance characteristics of the conductor traces are of sufficient magnitude to affect overall circuit performance. Since capacitance directly affects propagation delay in a conductor, Knaupp's equation estimates this delay as a function of the dielectric constant of the insulator surrounding the conductor:

$$T_{Pd} = 0.0333 \sqrt{\epsilon_r} \qquad (2)$$

$$= \text{Propagation Delay in ns/cm}$$

where ϵ_r = dielectic constant

Thus package materials with a high dielectric constant have high capacitance and large propagation delay. Table I gives the characteristics of several common electronic materials. With ϵ_r values from 3.0 for polyimide to 10.0 for 90% Al_2O_3, values of T_{Pd} from 0.06 to 0.1 ns/cm may be encountered.

The performance of a package as part of the electronic circuit is also significantly affected by the electrical resistance of signal traces. The buried conductors in multilayer ceramic package are typically tungsten metal. Up until recently, the sheet resistivity of these conductors was typically in the range of 12 to 15 milliohms/ square.

Heat Dissipation

The required dissipation of device-generated heat varies with power level and level of integration. In addition, the ability of packaging materials to conduct that heat also varies by several orders of magnitude among the candidates. Table I also gives the thermal conductivity of these materials.

Structural Considerations

Among the structural requirements of a package are the material strength, thermal stress resistance and dimensional tolerances imposed by assembly and use. Stresses are imposed by handling and assembly processes, shock and vibration in use, thermal gradients and the differences in thermal expansion coefficient of the materials comprising the package and devices. Strength of ceramic materials is generally determined by a flexure test. Data on the strength of several package ceramics is also included in Table I.

TABLE I
Properties of Electronic Materials

	90% Al_2O_3	Polyimide	BeO	AlN	Kovar/42	OFHC Cu	10Cu/W,	W,	Mo	Thin Film Cu
Density - g/cc	3.7		2.85		8.2	8.9	17.3			
Mechanical Properties										
Flexural Strength - MPa	275.0	138.0	240.0	300						
Hardness - GPa	12.7		12.0							
Young's Modulus - GPa	255.0		345.0	300						
Thermal Properties										
Coefficient of thermal Exp(x106/°C)	7.0		7.2	4.2	6.0/8.0	18.3	6.0	4.3	5.7	
Thermal Conductivity - cal/cm sec °C	.04	0.0005	0.60	.20	.04	.94	.50	.40	.35	
Specific Heat - cal/gm °C	0.2		.25		0.1	.09	.05		.06	
Electrical Properties										
Dielectric Strength - KV/mm	10	7.8	10							
Resistivity										
Volume Ωcm (μΩcm)	1014	1016	1017		(48/58)	(1.7)				
Sheet m Ω/sq								10		4/8
Dielectric Constant	10	3.2	6.5	10						
Dielectric Loss Angle @ 1MH$_z$ (x10-4)	18	18	2	.05						
Loss Factor (x10^{-4})	180		10							

Since a device and package are composed of many different materials there is a potential for mismatch and failure from excess thermal expansion induced stress. In addition, thermal gradients within a package will also produce thermal stress in proportion to the severity of the thermal gradient and the coefficient of thermal expansion (CTE) of any one material. The values of CTE for packaging materials are included in Table I.

The processes used to manufacture ceramic packaging include thick film technology, ceramic forming and firing, metal plating and brazing, and have recently been expanded to include thin film technology. Each of these processes imposes its own restrictions on the dimensions and tolerances which are available for size and configuration of the package. The allowable dimensions and tolerances on ceramics and metallizations will be discussed in a later section.

MANUFACTURING TECHNOLOGY

There are many technologies involved in ceramic package manufacturing: Ceramic forming and firing, screen printing, metalization, plating, brazing and thin film. A brief description of those that determine tolerances will aid understanding of the design guidelines to be discussed subsequently.

Multilayer Ceramic Packages

The process for manufacturing typical MLC packages is described in Figure 1. Several steps in this process are responsible for the limits on dimensions and tolerances.

Slip Preparation - A multi-layer Al_2O_3 ceramic body consists of Al_2O_3 plus up to 10% other oxides which are selected for their ability to permit firing at reasonable temperature, or to color the sintered body. The ceramic slip contains these oxides, plus binders and rheologic control agents in a suitable solvent. Control of particle size and size distribution, thorough mixing, removal of all gas bubbles, and control of viscosity are all required to permit casting of uniform and reproducible ceramic tape with consistent shrinkage during firing.

Tape Casting - Slip viscosity, belt speed and speed constancy, carrier surface smoothness, casting head height, doctor blade opening and drying control determine the final tape thickness, uniformity, quality and firing shrinkage. Since ceramic tape is continuously cast onto a moving carrier belt, there is a slight anisotropy in properties of the resultant tape. This is controlled and allowed for in subsequent operations.

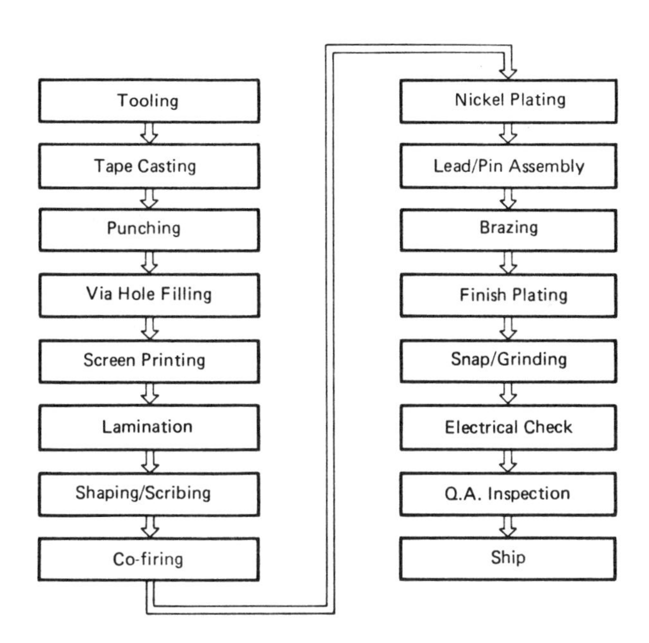

Figure 1. Multilayer Ceramic Package Process

Screen Printing - Conventional thick film screen printing technology is used to form the pattern of conductor traces on each layer of ceramic tape. A modified screen printing technique is used to fill via holes in the tape. The available dimensions and tolerances applicable to thick film printing therefore apply to this step.

Lamination - While registration of layers is accomplished using the same guide pin holes used in the printing step, this alignment is not perfect and must be allowed for by providing via cover dots larger than the actual via diameter.

Co-Firing - The sintering of both ceramic and tungsten powders is accomplished simultaneously. This co-firing operation is performed in a reducing atmosphere at temperatures in excess of 1550°C. Atmosphere control, and control of the firing time and temperature profile are required to produce sintering which is uniform in all locations of a part to avoid warpage, and to produce repeatable shrinkage for control on dimensional tolerances. Tolerances on fired parts are thus limited to ± 1%.

Thin Film Multilayer

Because of the limitations in tolerances, line density, and t_{pd}, the complementary technology of copper-polyimide thin film multilayer has been added to ceramic multilayer packaging. A typical thin film process is shown in Figure 2. Because this process is not subject to the shrinkage involved with ceramic sintering, tolerances are greatly improved. In addition, the photolithographic process is capable of two or more times the resolution possible with thick film screen printing used with ceramic multilayer. On the other hand, thin film technique is presently limited to five layers.

Ceramic Metallization

Moly-Manganese metallization can also be used to extend the capabilities of co-fired multilayer ceramic. Because this metallization is applied after firing of the ceramic, closer tolerances on pattern location are possible than with a co-fired metal. Tolerances are imposed by the capabilities of silk screen printing rather than by ceramic sintering. Adherence of the metallization, which is important for seal ring or pin brazing is comparable to that for co-fired tungsten metallization. Unfortunately, only one layer of this metallization is usually available.

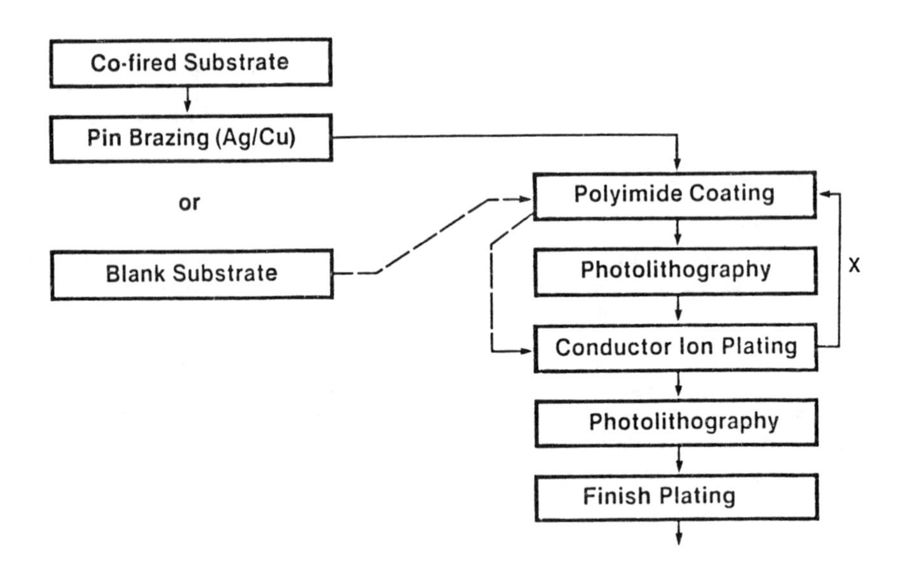

Figure 2. Thin Film Process Flow

DESIGN OPTIONS

The many package types, variety of materials and design options available present the package user with a great variety of options when incorporating a packaged device into a microelectronic system. Many of these options will be presented along with the decision factors required for selection.

Package Type

Initial decisions on package type involve how each device will communicate with other active and passive devices: i.e. shall that first level of packaging involve a single I.C., or should many I.C.'s and passives be packaged together. Multilayer ceramic technology, with or without addition of further metallization or thin film structure is capable of providing substrate or hermetic packaging for either case.

Devices can be individually hermetically packaged in leadless ceramic chip carriers (LCCC), ceramic leaded chip carriers (CLCC), or pin grid array (PGA). These packages may then be assembled either directly to a ceramic or other circuit board or indirectly through use of a ceramic "mother board" which will hold several devices normally packaged in LCCC's. The lead frame of such a mother board is then attached to a non-ceramic circuit board. The use of any of these individual packages, and mother boards, allows device testing at the individual chip level. The use of leaded packages, either CLCC or PGA, or of LCCC's on a mother board prevents problems of solder joint fatigue due to thermal expansion mismatch between ceramic packages and non-ceramic boards. Alternatively leadless packages can be soldered directly to ceramic circuit boards.

In order to minimize signal propagation delay, many circuits are packaged with active and passive devices on a single ceramic MLC substrate with seal ring and leads to form a hybrid package.

Figure 3 shows a variety of these single chip and hybrid packages.

Thermal Management

Increasing levels of integration imply increased operating power density; e.g. 3 watts for CMOS devices, 10 watts for Bipolar chips, and 10 or more watts for multichip configurations. Such power densities begin to approach the thermal loading imposed on space reentry vehicles. To provide the cooling necessary to prevent unacceptable circuit junction temperature rise requires some type of active or passive cooling. Active cooling involves forced air or water flowing over heat conductors designed into a package. Passive cooling may utilize air cooling fins or conductive heat sinks or spreaders similarly attached to the package.

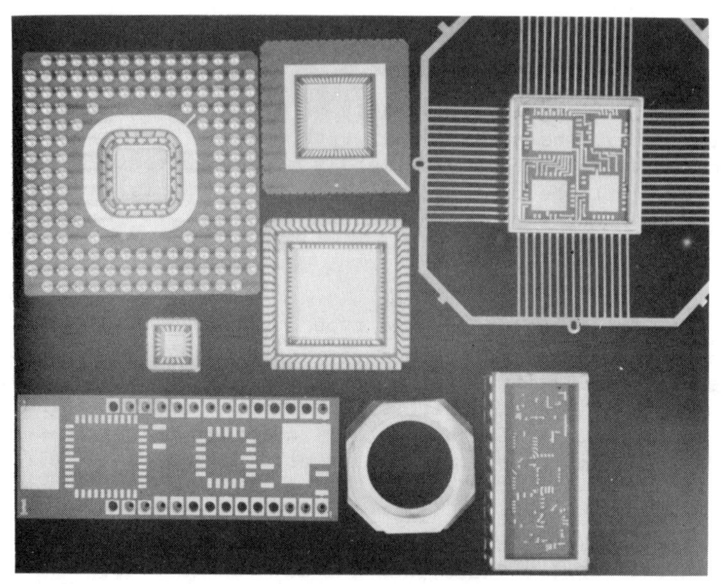

Figure 3. Typical MLC Packages

The choice of materials of package construction can therefore greatly affect the heat flow and resulting junction temperature rise. An examination of the thermal conductivity values shown in Table I illustrates the significance of this selection. Al_2O_3 has a thermal conductivity significantly greater than that for polymers (e.g. 0.04 vs 0.0005 cal/cm sec °C for polyimide) which is one of the significant reasons for the use of ceramic rather than plastic packaging. Even this level of conductivity is insufficient for high power density devices. BeO increases heat flow by an order of magnitude and has therefore found wide application in microwave packaging. However its cost and handling problems limit wider use in IC packaging. Where devices may be attached to an electrically conducting surface, copper/tungsten composite provides thermal conductivity equivalent to that of BeO without the handling concerns. Several other ceramic materials have been proposed, or are being evaluated as solutions to this problem. AlN, SiC and Si_3N_4 each have thermal conductivity which is significantly improved over that of Al_2O_3. However, purity level and process control in manufacture of these materials has yet to allow consistent properties and reasonable cost.

Electrical Performance

The electrical performance of a package involves selection of conductor and insulator materials, and conductor layout. Materials selection also involves the choice of package construction, to be discussed in the next section. For MLC packages, the present choice is usually 90 or 92% Al_2O_3, although research is progressing at many organizations on the application of glass ceramic or ceramic filled glass for this application. Equation (2) shows the influence of the ceramic dielectric constant on signal propagation delay. Thus if a ceramic with lower dielectric constant is utilized in the future, one benefit will be reduced t_{pd}. Where the requirement for conductor spacing dictates addition of thin film to MLC, Table I and Equation (2) show that an additional benefit is decreased t_{pd}.

Conductor electrical resistivity has a direct influence on the circuit thru the matching of impedance and overall voltage requirement. Significant progress has been made in reducing the resistivity of tungsten traces in MLC's from 15 to 10 mΩ/square, and further reductions are anticipated. Use of copper as the conductor in thin film layers on MLC's, provides resistivity in the range of 4 to 8 mΩ/square, although the reduced line width in these layers results in overall resistance equivalent to those traces in the MLC. Future use of copper as the conductor material in glass/glass ceramic type MLC's should provide further reduction of resistance.

The pattern of conductor traces in MLC's is presently laid out using computer aided design (CAD) with primary regard to efficient routing for minimum trace length, and producibility. Thus, this process has yielded optimum voltage drop results. However, to date, only minimal attention has been payed to the trace capacitance and inductance values beyond some general design guidelines. Increasing circuit speed and density has however increased the need for control of these impedance parameters. For this purpose, a high performance trace impedance analysis facility is being assembled. In addition, conductor trace impedance modeling is being developed at several institutions and will be incorporated in package design in the future.

Package Construction

Choice of package construction must be made with regard to:

1. Number of devices per package
2. Package I/O count
3. Package conductor trace density
4. Wire bond pad and lead geometric tolerance
5. Package to board interface

Package I/O count and interface type will determine the choice between leaded or leadless package, use of motherboards, and choice between leaded or PGA type package. All of these choices may be accomodated using conventional MLC technology. The tolerance reqirement on wire bond pads dictates a decision on use of an additional process such as metallizations or thin film. If the tolerances of $\pm$ 1% imposed by MLC cofiring are acceptable, then the more precise additive technologies must be used. The choice between thick film (Mo/Mn) metallization and thin film (Polyimide/Cu) is made on the basis of required trace density. Thick film technology cannot reduce line center spacing below 200μm, whereas the present thin film process is capable of 130μm spacing and future processes should reduce this to 70μm. Thus a doubling of conductor density could allow a significant increase in device density, the extent of which would depend on chip size, I/O count and choice of chip bond method (i.e. flip chip, wire bond, TAB). The limits of dimensions for conductors and vias available by each of these processes is given in Figure 4 and Table II.

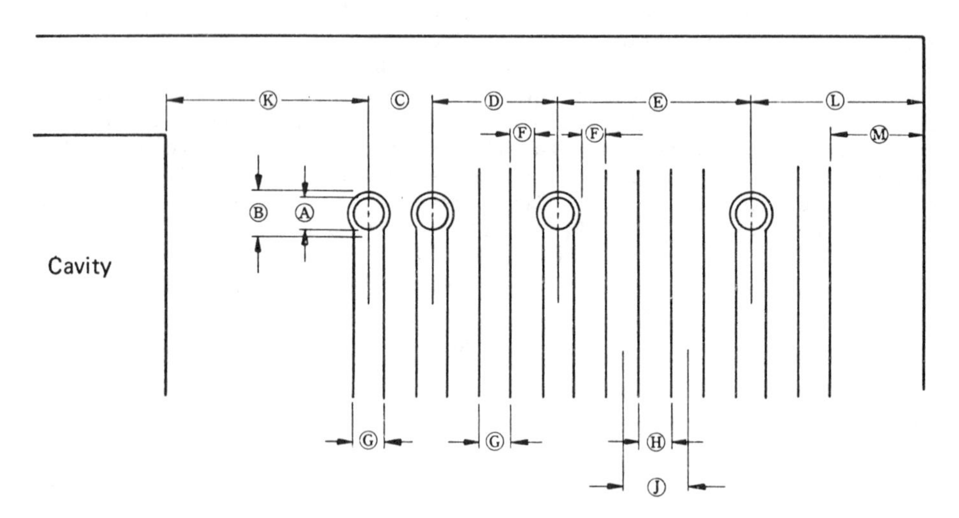

Figure 4. Key to Minimum Dimensions of Conductors and Vias (See Table II)

TABLE II
MINIMUM AVAILABLE CONDUCTOR AND VIA DIMENSIONS [2]

KEY	FEATURE	MULTILAYER CERAMIC			THICK FILM	THIN FILM	
		STANDARD	CUSTOM	HDCM		PRESENT	FUTURE
A	Via Hole [1]	0.203	0.203	0.102	0.640	0.080	0.035
B	Via Cover Dot [1]	0.508	0.381	0.152	0.80	0.120	0.050
C	Via Center Spacing	0.762	0.635	0.254	1.02	0.189	0.090
D	Via Center Spacing	1.270	1.016	0.508	1.05	0.310	0.160
E	Via Center Spacing	1.778	1.219	0.711	2.32	0.440	0.230
F	Via Cover Dot to Line Clearance	0.305	0.254	0.127	0.18	0.070	0.040
G	Line Width	0.254	0.102	0.102	0.51	0.050	0.030
H	Line to Line Clearance	0.254	0.102	0.102	0.25	0.080	0.040
J	Line Center Spacing	0.508	0.203	0.203	0.64	0.130	0.070
K	Cavity to Via Center	0.762	0.635	0.508			
L	Ceramic or Polyimide Edge to Via Center	0.889	0.762	0.508		0.500	0.500
M	Ceramic Edge to Line Clearance	0.762	0.508	0.254			
N	Polyimide to Ceramic Edge Clearance					0.200	0.200

1 Via dimensions are diameter for ceramic and thick film, and square for thin film.
2 Units are mm.

Summary

A wide choice of materials and design options are available for construction of ceramic packaging for high interconnect devices and circuits. The basic technology used in these packages is multilayer, cofired ceramic. High lead count and extensive hybrid circuit interconnection is provided by this method. A hermetic package is thereby provided which meets the majority of LSI or VLSI devices at minimum cost. If chip size or some other factor requires tighter tolerance on wire bond pads than is available by the cofire process, then an additional technology must be utilized. If this tolerance is the only issue, then post-fire metallization will provide the solution. If, however, circuit density or I/O count exceeds the capabilities of screen printing (for either co-fire or metallization) and a finer conductor spacing is required, then addition of the thin film multilayer to MLC package technology will provide the optimum design.

VLSI/VHSIC PACKAGING

P. L. Young, C. C. Chen, J. M. Cech, K. Li+ and M. M. Oprysko
Gould Research Center, 40 Gould Center, Rolling Meadows, IL 60008

ABSTRACT

In order to take full advantage of the high-performance VLSI and VHSIC devices that will become available in the near future, the packaging of semiconductor chips in a system becomes a very important issue. Present technology using single chip carriers mounted on a multilayer circuit board is inadequate for high speed systems. This paper describes the preliminary efforts at the Gould Research Center and elsewhere in the design and fabrication of a multi-chip module containing two layers of low-resistance thin film signal lines.

INTRODUCTION

In order to take full advantage of the high-performance VLSI and VHSIC devices that will become available in the near future, the packaging of chips in a system becomes a very important issue. The package must be capable of (i) minimizing the distortion and delay of pulses propagating between chips, (ii) handling a large number of signal lines, and (iii) enhancing the dissipation of the heat generated by high-performance chips. Present technology using single chip carriers mounted on a multilayer circuit board is incapable of satisfying these stringent requirements. Even the more advanced packaging concept used in the IBM 3081 computer cannot meet all the future VLSI/VHSIC packaging requirements without significant improvement in line resolution and in power handling capability.

One potential approach to increase the packing density of the chip module is to fabricate an integrated circuit that covers the entire surface of a silicon wafer, hence the name wafer scale integration (WSI). However, in spite of research efforts by several organizations, the promise of producing WSI chips with reasonable yield remains unfulfilled. Based on computer model calculations[1], one cannot be overly optimistic that monolithic WSI will become a proven fabrication technique in the next five years.

On a less ambitious scale, a hybrid approach can be developed so that as many as several hundred chips may be attached to a substrate containing very high density interconnecting lines. The hybrid approach to WSI, using thin-film interconnecting lines, can ultimately offer a performance level comparable to the monolithic approach. The hybrid approach also offers the freedom to incorporate, on the interconnect substrate, the most advanced semiconductor chips that will be available in the future. Hence, the expected circuit density will depend primarily on the status of IC fabrication at the time when such a module will be implemented. However, it has been recognized that standard thin film technology

+Present address: Hypress Inc., Elmsford, N.Y.

is inadequate in fabricating such a module. Detailed discussions of the difficulties in fabricating a multichip thin film module and methods to overcome these difficulties will be presented. Based on these preliminary investigations, we concluded that the hybrid approach to WSI can become a viable manufacturable technique to satisfy the future requirements in the computer and military electronics market. At the recent IEEE International Conference on Computer Design (ICCD), Neugebaur[2] compared several packaging alternatives including printed wiring board, thick film multilayer hybrid on ceramic, thin film multilayer hybrid and WSI. His conslusion was that the thin film approach had the highest overall merit, in agreement with our work.

ELECTRICAL REQUIREMENTS

Let us first consider the electrical requirements of the chip interconnection lines and the resultant constraints that will be imposed on the fabrication technology.

The electrical requirements of a high performance multi-chip module will vary considerably depending on the properties of the devices used in the actual system. However, it is instructive to use some simple models to derive the basic configuration for thin-film modules. Davidson[3] considers the design of a high speed computer package using lossless transmission lines and estimates the total delay as well as the noise tolerance as a function of the board characteristic impedance for a single chip module. When the board impedance is low, the noise tolerance is small and the total delay is large. At high board impedance, the noise level rapidly exceeds the noise tolerance level. An acceptable design space is to hold the board impedance to between 40 and 100 ohms.

In order to increase the wiring densities to several hundred lines/cm/layer, it has been suggested[4] that fine resolution thin-film signal lines should be used. For these lossy transmission lines, Ho et al[4] showed that no termination resistors are necessary if the total line resistance is in the range

$$\frac{2Zo}{3} < \frac{R}{L}\ell < 2Zo$$

where Zo is the characteristic impedance of the transmission line, R/L is the line resistance per unit length, and ℓ is the length of the line. Consider a 10 x 10 cm chip module. The distance between two opposite corners of the module is 20 cm, if the interconnect line runs along the edges of the module. For a copper conductor system, the resistivity at room temperature is 2×10^{-6} ohm-cm. In order to propagate pulses with small distortion in such a line with Zo = 50 ohms, the required resistance per unit length is calculated to be 5 ohms/cm. The cross sectional area for such a thin-film conductor is calculated to be 40 μm^2. To obtain a transmission line with this cross-sectional area and with a characteristic impedance of 50 ohms, the conductor lines should be 8 μm wide, 5 μm thick, with 6 μm thick polyimide films separating the signal and ground planes. Such a two signal layer module has a wiring density of 800 lines/cm, sufficient to interconnect the future VLSI/VHSIC devices. A thinner conductor, say 1 μm thick, will require a much wider line and a much thicker dielectric to achieve the same value of Zo. In that case, the wiring density is greatly reduced and is not desirable. It is this requirement on the thickness of the conductor lines that poses a major obstacle for the fabrication of multilayer thin film modules. This issue will be discussed in detail in the next section.

FABRICATION

In order to fabricate a substrate containing a high density of 5μm thick conductor lines, a number of problems not normally encountered in conventional thin-film technology exist. Here, the most crucial ones will be discussed and solutions proposed in the literature, and by the present work will be described.

The most obvious difficulty can be seen in Figure 1. In depositing several microns thick second-level metal over the first level, we face a very severe problem in step coverage. In the region where a second-level conductor crosses the edge of a first-level conductor, the metal line will be much thinner, leading to possible line breakage. If the dielectric layer is not thick enough there is also a possibility that conductors in adjacent layers may be shorted. Ideally, the step coverage problem may be overcome by usihng a planarization process (Figure 2). Here the dielectric film completely covers the first-layer conductor and forms a planar surface for the second layer. As a result, the step coverage difficulty is completely eliminated, and the yield for producing thin-film modules is enhanced. In addition, the variation in film thickness and hence the variation of characteristic impedance is minimized, a crucial requirement for any high performance interconnect module.

The desirability of planarization on the chip level has been recently discussed by Moriya et al.[5]. However, the metal conductors on the chip level are only a few tenths of a micron thick and recent advances in thin-film deposition and bias sputtered etching of insulators are sufficient to ensure planarization. In the present case, conductors are 5μm thick, making the task of planarization far more difficult.

One method that has been used to planarize the dielectric over the thin film signal lines is to cover the conductors with a 25 μm thick layer of polymide.[6] Good electrical properties have been reported on propagation of high speed pulses on these thin-film lines. However, the reported wiring density is only 80 lines/cm/signal layer, sufficiently less than future requirements. A second method to achieve planarization is to employ photolithographic techniques to define and etch the area for conducctor lines in a polyimide film. Then, the metal conductors are deposited via an additive process, either by lift-off, by electroplating or by electroless plating. Results are shown in Figure 3. Copper conductor lines, 10 μm wide and 5 μm thick, are completely planarized using a 5 μm thick layer of polyimide overcoating. The second signal layer can be fabricated on top of the first layer without any step coverage problems. (Fig 4.) With these results, a wiring density of 400 lines/cm/layer has been achieved.

DEFECT REPAIR

Perhaps the second major obstacle in making thin-film multi-chip modules is to obtain acceptable yields. A single defect on the substrate may be fatal to the whole package. "Hence, achieving high yield in fabricating a thin-film module with a substrate size of 10 cm x 10 cm and a minimum feature of 10 μm surely exceeds the capability of present integrated circuit processes practiced in stringent clean room environment".[4]

This obstacle may be overcome using a laser photolytic or pyrolytic deposition technique which can deposit micron size metal lines on either silicon or glass substrates. A schematic diagram of the laser apparatus is shown in Figure 5. At present, this technology has been successfully used to repair micron size clear defects on photomasks.[7] It can be easily extended to repair defective conductors in thin-film modules. Furthermore, the same technique can be used to modify and rework interconnect patterns after the chips have been mounted on the module.

THERMAL MANAGEMENT

For any multi-chip module approach using high speed device technology to be successful new techniques must be implemented to efficiently remove the heat generated by the semiconductor devices. A model based on finite difference method, has been developed to predict the internal thermal resistance of multi-chip modules. The equation which governs the heat transfer characteristic is given by

$$\frac{\partial^2 T_i}{\partial x^2} + \frac{\partial^2 T_i}{\partial y^2} + \frac{\partial^2 T_i}{\partial z^2} = 0$$

where Ti denotes the temperature in the ith section of the module. Using the boundary conditions given in Table 1, a three dimensional model (Fig. 6) has been developed to predict the internal thermal resistance of a multi-chip module.

Consider a semiconductor chip .6 cm x .6 cm occupying an area 1.6 cm x 1.6 cm. The semiconductor chip is bonded directly to a silicon substrate containing the thin-film signal planes. The back side of the silicon substrate is attached to a copper heat sink.+ Based upon reasonable estimate of the bond thicknesses, the internal thermal resistance of this package is calculated to be less than $1^oC/W$. Therefore, it is expected that heat removal from high speed device can be adequately managed in a multi-chip configuration.

+We are presently experimenting with different polymers for bonding the silicon substrate to the copper heat sink. Experimental results will be published when this study is completed.

SUMMARY

Packaging of high speed semiconductor devices require high interconnect densities which can be achieved on thin-film multi-chip modules. The required electrical properties on the thin-film conductors impose difficulties in the fabrication of these modules. Methods used by others and by us to overcome these difficulties were described. Finally, computer simulation suggests thermal management can be satisfied by bonding semiconductor chips directly to silicon substrates containing thin-film signal lines.

REFERENCES

1. S. M. Hu, IEEE Electron Devices Issue No. 69, 8 (1984).
2. C. A. Neugebaur, ICCD Proceedings, 115 (1984).
3. E. E. Davidson, IBM J. Res. Develop. 26, 340 (1982).
4. C. N. Ho et al, IBM J. Res. Develop. 26, 286 (1982).
5. T. Moriya et al, IEDM Technical Digest, 550 (1983).
6. R. J. Jensen, J. P. Cummins and H. Vora, IEEE Trans. CHMT-7, 384 (1984).
7. M. M. Oprysko, M. W. Beranek and P. L. Young, IEEE Elec. Dev. Lett. EDL-6, 344 (1985).

Table I

At $z = 0$

$$-k_1 \frac{\partial T_1}{\partial z} = \begin{cases} q & \text{for } o < |x| < \ell_x,\ o < |y| < \ell_y \\ o & \text{for } \ell_x < |x| < L_x,\ \ell_y < |y| < L_y \end{cases}$$

At $z = z_1$ and z_2

$$\left.\begin{aligned} -k_i \frac{\partial T_i}{\partial z} &= -k_{i+1} \frac{\partial T_{i+1}}{\partial z} \\ T_i &= T_{i+1} \end{aligned}\right\} \quad \begin{aligned} &\text{for } o < |x| < L_x,\ o < |y| < L_y \\ &i = 1 \text{ and } 2 \end{aligned}$$

At $z = z_3$

$$-k_3 \frac{\partial T_3}{\partial z} = h\,(T_3 - T_a) \qquad \text{for } o < |x| < L_x,\ o < |y| < L_y$$

At $|x| = 0$ and L_x

$$\frac{\partial T_i}{\partial x} = o \qquad \text{for } o < z < z_3$$

At $|y| = o$ and L_y

$$\frac{\partial T_i}{\partial y} = o \qquad \text{for } o < z < z_3$$

where q is the heat flux from the die, h is the effective heat conductance of the heat sink, and T_a is the cooling air temperature. The coordinate system is shown in Fig. 6.

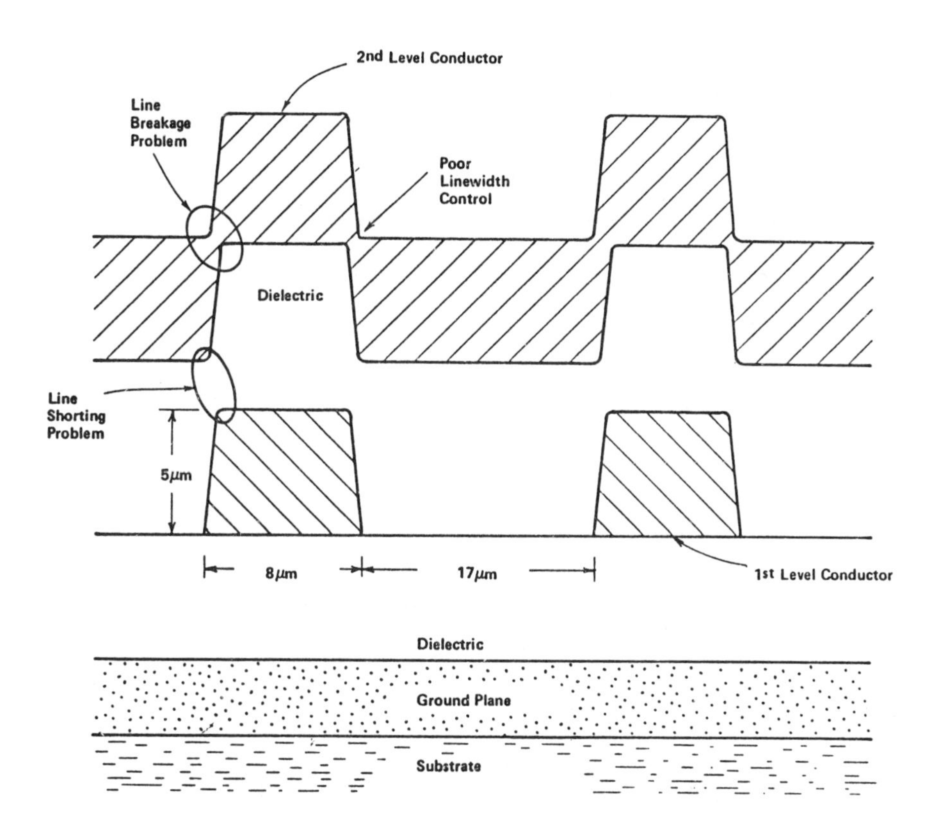

FIG 1. PARTIAL CROSS SECTIONAL VIEW OF A NON-PLANARIZED THIN-FILM MODULE SHOWING X and Y SIGNAL LINES

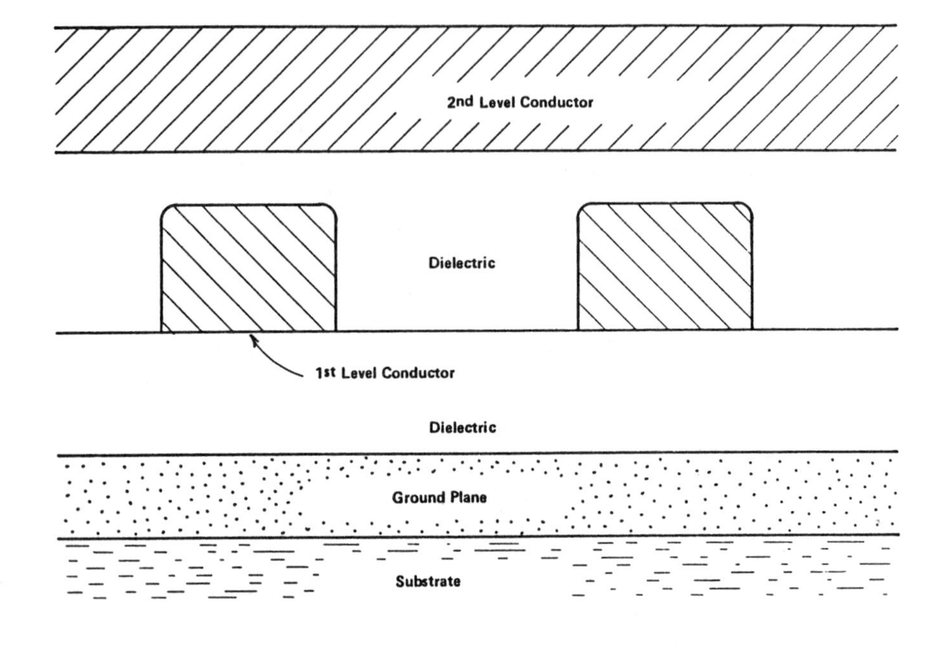

FIG 2. PARTIAL CROSS SECTIONAL VIEW OF A PLANARIZED THIN FILM MODULE SHOWING BOTH X and Y SIGNAL LINES

FIG. 3 SEM MICROGRAPH SHOWING 10μm WIDE BY 5μm HIGH COPPER LINES EMBEDDED IN POLYIMIDE. NOTE THE PLANAR SURFACE OF THE POLYIMIDE, AND THE NEAR VERTICAL WALL OF THE COPPER CONDUCTORS

FIG. 4 SEM MICROGRAPHS SHOWING TWO LAYERS OF PLANARIZED THIN-FILM INTERCONNECTING SIGNAL PLANES. NOTE THE PLANAR SURFACE OF EACH LAYER AND THE NEAR VERTICAL WALL OF THE COPPER CONDUCTORS.

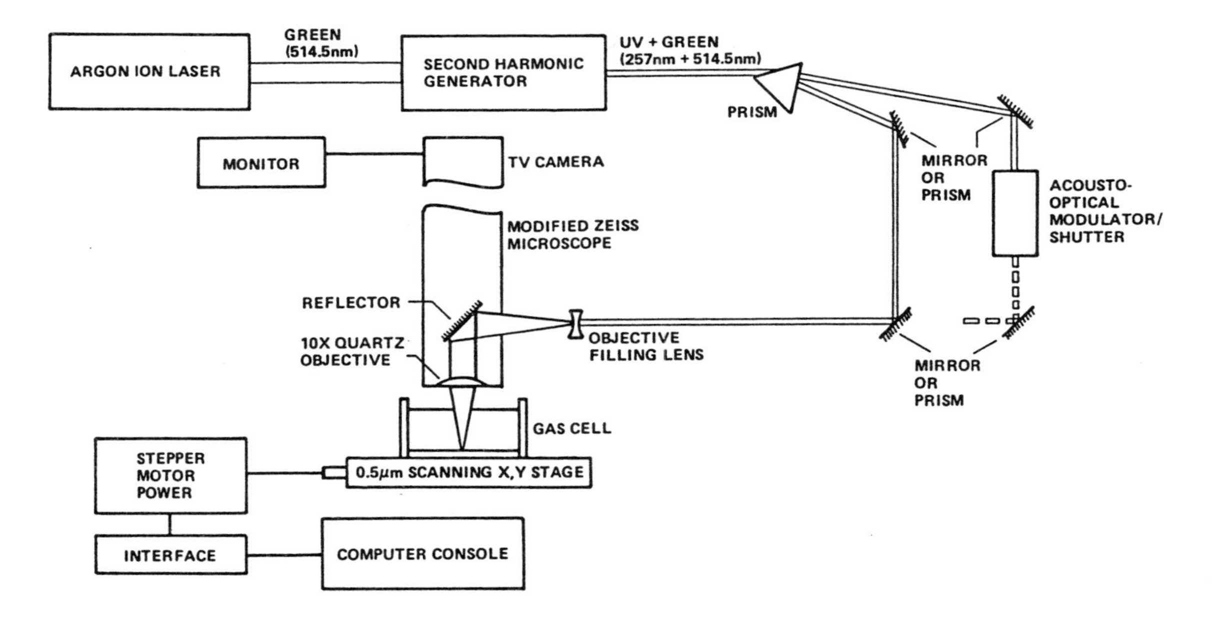

FIGURE 5 LASER MICRO DEPOSITION APPARATUS

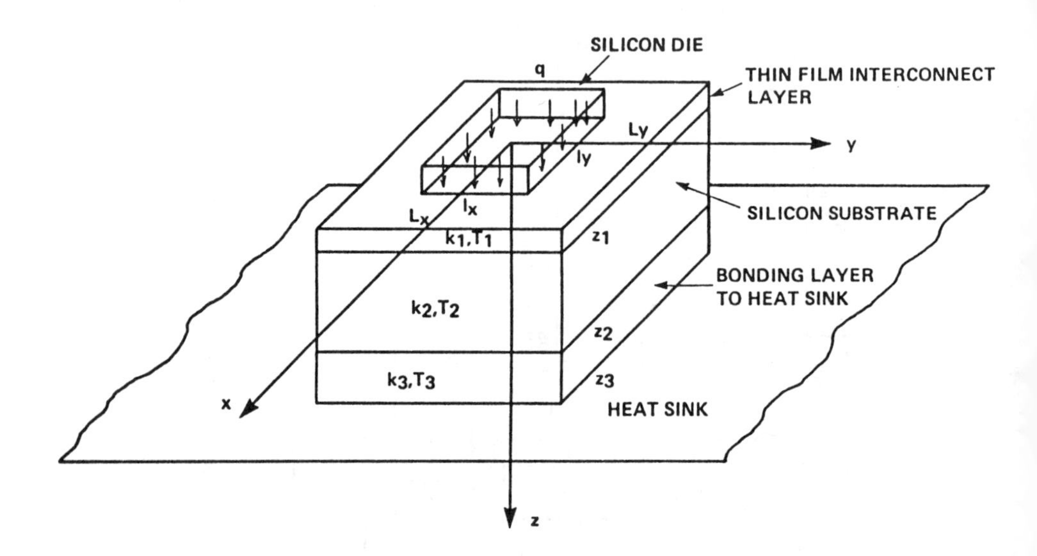

FIGURE 6 THREE-DIMEMSIONAL RECTANGULAR CO-ORDINATE SYSTEM FOR A MULTICHIP MODULE

CHAPTER VI

VERY LARGE SCALE INTEGRATION (VLSI) ISSUES

MATERIALS AND TECHNOLOGIES FOR MULTILEVEL METALLIZATIONS IN VLSI

A. N. Saxena
Gould AMI Semiconductors
3800 Homestead Road
Santa Clara, CA 95051

Abstract

In very large scale integrated circuits (VLSIC's), the materials and technologies, in particular deposition and patterning, play an ever increasing important role in multilevel metallizations. This is evident from the fact that the largest area on a VLSIC chip is occupied by multilevel metallizations. This is expected to be true also for chips in future VLSIC's, usually referred to as ultra large scale integrated circuits (ULSIC's). The chip sizes in VLSI and ULSI are not limited by device scaling; instead, they are goverened by multilevel metallizations. The materials required for multilevel metallizations are in two categories: metals (conductors) and dielectrics (insulators). Both of these types of materials need to be chosen and used properly for successful multilevel metallizations. Associated technologies to use these materials are also essentially in two categories: deposition and patterning. A brief review of the most widely used materials and associated technologies, as well as their future directions, will be presented in this paper. It will also be shown that improved vacuum technologies are needed, which can significantly impact both the chemical vapor deposition (CVD) and physical vapor deposition (PVD) processes, and PVD can regain some of the territory lost to CVD.

Refractory Metals and Silicides for Very Large Scale Integration Applications

Krishna C. Saraswat
Center for Integrated Systems, Stanford University, CA 94305

Continuous advancements in technology have resulted in integrated circuits with smaller device dimensions, and larger area and complexity. The overall circuit performance has depended primarily on the device properties. However, the parasitic resistance and capacitance associated with interconnections and contacts, as shown in Fig. 1(a) for an MOS transistor, are now beginning to influence the circuit performance and will be the primary factors in the evolution of submicron VLSI technology. Fig. 1(b) - 1(d) show examples of contact resistance, shallow junction series resistance and RC time delay of interconnections, respectively. Results of theoretical modeling indicate that below 1 μm minimum feature size the impact of these parasitics will seriously hurt the circuit and system performance [1]. The RC time delay, the IR voltage drop, the power consumption and the crosstalk noise due to these parasitics will become appreciable. Thus even with very fast devices the overall performance of a large circuit could be seriously affected by the limitations of interconnections and contacts.

During the last few years a good amount of work has been done on new and innovative materials, device structures and fabrication technology to overcome these problems. Fig 2 shows the past, present and the future status of the technology. It is evident that refractory metals and silicides are playing increasingly important roles. It must also be pointed out that these materials will not replace aluminum but compliment it.

In order for a certain conductor to be used to form multilayer interconnections, several requirements which are imposed by fabrication technology and required circuit performance must be met.

The main requirement is good conductivity, because it can substantially improve the resistance and delay times of the electrical interconnection lines used for VLSI structures. In general, several other requirements are imposed on interconnection materials by fabrication technology. In a multilayer interconnection structure, the layers incorporated early in the process sequence might be subjected to several fabrication steps, that layers incorporated later might not be subjected to. The most rigorous set of requirements are, low resistivity, ease of deposition of thin films of the material, ability to withstand the chemicals and high temperatures required in the fabrication process, good adhesion to other layers, ability to be thermally oxidized, stability of electrical contacts to other layers, ability to contact shallow junctions, good MOS properties, resistance to electromigration and ability to be defined into fine patterns.

The materials which have been used or proposed for forming interconnections can be broadly classified into four categories: heavily doped polysilicon, low temperature metals, high temperature refractory metals and metal silicides. Table [1] compares properties of some of these materials showing their compatibility with present silicon fabrication technology.

For a long time, aluminum has been used to form the metal interconnects, however as device dimensions are scaled down, its reliability is becoming a major issue. Some of the problems are, electromigration, high solubility of silicon leading to poor contact reliability to shallow junctions, hillock formation causing electrical shorts between successive layers fo A1, and corrosion. Many of these problems have been solved by using alloys of aluminum with Si, Cu and recently Ti [2].

With the advent of silicon-gate MOS technology, polycrystalline silicon has been extensively used to form gate electrodes and interconnections. The success of the silicon-gate MOS technology can be largely attributed to the use of polycrystalline silicon as an additional layer of interconnections. In many applications two or even three layers of polycrystalline silicon have been used. From numbers shown in Table [1] and Fig. 1(d) it is evident that the resistivity of polycrystalline silicon is too high and is beginning to limit the performance of large circuits with small dimensions. In all other respects, it is a very nice material to work with because basically it is silicon and therefore highly compatible with silicon technology.

During the last few years, use of refractory metal silicides have been heavily investigated for this application and for silicidation of diffusions and the results have been very exciting. From Table [1] it can be seen that silicides of tungsten (W), molybdenum (Mo) and tantalum (Ta) have reasonably good compatibility with IC fabrication technology. They have fairly high conductivity, they can withstand all the chemicals normally encountered during the fabrication process, thermal oxidation of their silicides can be done in oxygen and steam to produce a passivating layer of SiO_2, their contacts to shallow p-n junctions are reliable, and fine lines can be etched by plasma etching these materials. The formation of thin films of silicides is not as straightforward as that of aluminum and polycrystalline silicon, but it can be done by a variety of deposition techniques. These properties suggest that the silicides of W, Mo and Ta can be used in all of the applications where polycrystalline silicon has been used so far. While $TiSi_2$ is very attractive for this application, its etch rate in hydrofluoric acid (HF) is very fast. If the use of HF can be avoided during the fabrication by employing dry etching techniques, $TiSi_2$ can also be used successfully.

Several problems remain to be solved before silicides can be used on a routine basis in production. Deposition techniques will have to be much more controllable and reproducible than those being used today. CVD technology offers the most promising results [3]. Silicide films as deposited by current deposition techniques have very low conductivity, and additional high temperature annealing is required to increase it. These temperatures might be too high for the VLSI technology of the the future, since these deposited films can have tensile or compressive stress in them the type and magnitude of which change as a function of processing temperature. The magnitude of this stress can be enough in some cases to cause cracks and poor adhesion. This poses serious yield and reliability problems.

During the early seventies, several attempts were made to develop refractory metal (tungsten and molybdenum) gate technology. However, in comparison to silicon gate technology it had no apparent advantage. The minimum feature size at that time was around 10 µm and the chip area was rather small. Therefore the RC time delay was not a big issue, and the industry's interest in refractory metal gate technology gradually cooled off. As circuit complexity and size grew, and the inadequacies of polycrystalline silicon and Al started surfacing, interest in refractory metals was rejuvenated. During the last few years, W and Mo have been successfully used to form gate electrodes and interconnections in several VLSI applications and it appears that this trend will grow.

These materials have very attractive properties for multilayer interconnection applications. Their resistivities are only slightly higher than that of Al, their melting points are much higher than silicon, the thermal expansion coefficients are closer to that of silicon (see Table 1) and they are highly resistant to electromigration. They are inert to many chemicals allowing the use of patterned films of these materials as etching masks for both SiO_2 and Si_3N_4. They are easy to etch using the plasma technology and thus fine lines can be defined with relative ease. At temperatures below the silicide formation temperature (~600°C) the solubility of silicon is negligible and thus the stability of contacts is excellent.

Deposition of the thin films of Mo and W has been investigated by electron beam evaporation, sputtering and recently by chemical vapor deposition. The films deposited by CVD generally have such superior properties as low contamination, lower resistivity, better step coverage, lower stress and better compatibility with batch production techniques. Selective CVD of W appears to be a very promising technology for many VLSI applications [4].

The major disadvantage of W and Mo is that their oxides are volatile at high temperatures (~ > 400°C). Therefore these metals can be used only when all of the steps where exposure to high temperature oxidizing ambient might occur are over, but high temperature steps in inert ambient are still remaining. Recent work involving the use of H_2 rich H_2O to avoid oxidation of W yet oxidize Si appears to be very promising and may provide the necessary ingredients of the submicron MOS technology [5].

References

[1] K.C. Saraswat and F. Mohammadi, "Effect of Interconnection Scaling on Time Delay of VLSI Circuits", *IEEE Trans. Electron Dev.*, Vol. ED-29, April, 1982., pp. 645-650.

[2] D.S. Gardner, T.L. Michalka, K.C. Saraswat, T.W. Barbee Jr., J.P. McVittie and J.D. Meindl, "Layered and Homogeneous Films of Al and Al/Si with Ti and W for Multilayer Interconnects" *IEEE Trans. Electron Dev.*, Vol. ED-32, February, 1985, pp. 174-183.

[3] K.C. Saraswat, D.L. Brors, J.A. Fair, K.A. Monig and R. Beyers, "Properties of Low Pressure CVD Tungsten Silicide for MOS VLSI Interconnections", *IEEE Trans. Electron Dev.*, Vol. ED-30, November, 1983, pp. 1497-1505.

[4] K.C. Saraswat, S. Swirhun and J.P. McVittie, "Selective CVD of Tungsten for VLSI Technology", *VLSI Science and Technology*, Electrochemical Society, 1984, pp. 409-419.

[5] S. Iwata, N. Yamamoto, N. Kobayashi, T. Terada and T. Mizutani, " A New Tungsten Gate Process for VLSI Applications," *IEEE Trans. Electron Dev.*, Vol ED-31, pp.1174-1179, September 1984.

PROPERTIES OF INTERCONNECTION MATERIALS

Silicide	Thin Film Resistivity ($\mu\Omega$ cm)	Eutectic Temp. with Si (°C)	Process Compatibility	Reaction with Al (°C)	Thermal Expansion Coefficient (ppm/°C)
WSi_2	30-70	1440	Good	500	12.5
$MoSi_2$	40-100	1410	Good	500	8.25
$TaSi_2$	35-55	1385	Etches slowly in HF	500	8.8-10.7
$TiSi_2$	13-16	1330	Etches in HF	450	12.4
$CoSi_2$	18-25	1195	Etches in HF	400	10.14
PtSi	28-35	830	High temp.?	250	---
$Si(N^+)$	500	---	---	577	3
W	8-10		POOR OXIDATION RESISTANCE	500	4.5
Mo	8-10		POOR OXIDATION RESISTANCE	500	5.0

Fig. 1

(a)

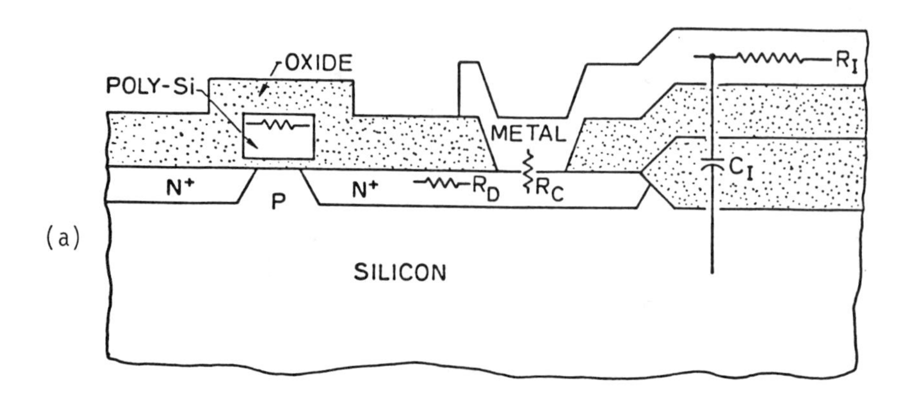

Fig. 1(a) A cross section of an MOS transistor with parasitic elements associated with poly-Si, shallow diffusions, contacts and interconnections.

(b)

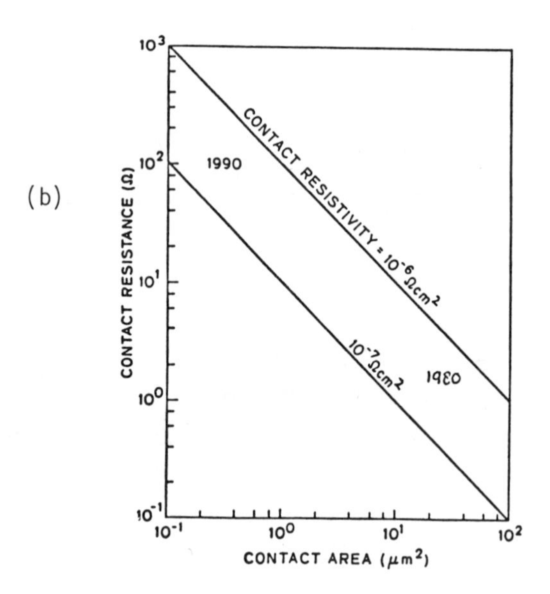

Fig. 1(b) Contact resistance vs contact area

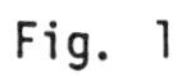

(c)

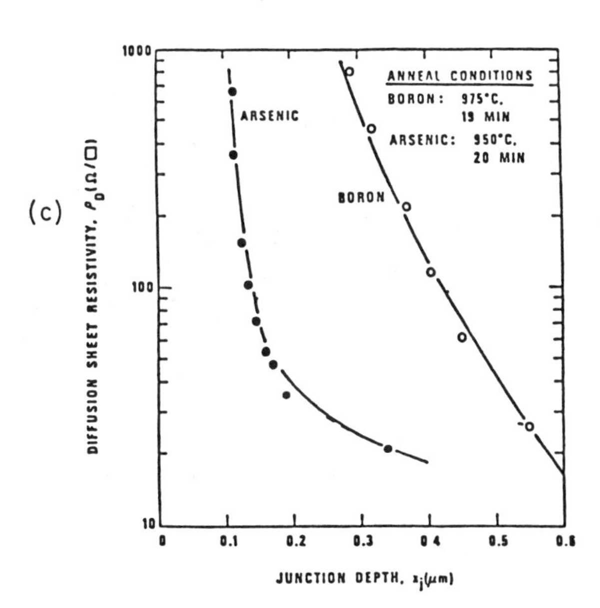

Fig. 1(c) Diffusion sheet resistance vs junction depth

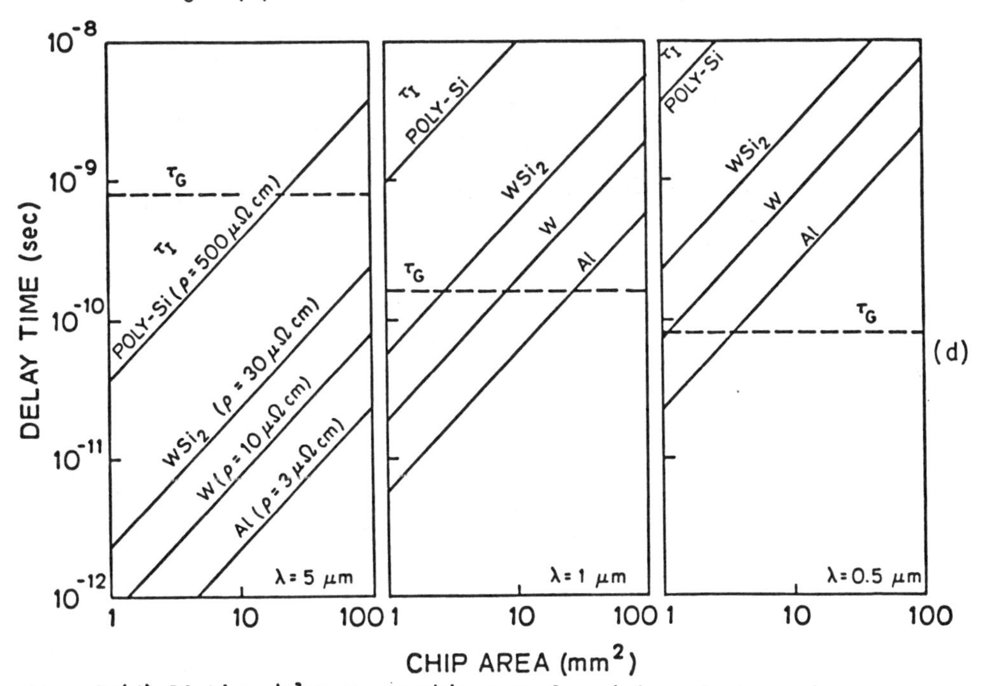

(d)

Fig. 1 (d) RC time delay τ_I vs chip area for minimum feature size, λ of 5,1 and 0.5 μm. Also shown is the intrinsic device delay τ_G

Fig 2

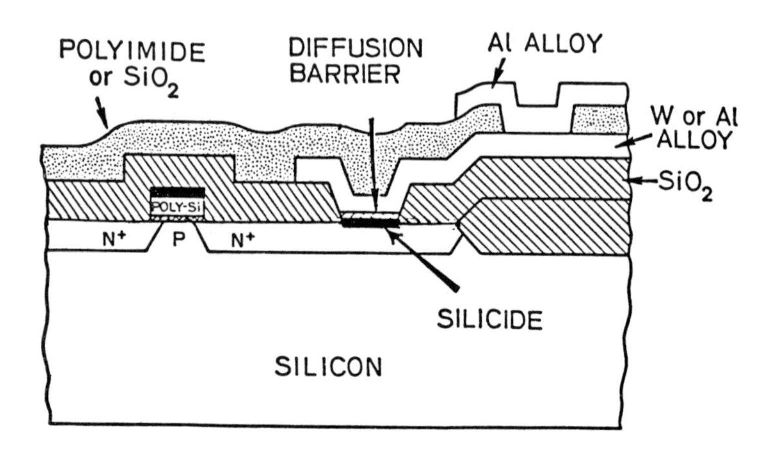

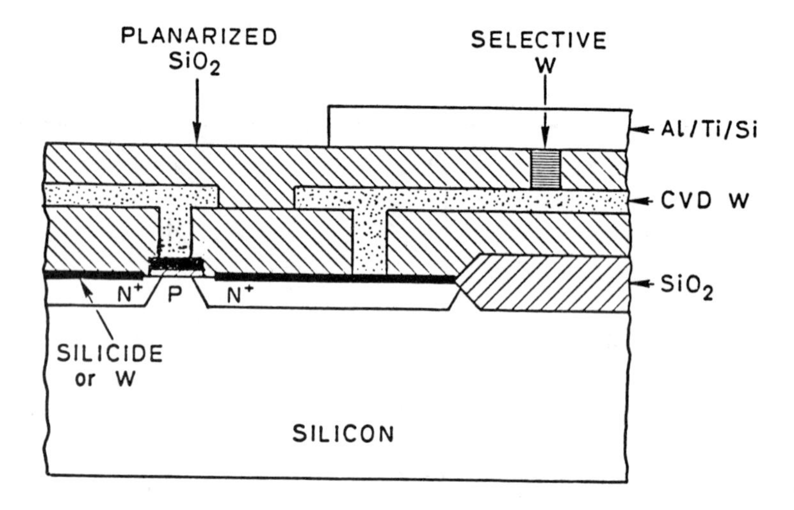

Fig. 2. Use of refractory metals and silicides to solve the problems created by parasitics shown in Fig. 1(a) (a) present technology (b) Future technology

DIELECTRICS FOR VERY LARGE SCALE INTEGRATION (VLSI)

R. Singh
School of Electrical Engineering and Computer Science
University of Oklahoma
Norman, OK 73019

ABSTRACT

Dielectric films are an integral part of the integrated circuits. Key applications can be classified as, (i) dielectrics for growing semiconductor films e.g., silicon on insulator, (ii) layer of dielectric films which do not appear in the finished device e.g., dielectrics for diffusion and ion implantation masks, and (iii) permanent dielectric films which play a key role in the electrical operation of the circuit e.g., gate dielectric in a MOSFET. In this paper a review is presented of the dielectrics with main emphasis on the permanent dielectric films. Techniques for forming dielectric films reviewed are evaporation, sputtering, plasma deposition, thermal process and various types of chemical vapor depositions. A comparative study of alternative deposition techniques for various dielectric films is presented. Materials and techniques that have potential for achieving performance and reliability goals of submicron integrated circuits are evaluated in depth.

I. INTRODUCTION

Since the invention of integrated circuits in 1959, tremendous progress has been made in the performance, chip density and power dissipation of integrated semiconductor circuits. This phenomenal growth of system complexity has been achieved by drastically reducing the lateral and vertical dimensions of the individual building blocks. In the last 26 years, the minimum device dimension has reduced from "mills" to about a micron. The thickness of the dielectrics used in isolation and as gate dielectric material is also being reduced[1].

Dielectrics both as substrate and in thin film form are an integral part of the integrated circuits. Key applications of the dielectrics can be classified as, (i) dielectrics used in the growth of semiconductor films e.g. silicon on insulator, (ii) layer of the dielectric films which do not appear in the finished device, e.g., dielectrics used as diffusion and ion implantation masks, and (iii) permanent dielectric films which play a key role in the electrical operation of the circuit e.g. gate dielectric in a metal-oxide-semiconductor field effect transistor (MOSFET). In this paper a review is presented of the dielectrics with main emphasis on the permanent

0094-243X/86/1380289-18$3.00

dielectric films used in the integrated circuits. The main thrust of this paper is for Si MOSFET, although the majority of the discussion applies to Si bipolar transistors also and may be applicable to non-silicon integrated circuits. Recent work on the dielectrics for compound semiconductor based integrated circuits is described in Ref. 2. It is beyond the scope of this article to cover all the details of preparation and evaluation of various dielectrics. However, we shall cover the highlights and give sufficient reference for the reader interested in a more detailed description. Techniques for forming dielectric films reviewed are evaporation, sputtering, plasma deposition, thermal processes, ion beam cluster and various types of chemical vapor depositions. Materials and techniques that have potential for achieving performance and reliability goals of submicron integrated circuits are evaluated in depth.

II. DIELECTRICS AS SUBSTRATE

Dielectrics as substrate have been used for growing Si and other semiconductor films. One such key example is silicon on Insulator (SOI) which refers to a generic class of materials where a crystalline silicon film is supported on an insulating substrate. In addition to 3 dimensional integration, and radiation hardened integrated circuits, there is a growing interest in SOI for other fields such as large area display and high voltage applications[3]. In the past sapphire (Al_2O_3) or spinal ($MgAl_2O_4$) were used as the insulating material. Currently, explored key SOI techniques include beam recrystallization and implanted buried oxide[3]. In case of beam recrystallized SOI, the dielectric is mostly an oxide layer (~ 1 μm) thermally grown on top of a single crystal silicon wafer. The implanted buried oxide or separation by implantation of oxygen procedure requires the implantation of large doses of oxygen (~ 1-2 X 10^{18} atoms per cm^2) in order to form a buried silicon dioxide (SiO_2) layer. In order to preserve the crystalline nature of the surface layer, the implants must be performed at elevated temperatures. There are several recent reviews in the literature on the subject of SOI[3-6].

III. TEMPORARY DIELECTRIC FILMS

There are a number of dielectric films that are used temporarily in the fabrication of integrated circuits and do not appear in the finished device. Such films are employed for their chemical or physical properties and their electrical behavior is not important in the finished device. Following are the major applications of temporary dielectric layers, (i) masking structure against oxidation, (ii) gettering layers for impurities, (iii) solid state diffusion sources, (iv) diffusion and ion implant capping layers, (v) photolighographic masks for etching, (vi) taper etching control layers, (vii) sacrificial planarization media, (viii) inorganic etch masks, (ix) x-ray transparent membranes for x-ray masks and (x) components for multilayer resist structures. For most applications of temporary layers Ref. 7 is an excellent source of information.

IV. PERMANENT DIELECTRIC FILMS

The main focus of this paper is on the "permanent" dielectric films which are present in the finished device or circuit and whose dielectric properties critically determines the electrical performance of the device or circuit. Following are the major applications of the permanent dielectric films, (v) final dielectric layer, (iv) low temperature intermetal dielectrics for multilevel metallization, (iii) high temperature intermediate dielectrics (between gate and metallization layer), (ii) field oxide for isolation and (i) gate dielectrics. In this section different techniques of dielectric formation are reviewed.

Formation of Dielectric Films

Basically, the dielectrics can be divided into two classes: the native dielectrics and the deposited dielectrics. The term native dielectric is given to films which are formed by chemical reaction (e.g., oxidation, nitridation, etc.) of the substrate semiconductor material. Dielectrics formed without any chemical reaction with the semiconductor substrate are grouped in the deposited category[8]. It is impossible to discuss all techniques in detail in a short paper. For completeness, we will briefly review most of the formation techniques and cite proper references. For details, see Ref. (9-12).

A. NATIVE DIELECTRICS

(a) Thermal Oxides

Dry oxidation is usually carried out from 900°C to 1150°C at atmospheric pressure. Such oxides have excellent properties of $Si-SiO_2$ interface. Modified form of dry thermal oxidation include use of dry O_2-HCl mixture in place of O_2. Recently Morita, et al[13], have added NF_3 gas to a dry oxygen atmosphere resulting into enhanced oxidation rate. In case of wet oxidation water vapor is present in the oxidizing ambient and is done at about 900-1000°C. Wet oxidation occurs faster than dry oxidation and usually can be performed at a lower temperature.

(b) High and Low Pressure Oxides

For high pressure oxidation, a pressure of about 5 to 20 atmosphere and temperature between 800 and 1000°C is usually required[14-15]. This method can be used for faster oxidation at a lower processing temperature with lesser impurity diffusion and oxidation induced stacking faults.

Low pressure oxidation is usually done at a pressure of about 0.2 to 2.0 Torr[16]. For this method, the oxide growth rate is slower which makes the method useful if the oxide thickness is very critical.

(c) Optical Induced Low Temperature Thermal Oxides

In order to reduce the processing temperature, several beam technology are currently being pursued to form oxide layer on Si[10]. The most promising beam technology is the rapid thermal processing for low temperature thermal oxidation[17]. In this method an incoherent radiation source of light such as an array of tungston-halogen lamps, arc sources or electrically heated graphite sources are used to generate high intensity light which produces rapid heating and cooling rates with steady state temperature in the range of 400-1400°C.

(d) Thermal Silicon Nitrides and Silicon Oxynitrides

Thin silicon nitride dielectric can be grown by exposing a silicon surface to NH_3 gas at a temperature of about 950-1300°C. This method gives a thickness within 100 A° because the growth is limited by the diffusion of nitrogen radicals. Recently rapid thermal processing has been used for the rapid thermal nitridation of thermally grown oxides.

A silicon oxynitride film is formed by nitriding a thermal oxide film in NH_3 at a temperature of about 900-1200°C. Nitrogen gas can also be used to nitride a sample but is usually not used because NH_3 produces films more controllably[10].

B. DEPOSITED DIELECTRICS

Most of the deposited dielectric fabrication techniques are based on one of the following methods (a) physical methods, and (b) chemical methods (c) physical-chemical methods. Key physical methods are evaporation, sputtering, ion implantation and ionized cluster beam. In case of chemical methods, spin on, spray on, atmospheric pressure and low pressure CVD are the key techniques. Reactive evaporation, reactive sputtering, plasma oxidation, plasma nitridation are the key techniques based on physical - chemical methods.

Various types of chemical vapor deposition (CVD) techniques are very important for the deposited dielectrics. For this reason further discussion of the deposited dielectrics is divided into two categories, (a) non-CVD method, and (b) CVD methods.

(a) Non CVD Methods

Vacuum evaporated and reactive evaporated films usually are of poor quality for permanent dielectrics applications and will not be used to any large extent for VLSI. Therefore, we will not discuss these films any more in this paper. The most important application of ion implantation for the fabrication of dielectrics is implanted burried oxide as described earlier in Section II. The most important issue in the deposition of dielectric films by sputtering techniques is the control of process generated particles[18]. Recently Mogami, et al.[19] have successfully used two step RF bias sputtering for the deposition of SiO_2 for surface planarization.

Plasma oxidation, utilizing a highly activated oxygen plasma is a promising low temperature technique for the deposition of dielectric films. There are basically two types of plasma oxidation. In the case of plasma anodization, the specimen to be oxidized is positively biased. On the other hand no external bias is applied in the case of plasma oxidation. Both types of plasma oxidations are subject of current investigation[20-21].

Ionized cluster beam (ICB) technique is a promising low temperature technique for the deposition of dielectrics[22]. In this technique, the dielectric material to be deposited is heated and vaporized inside a crucible. Inside the crucible high pressure is built up, and the material is then ejected through a small nozzle into the high vacuum. Prior to deposition, the ejected species are bombarded by an electron beam. Limited information is available in the literature to judge the potential of this technique for VLSI.

(b) CVD Methods

In case of chemical vapor deposition method, key process variable are temperature, pressure, input concentration and flow rates. The important CVD processes include atmospheric pressure (APCVD), low pressure CVD (LPCVD), plasma enhanced CVD (PECVD), photochemical CVD (PHCVD). The chemical aspects, thermodynamics, kinetics, transport phenomena, reactor design and the film growth aspects of CVD technique are described in detail in Ref. 10 and 23. In a recent publication Kern[9] discussed current design of key CVD reactors.

Atmospheric pressure CVD technique is the simplest technique and is used for the deposition of SiO_2, phosphosilicate glass (PSG), borosilicate glass (BSG) etc.

Contamination is the main problem in APCVD and requires continuous cleaning. In order to eliminate contamination as much as possible and to also have more control over the stoichiometry of the dielectric films, low pressure CVD (pressure about a few hundred milli torr) is mostly used. Depending on the processing requirement most of the LPCVD processes are done between 300 and 1000°C. In a LPCVD system low temperature oxides are deposited by the pyrolysis of silane in the presence of oxygen at a temperature of about 700-800°C. LPCVD is also used for the deposition of silicon nitride by a silane-ammonia or a silicon tetrachloride-ammonia reaction at a temperature of about 700-800°C[24].

Instead of thermal energy, glow discharge or plasma can be used to drive LPCVD reactions. This technique is called plasma enhanced CVD (PECVD). Since the PECVD system is not thermally activated, one has much more control of the reaction. As compared to LPCVD system films deposited by PECVD system tend to be less stoichiometric. The deposition is carried out at much lower substrate temperature and with high deposition rate as compared to standard LPCVD process. Silicon nitride and silicon dioxide deposited at a substrate temperature of about 300°C and 200°C respectively are the most important dielectrics deposited by PECVD.

Intense source of ultra violet light is used in case of photochemical CVD (PHCVD) process. The reaction can occur at low pressure either by mercury sensitization[25] or by direct photolysis[26]. As compared to PECVD method, bombardment effects to the substrate by energetic particles do not happen in PHCVD process. Chemical reaction in the direct photolysis is possible only if reactant gas molecules absorb resonant energy from electromagnetic radiation of specific wavelength. Direct photolysis of SiO_2 using a low pressure mercury lamp (1849A°, 2537A°) is reported by Takahashi et al.[26]. Okuyama et al.[27] published a paper on the direct pyrolysis of SiO_2 utilizing a lamp having much vacuum ultra violet light. In case of mercury sensitization, reactant molecules are excited by collisions with UV photosensitized mercury atoms that have resonantly absorbed radiation energy at 2537A°. The mercury atoms in the reactant gas mixture act as an efficient kinetic energy transfer catalyst. The role of the mercury residual contamination in the deposited films is not clear at the present time. Photochemical CVD process is a promising technology, although limited information exists in the literature to access the role of this technique for VLSI applications.

V. PROPERTIES OF DIELECTRIC FILMS

Oxides, nitrides oxynitrides, silicates and polymides are the important dielectric materials that are currently being investigated for current and future VLSI technology. Major material used at present are listed in Table I.

Table I. Major Permanent Dielectric Materials Used in VLSI

Oxides	Nitrides	Oxynitrides	Silicates	Organic Polymers
Silicon dioxide, Plasma silicon dioxide, Aluminum oxide, Titanium dioxide, Tantalum pentaoxide	Silicon nitride, Plasma silicon nitride	Silicon oxynitride, Plasma silicon oxynitride	Phospho-silicate glass, Borophospho-silicate glass, Arseno-silicate glass	Polymides

It is not possible to cover all the dielectric materials in this paper due to limited space and only silicon oxide, silicon nitride and oxynitrides, and silicates are covered in detail. As a high dielectric constant material, there is great interest in $Ta_2 O_5$, TiO_2 etc. In addition CaF_2, BaF_2, SrF_2 etc. are also being investigated as gate dielectric material and for growing semiconductor films.

(a) Silicon Dioxide

1. Thermal silicon dioxide

Thermally grown silicon dioxide has been the conventional dielectric material used in silicon integrated circuits for a number of applications. The physical and chemical properties of silicon dioxide formed by thermal oxidation of silicon are not much affected by moisture or moderate heat treatment. Due to excellent thermal stability, thermally grown silicon dioxide is used as a standard to compare silicon dioxide formed by other techniques.

Thermal oxides grown at lower temperature (< 900°C) are currently being explored for the passivation of silicon surface and for forming gate and field dielectrics in MOS integrated circuits. Low temperature thermal oxidation issues important to VLSI are understanding the growth kinetics and reliability of thin oxides. Recently we have developed a new model of Si oxidation in pure dry oxygen based on the internal electrif field created by the oxide charges and the stress dependent diffusivity of oxidant due to viscoelastic behavior of the oxide[28]. In addition to the explanation of growth kinetics results within an accuracy of $\pm$ 3% this model explains the oxidation under external electric field. An added feature of this model is the explanation of multistep oxidation process and its correlation with the dielectric breakdown and interface characteristics. In a recent publication Bhattacharya[29], et al. studied thin oxides grown at 900°C and correlated the Si/SiO_2 interface roughness with the undesirable electrical properties. The quality of the films is greatly improved by adding an intermediate annealing step to the oxidation process.

Rapid thermal oxidation (RTO) is a promising technology for the fabrication of thin silicon dioxide films. Breakdown voltage and interface charcteristics of rapid thermal oxides are better or compatible to the thin oxides formed by other techniques. The oxide films show uniformities of better than 2% across the 3 inch wafer in the thickness range of 40-130A°[19].

(2) Plasma oxide

In case of plasma oxidation, silicon is partly consumed by surface reaction with oxygen. The growth kinetics of plasma oxides is complex[20]. The exact role of the electric field and the ionic species is not clear at the present time and no model has been generally accepted. As shown in Table II, the properties of plasma oxides grown in the pressure range above 10m Torr and at 500°C are comparable to the thermal SiO_2 grown at 1000-2200°C[19].

Table II. Comparative study of plasma oxides grown at 500°C and thermal oxides grown at 1100°C

Properties	Plasma Oxide	Thermal Oxide
Etch rate in 1:9 BHF (nm/min)	74-76	75
Refractive Index	1.461-1.465	1.462
Stress (dynes/cm^2)	1.5-1.6x10^9	3.1-3.4x10^9
Fixed charge (No./cm^2)	2-6x10^{10}	2x10^{10}
Interface States (No./cm^2-eV)	2-6x10^{10}	2x10^{10}
Retention time (sec.)	~ 100	>500
Breakdown strength	4.8	10

(MV/cm)		
Boron depletion	Absent	Present
Bird's beak effect	Absent	Present
Oxidation-induced defects	Absent	Present

Haneji et al.[21] recently reported that by adding about 1.5% of chlorine to oxygen plasma, the anodization rate of Si is enhanced and the interface trap density is reduced from $7 \times 10^{11}/cm^2$-eV to 5×10^{11} cm^2-eV at the midgap of silicon. The results of microwave plasma oxidation were recently reported by Kimura et al.[30]. The plasma oxidized SiO_2 at temperatures greater than 500°C are identical to the thermal oxides. Based on IR absorption and etch rate measurements it is fair to say that the physical properties of plasma oxides grown at temperature greater than 500°C are comparable to the high temperature thermal silicon dioxides[30].

(3) Sputtered silicon dioxide

As we mentioned earlier there are several issues related to the reliability and particle contamination of sputtered dielectric films. However, there is some interest in the use of sputtering for depositing a planer layer of silica, as an interlayer dielectric layer[18,31]. In this process silica is sputtered in an RF planar magnetron system and simultaneously RF bias is applied to the wafers. The physical properties of sputtered silicon dioxide are comparable to the thermal oxide[31].

(4) Miscellaneous techniques

In case of high pressure oxidation, the rate of transport of oxidant and the amount of oxidant arriving at the Si-SiO_2 interface is enhanced resulting into faster growth rates when compared to thermal oxides[32]. Oxidation induced stacking faults are reduced as compared to wet oxidation at atmospheric pressure. The charge levels in high pressure oxidation are comparable to one atmospheric oxidation[33]. High pressure oxidation does not introduce measurable contamination defects[33]. The length of bird's beak is drastically reduced in high pressure oxidation[34]. The major problem of this technique is the nonuniformity of temperature in systems operating at high pressures[32].

(5) CVD Silicon dioxide

Dielectrics deposited by CVD techniques meet the electrical, mechanical, and protective requirements for a number of device applications. The key parameters to be considered in the optimum deposition technique are uniform deposition, deposition rate, density, refractive index, etch rate dielectric constant, dielectric strength, step coverage and thermal stress. Low pressure CVD (LPCVD) and plasma enhanced CVD (PECVD) are the two important CVD techniques for the deposition of SiO_2. The PECVD SiO_2 films contains 5-10 atomic % hydrogen. The main problem with PECVD is that surface state charge for the plasma oxide is generally high (e.g. $\geq 10^{12}/cm^2$-ev) possibly due to radiation damage from bombardment of the wafer with neutral and charged particles. However, high quality films are obtained by this technique. In Table III, the deposition condition and some properties of PECVD, low temperature CVD and thermal SiO_2 are shown[12].

Table III. Comparison of PECVD, low temperature oxide and thermal oxide

Reaction	Deposition temperature (^oC)	Oxidant $/SiH_4$	Auger SiO_xN_y	RI	Etch rate (A/sec)
PECVD					
SiH_4+O_2	300	3	-	1.46	45
$SiH_4+N_2O+O_2$	300	55	$SiO_{1.94}N_{0.06}$	1.50	30
$SiH_4+N_2O+O_2$	380	55	$SiO_{1.94}N_{0.06}$	1.50	24
$SiH_4+N_2O+O_2$	380	36	$SiO_{1.91}N_{0.10}$	1.52	19
Thermal Oxide	900	-	SiO_2	1.46	17
LTO					
SiH_4+O_2	450	1.5	SiO_2	1.46	60

In Table IV we have shown a comparison of PECVD, LPCVD and APCVD silicon dioxide based on the results of Ref. 35 and 36.

Table IV. Comparison of PECVD, APCVD and LPCVD Silicon dioxide

	PECVD	APCVD	LPCVD
Gases	SiH_4+N_2O	SiH_4+O_2	$SiH_2Cl_2+N_2O$
Deposition temperature (^oC)	350	430	900
Deposition rate (A^o/min)	600	1000	120
Refractive index	1.5	1.45	1.45
Density (gm/cm^3)	2.0	2.15	-
Dielectric constant	4.6	4.2	4.2
Step coverage	conformal	poor	conformal

Resistivity (ohmxcm)	10^{16}	10^{16}	10^{16}
Stress (dynes/cm^2)	$2x10^9$	$2x10^9$	$3x10^9$
Dielectric strength (V/cm)	$3-6x10^6$	$8x10^6$	$10x10^6$
Etch rate (A°/min)	400	60	30
Thermal stability	Losses H	Densifies	Losses Cl

In a recent publication Smolinsky and Wendling[37] have studied the temperature dependent stress of silicon dioxide films prepared by a variety of CVD methods. No correlation was observed between the etching rate and the thermal stress.

(b) Silicon Nitride and Oxynitride

Silicon nitride layer act as an excellent diffusion barrier against practically most materials of concern in silicon devices. Silicon nitride grwon by rapid thermal processing is very promising for gate dielectric applications. For other applications LPCVD and PECVD are the key deposition technologies. Table V compares the physical and chemical properties of PECVD Si_3N_4 to those of high temperature LPCVD[12]. In the following paragraphs we have discussed the results of recent investigations.

Table V. Properties of PECVD silicon nitride compared to those of high-temperature CVD nitride.

Property	HT-CVD-NP 900°C	PE-CVD-LP 300°C
Composition	Si_3N_4	$Si_xN_yH_2$
Si/N ratio	0.75	0.8-1.0
Density	2.8-3.1 g/cm^3	2.5-2.8 g/cm^3
Refractive index	2.0-2.1	2.0-2.1
Dielectric constant	6-7	6-9
Dielectric strength	$1x10^7$V/cm	$6x10^6$V/cm
Bulk Resistivity	$10^{15}-10^{17}$ ohms/cm	10^{15} ohms/cm
Surface resistivity	$>10^{13}$ ohms/sq.	$1x10^{13}$ ohms/sq.
Stress at 23°C on Si	$1.2-1.8x10^{10}$ dyn/cm^2 (tensile)	$1-8x10^9$ dyn/cm^2 (compressive)
Thermal expansion	$4x10^{-6}$/°C	$>4<7x10^{-6}$/°C
Color, transmitted	None	Yellow

Step coverage	Fair	Conformal
H_2O permeability	Zero	Low-none
Thermal stability	Excellent	Variable>400°C
Solution etch rate		
HFB 20-25°C	10-15A/min	200-300 A/min
49% HF 23°C	80A/min	1500-3000A/min
85% H_3PO_4 155°C	15A/min	100-200A/min
85% H_3PO_4 180°C	120A/min	600-1000A/min
Plasma etch rate		
70% CF_4/30%O_2,		
150W, 100°C	200A°/min	500A°/min
Na^+ penetration	<100A°	<100A°
Na^+ retained		
in top 100A°	>99%	>99%
IR absorption		

Plasma enhanced CVD deposited Si_3N_4 contains significant concentration of hydrogen[38] (>10^{21} cm^{-3}) and is partly responsible for the device performance and reliability. Very recently Maeda and Nakamura[39] first time reported the hydrogen chemical bonding for PECVD Si_3N_4 deposited by SiH_4 and NH_3 as the reactant gases. The absorption bands around 2180 cm^{-1} corresponding to silicon-hydrogen stretching vibration are separated into three Gaussian profiles at 2120, 2180 and 2255 cm^{-1} and are assigned to Si-H, Si-H_2 and/or (Si-H_2)$_n$ chains and Si-H_3 groups respectively. The Si-H_2 group results into the most thermally stable bonding configuration and N-H_2 bands are more thermally stable than N-H bonds. Maeda and Nakamura[39] also concluded that N-H bonds form the hydrogen bonding in the films.

In place of SiH_4-NH_3 mixture, SiH_4-N_2 gas mixture is being investigated in order to get Si_3N_4 PECVD films with less hydrogen concentration and/or stabler hydrogen bonding. Zhou, et al.[40] recently reported the results of PECVD silicon nitride films deposited by SiH_4-N_2 gas mixture. Such films exhibit (i) less hydrogen, (ii) higher thermal endurance, (iii) higher density, and (iv) smaller etching rates as compared to the films deposited from SiH_4-NH_3 gas mixture. These films have lower hydrogen concentration. The resistivity, breakdown strength and interface state density are 10^{15} ohm x cm, 9 x 10^6 V/cm and 5 x 10^{11}/cm^2 eV respectively.

Fluorinated Si_3H_4 films using a gas mixture of SiF_4, N_2 and H_2 gases are also being investigated in order to get a stable film with less hydrogen[41]. Such films contains about 25 at % fluroine and the N-H bond remains stable up to 640°C. Fujta et al.[42] reported resistivity of 7 x 10^{16} ohm x cm, breakdown strength of 10 x 10^6 V/cm and low deep trap density. Recently authors of Ref. 42 used SiF_2 gas instead of SiF_4 gas for the plasma deposition of silicon nitride. Such films have following properties as compared to SiF_4 films, (i) higher deposition rate for the same N/Si ratio, (ii) more nitrogen, less fluorine and less oxygen, and (iii) deposition without H_2 gas.

In addition to PECVD and LPCVD other techniques being explored for the deposition of silicon nitride are photo CVD[43], electron beam assisted CVD[44] and sputtering[45]. Photo CVD silicon nitride films are less dense than Si_3N_4 films deposited by high temperature CVD. As compared to PECVD films, photo CVD films have basically similar structure properties and better electrical properties[43].

(c) Silicates

Phosphosilicate glass (PSG) films are widely used in integrated circuit as reflow glass layers, interlevel dielectrics in multilevel metallization systems and passivation over metallization. Conventionally PSG layers have been deposited by APCVD method by using SiH_4 and PH_3 as the reactant gases. Such films have undesirable step coverage and insufficient mechanical strength. Alternative techniques like PECVD are being explored[46]. A comparison of the characteristics of PECVD and APCVD PSG is described in Ref. 46.

There are several studies in the literature dealing with the chemical state of P in PSG[46-50]. Houskova et al.[50] found that plasma, plasma enhanced and low temperature low pressure PSG and borophosphosilicate glass (BPSG) films contain phosphorous pentaoxide, phosphorous trioxide and phosphine. The fourth compound has not yet been identified undoubtly.

Modified PSG with lower glass transition temperature have been investigated by Nassau, et al.[51] Oxides of boron, germanium or arsenic were bulk fused to phosphosilicate matrix. Addition of boron was found to be most effective in lowering the transition temperature of the glass. Atmospheric CVD[9], LPCVD[52] and PECVD[53] are being investigated for the deposition of BPSG. Ashwell et al. recently[54] reported the properties of arsenosilicate (ASG) as an interlayer dielectric material. Rapid isothermal annealing using laser beam or high intensity lamps have been used to reduce the exposure time of silicon to high temperature during the processing of silicates. There are several reports in the literature that shows that incoherent lamp annealing is a potential low temperature processing technique for the reflow of silicates[54-58].

(d) Polymides

Polymide films have interesting features that make them attractive for integrated circuits. Due to space limitation this section is not discussed in detail, however current references[59-61] are an excellent source of information.

VI. APPLICATIONS OF PERMANENT DIELECTRIC FILMS

A. Isolation and Gate Dielectrics

Both for field and gate dielectrics usually native oxides are generally used. High pressure oxides are favorable for field oxides. From materials and processing point of view thin gate dielectrics have very stringent requirements. Current trends are in the direction of low temperature processing and rapid thermal processinng is a promising technique. In addition to the structural imperfections, hot carrier related effects poses serious concern in achieving desired reliability of thin gate dielectrics based submicron devices and circuits[62]. Microscopic understanding of the insulator - Si interface is highly desirable to make further advancement in reducing hot carrier effects. At the present time, there is no appropriate method to measure the breakdown strength of ultra thin gate dielectrics. Thin gate dielectrics may degrade during subsequent processing steps in a VLSI fabrication sequence and optimization of process parameters is highly desirable. Tight control of surface contamination and high purity gases and chemicals will be required for thin gate dielectrics. In our opinion development of a high dielectric constant gate material with interface characteristics compatible to Si-SiO_2 interface can improve the performance and reliability of the next generation of MOSFET bases IC's. Presently, there is great interest in thin oxides of polysilicon for memory devices[63].

B. Low Temperature Interlevel and High Temperature Intermetal Insulation

Multilevel metallization is a key technology issue in VLSI. In case of high temperature metallization temperature can be greater than about 500°C but should be lower than about 800 ± 900°C. For low temperature metallization the deposition temperatue should be lower than 500° preferably about 300°C. Saxena and Pramanik[78] have written an excellent review article on the VLSI multilevel metallization. Both for low and high temperature metallization physical, chemical and electrical characteristics of dielectrics are specified in Ref. 78. In addition the key materials being investigated recently are compared in three categories of good, fair and poor.

For high temperature metallization PSG deposited by APCVD or LPCVD is used. For some applications PSG is being replaced by BPSG. Rapid thermal annealing is a promising low temperature technique for oxide reflow. Planerized silicon dioxide deposited by magnetron sputtering is also being evaluated for high temperature metallization. Polymides have shown promise and are being investigated for submicron technology.

For low temperature metallization PECVD SiO_2 and $SiO_xH_yH_2$ deposited at about 300°C are the best dielectric material. It is possible that in the future this material deposited by Photo CVD will be used for low temperature metallization. Polymides can be used as an interlevel dielectric material, but due to a number of technical problems it is doubtful if they will widely accepted for submicron devices[6].

C. Final Dielectric Layer

Protective coatings are used to cover the entire chip except the bond pads and grid lines of the top level metallization. The protective dielectric layer should have following key properties[9].

(1) The dielectric film should be deposited by a low temperature deposition technique compatable with aluminum metallization.
(2) The dielectric film should prevent aluminum corrosion of interconnect lines. In addition the film should be physically hard, rugged and have perfect integrity.
(3) The final dielectric layer should protect the chip interior from moisture, and deleterious impurities such as alkalines.
(4) The intrensic stress of the film should be of compressive nature.

Key materials being investigated are PSG, BPSG, $SiO_xN_yH_z$, SiN_xH_z and polymide. The most likely candidate is SiN_xH_y because of its impurity barrier effectiveness and compressive stress. Upon heating these films, the hydrogen contents decreases. Kern[6] has shown that these films can endure heating to 500°C without any undesirable features. In addition silicon nitride deposited by other techniques[56-58] contains less hydrogen and a stronger N-H bonding and may prove better than PECVD silicon nitride deposited by a gas mixture of SiH_4 and N_2.

CONCLUSION

In this paper we have reviewed the dielectrics for VLSI. Dielectrics used in the growth of semiconductor films and temporary dielectric films are briefly discussed and most of the paper deals with permanent dielectric films. Technique for forming dielectric films reviewed in this paper are evaporation, sputtering, plasma deposition thermal processes, ion beam cluster and various type of CVD. For final dielectric layer PECVD deposited SiN_xH_y is an excellent candidate for integrated circuits based submicron technology. More work is needed to establish photo CVD as a potential candidate for VLSI technology. Although, not discussed at great length in this paper thin gate dielectrics will be the focus of main research in the future.

REFERENCES

1. R. Singh, Proc. of the 1984 Intl. Symp. on microelectronics, ISHM, Montgomery, Alabama, 1984, p. 386.
2. Proc. of the Symposium on Dielectric Films on Compound Semiconductors published by Electrochemical Society (in press).
3. H. W. Lam, Tech. Digest International Electron Device Meeting, (IEDM) 1983, published by IEEE, New York, p. 348.
4. K. W. Lam, A. F. Tasch, Jr. and R. F. Pinnizzotto, VLSI Electronics Microstructure Science, Ed. by N. G. Einspruch, Vol. 4, published by Academic Press, 1983, p. 1.
5. N. Nakano, Tech. Digest IEDM, 1984, published by IEEE New York, p. 792.
6. H. W. Lam, Proc. 32nd National AVS meeting, J. Vac. Sci. Technol., 1985 (in press).
7. S. M. Sze, Editor, VLSI Technology, McGraw Hill, New York, 1983, p. 93
8. R. Singh, K.C. Taylor, G.A. Chaney, K. Rajkanan and G.S.R. Krishnamurthy, Proc. of the Second International Symposium on VLSI, Electrochem. Soc. 1984, p. 288. York, 1983 , p. 93.
9. W. Kern, Semiconductor International, 8 (7), 122 (1985).
10. W. Kern and V. S. Ban, in Thin Film Processes, edited by J. L. Vossen and W. Kern, Academic Press, New York, 1978, p. 257.
11. W. A. Pliskin and R. A. Gdula, Hand Book on SemiConductors, Vol. 3, edited by S. P. Keller, North-Holland, New York, 1980, p. 641.
12. T. B. Gorczyca and B. Gorowitz, VLSI Electronics Microstructure Science, Vol. 8, edited by N. G. Einspruch and D. M. Brown, Academic Press, 1984, p. 69.

13. M. Morita, T. Kubo, T. Ishihara and M. Hirose, Appl. Phys. Lett. 45, 1313 (1984).
14. M. Hirayama, H. Miyoshi, N. Tsubouchi and M. Abd, IEEE Trans. Electron Devices ED-29, 503 (1982).
15. L. N. Lie, R. R. Razouk, and B. E. Deal, J. Electrochem. Soc. 129, 2828 (1982).
16. A. C. Adams, T. E. Smith and C. C. Cheng. J. Electrochem. Soc. 127, 1787 (1980).
17. A. Gat, and J. Nulman, Semiconductor International 8 (5) 120 (1985).
18. P. Burggraaf, Semiconductor International, 7, (11), 70 (1985).
19. T. Mogami, M. Morimoto, Hokabayashi and E. Nagasawa, J. Vac. Sci. Technol. B3(3), 857 (1985).
20. K.K. Ng and J.R. Ligenza, J. Electrochem. Soc. 131, 1968 (1984).
21. N. Haneji, F. Arai, K. Asada and T. Sugano, IEEE Trans. Electron Devices ED-32, 100 (1985).
22. J. Wong, T.M. Lee and S. Mahta, J. Vac. Sci. Technol. B3(1), 453 (1985).
23. W. Kern, and R.S. Rosler, J. Vac. Sci. Technol. 14, 1082 (1977).
24. P.H. Singer, Semiconductor International 6, (5) 72 (1984).
25. J.Y. Chen, R.C. Henderson, J.T. Hall and J.W. Peters, J. Electrochem. Soc. 131, 2146 (1984).
26. J. Takahashi and M. Tabe, Jap. J. App. 24. 274 (1985).
27. M. Okuyama, Y. Toyoda and Y. Hamakawa, Jap. J. App.
28. R. Rajsuman and R. Singh, to be published in Applied Physics Letters and unpublished results.
29. A. Bhattacharya, C. Vorst and H. Carim, J. Electrochem. Soc. 132, 1900 (1985).
30. S. Kimura, E. Murakami, K. Miyake, T. Warabisako, H. Sumani, and T. Tokuyama, J. Electrochem. Soc. 132, 151 (1985).
31. P. Chang, Semiconductor International 7, No. 11, 79 (1984).
32. E.A. Irene, Semiconductor International, 5, (4), 99 (1983).
33. L.C. Katz in Silicon Processing, Ed. D.C. Cupta, ASTM, Philadelphia, PA, 1983, p. 238.
34. E. Bussman, Semiconductor International, 5, (4), 162 (1983)..
35. W. Kern, Semiconductor International 4 (3) 89 (1982).
36. Ref. 9, p. 116.
37. G.S. Molinsky and T.P.H.F. Wendling, J. Electrochem. Soc. 132, 950 (1985).
38. D.W. Hess, J. Vac. Sci. Technol. A2(2), 244 (1984).
39. M. Maede and H. Hakamura, J. App. Phys. 58, 484 (1985).
40. N. Zhou, S. Fujita, and A. Sasaki, J. Electron. Mat. 14, 55 (1985).

41. S. Fajita, H. Toyoshima, T. Ohishi and A. Sasaki, Jap. J. Appl. Phys. 23, L 144 (1984).
42. S. Fujita, H. Toyoshima, T. Ohishi and A. Sasaki, Jap. J. App. Phys. 23, L 268 (1984).
43. K. Hamano, Y. Namazawa, and K. Yamazaki, Jap. J. Appl Phys. 23, 1209 (1984).
44. D.C. Bishop, K.A. Emery, J.J. Rocca, L.R. Thompson, H. Zarnani, and G.J. Collins, Appl. Phys. Lett. 44, 598 (1984).
45. T. Serikawa and A. Okamoto, J. Electrochem Soc. 131, 2928 (1984).
46. A. Takamatsu, M. Shibata, H. Sakai and T. Yoshimi, J. Electrochem. Soc. 131, 1865 (1984).
47. A.J. Learn, J. Electrochem Soc. 132, 405 (1985).
48. O.K.T. Wu and A.N. Saxena, J. Electrochem. Soc. 132, 933 (1985).
49. R.A. Levy, S.M. Vincent and T.E. McGahen, J. Electrochem. Soc. 132, 1412 (1985).
50. J. Houskova, K.N. Ho, and M.K. Blazs, Semiconductor International 7 (5), 236 (1985).
51. K. Nassau, R.A. Levy and D.L. Chadwick, J. Electrochem. Soc. 132, 409 (1985).
52. T. Foster, G. Hoeye, and J. Goldman, J. Electrochem. Soc. 132, 505 (1985).
53. J.E. Tong, K. Schertenlib and R.A. Carpio, Solid State Technol. 25 (1), 161 (1984).
54. G.W.B. Ashwell and S.J. Wright, Semiconductor International 7 (1), 132 (1985).
55. T. Hara, H. Suzuki and M. Furakawa, Jap. J. Appl Phys. 23, L452 (1984).
56. J. Kato, and Y. Ono, J. Electrochem. Soc. 132, 1730 (1985).
57. W. Kern and R.W. Smeltzer, Solid State Technol. 26 (6), 171 (1985).
58. R. Iscoff, Semiconductor International 6 (10), 116 (1984).
59. R.K. Sadhir, and H.E. Saunders, Semiconductor International 7 (3), 110 (1985).
60. H.J. Newhaus, D.R. Day and S.D. Senturia, J. Electronic Mat. 14, 379 (1985).
61. T. Yokoyama, N. Kinjo, Y. Wakashima, and Y. Miyadera, IEEE Trans. Electrical Insulation EI-20, 557 (1985).
62. R. Singh, G. Ganesh Kumar and K.C. Taylor, Proc. of the Third Intl. Symp. on VLSI, Electrochemical Soc. 1985, p. 349 .
63. S. Chiao, W. Shish, C. Wang and T. Batra, Semiconductor Intl. 7 (4), 156 (1985).
64. A.N. Saxena and D. Pramanik, Solid State Technology, 25 (12), 93 (1984).

Metallization and Interconnects for Post - Mb DRAM Applications

by

A. K. Sinha
AT&T Bell Laboratories
555 Union Boulevard
Allentown, PA 18103

ABSTRACT

It is shown that the combination of ever higher integration levels and ever finer scaling generates several device technology concerns relating to (a) parasitics, (b) defect density, and (c) generic reliability, that require implementation of improved interconnect materials and processes. Thus, the RC time constant delay, now believed to be the most important limiting factor for VLSI circuit performance, has led to the development of poly-silicide and refractory metal gate level interconnects as well as two-level Al interconnects. The parasitic source/drain sheet resistance of small geometry MOSFET's has been reduced using self-aligned siliciding of entire source and drains. Defect density control is generally the underlying motivation for such developments as high-precision pattern transfer processes, Al process controls to reduce hillocks, voids, cracks, Si nodules, etc., and CVD processes for better step coverage. Finally, generic reliability issues associated with interconnects/contacts include electromigration, corrosion and α-particles. These have led to specific improvements in the metallurgy of the interconnects, which will be reviewed.

0094-243X/86/1380307-14$3.00

INTRODUCTION

The power of VLSI circuits is, to a large measure, derived through an extensive network of fine-line metallization interconnects used to wire up various components of the chip. For example, the 1 Mb DRAM[1] utilizes three levels of conductors to interconnect over one million each of the transistors and capacitors on a chip. As shown in Fig 1, the first conductor level consists of doped poly-Si capacitor field plates, the second level utilizes $TaSi_2$/n+ poly-Si for gates and word lines whereas the third level is made up of Si-doped Al bit lines.

Intra-chip interconnects are expected to continue to dominate the area of post-Mb VLSI/ULSI (Very Large Scale Integration / Ultra Large Scale Integration) circuits. In order for these circuits to be practical, the metallization technology must respond to two distinct industry-wide trends, shown in Fig. 2:

a. Ever higher integration levels (LSI → VLSI → ULSI), i.e. larger chip dimensions, longer and more numerous interconnections for exponentially increasing transistor counts that provide more functions per chips and hence lower system costs.

b. Ever finer scaling of feature sizes (5μ → 3μ → 1μ) and other consequences of scaling, e.g. shallower junction depths, thinner oxides, etc. to enable superior device performance.

The combination of larger chip dimensions and finer scaling generates several technology concerns that must be addressed by a variety of measures, including an optimum choice of metallization processes. These concerns are related to parasitics, defect density and generic reliability. Control of these three factors has turned out to be the main driving force for many technological improvements in VLSI circuits.

PARASITICS

A natural consequence of device scaling is to increase the parasitic capacitance and resistances associated with the circuit. The RC time constant produces performance-limiting delays along fine-line runners, whereas the source/drain series resistance can severely limit the current drive of small geometry MOSFET's. These parasitics threaten to offset performance advantages of small geometry transistors.

Indeed, the RC time constant delay is replacing electromigration as the factor most likely to limit continued progress towards fine-line VLSI. The concern is especially acute in microprocessor (and logic) circuits (see Fig. 3) which contain critical interconnect paths between superblocks of ALU, PLA's, ROM's, etc.

The RC time constant or rise time for a distributed transmission line of length y is a strong function of the interconnect length: $\tau \approx 0.5\ RCy^2$. At near micron and submicron design rules, the capacitive coupling between adjacent lines and fringing capacitance to the substrate become dominant components of the total capacitance C. Figure 4 shows that the rise time of even highly conductive, long Al lines can exceed the switching delay of fine-line MOSFET's.[2] The investigation of poly-silicides and refractory metal gates is motivated mainly by a desire to control this parasitic for gate-level interconnects.

The sensitivity of CPU clock rate to the sheet resistance of poly-silicide interconnects is dramatically illustrated by the results shown in Fig. 5 for a state-of-the-art 32 bit microprocessor. The expected speed improvements due to decreasing channel lengths and hence, faster-switching transistors are boosted even further if the sheet resistance of poly-silicide interconnects is reduced from 2.5 to 1.0 ohm/sq. by going from $TaSi_2$/poly-Si to $TiSi_2$/poly-Si. An analysis of critical path delay for such 32 bit microprocessors showed that successive generations of scaled transistor geometries reach a point of diminishing performance returns at micron levels, unless the gate-level interconnect material is made simultaneously more conductive (Fig. 6).

Poly-silicides (1 - 5 ohm/sq.) were a preferred means of reducing the RC time constants because they provided an excellent processing retrofit to the existing poly-Si gate (30 - 60 ohm/sq.) structures.[3] The added process complexity associated with implementing more conductive interconnects is likely to increase with refractory metal gates such as W and Mo that do provide a lower sheet resistance of 0.1 ohm/sq. but do not provide the chemical inertness including oxidation resistance of poly-Si and poly-silicides.

The best RC time constant performance can be obtained using multilevel Al interconnects[4] for logic circuits (Fig. 7). Two-level Al enables use of shorter interconnect lengths, somewhat relaxed design rules and yet greater device density for VLSI chips. However, the use of Al at both levels severely limits the choice of interlevel insulator materials and processing temperatures (<450°C). As the geometries get tighter, and structures more vertical, with multiple levels of poly-Si and metals, it would become necessary within the framework of above limitations to solve the issues of topography, step coverage, etch stops and defects.

Another effect of scaling transistor geometries below the one micron level is to increase the parasitic series resistance associated with shallow diffused source and drains, and small geometry metal-to-Si contacts. At values above several hundred ohms, the series resistance can reduce the amount of saturation drain current drive of fine-line transistors, and thereby increase the timing intervals required to charge critical nodes in the circuits.

The most promising approach to solve the series resistance problem involves siliciding the entire source and drain areas in a self-aligned manner (SALICIDE).[5] The sheet resistance is then reduced to 5 ohm/sq. range from several hundred ohm/sq. for shallow junction p+ source/drains. The small geometry contact resistance can be also drastically reduced since the contact interface now consists of metallic Al and metallic silicide layers.

DEFECTS

The defect density, D_o, can cause yield drop-off of large area VLSI circuits at the rate of $\sim 1/D_o^2$. For advanced VLSI process technologies, metallization defects can dominate those due to weak dielectrics and juctions. Defect density control is often the underlying motivation for such diverse developments as high-precision pattern transfer processes, controlled Al processing to reduce hillocks and excessive Si precipitates, and CVD processes for improved step coverage.

In Fig. 2, the dotted end points for foreseeable chip dimensions and feature sizes correspond to the maximum exposure field diameter and lens resolution respectively that are expected with conventional step and repeat lithography. Intra-field distortions limit the width of exposure fields. Registration accuracy and process margins for depth of focus/exposure become critical limitations at high resolution levels.

For patterning fine-line interconnects, optical lithography has been pushed to new levels of refinement that have resulted in a routine capability for 0.9 μm lines and spaces with good margins and low defect density. As illustrated in Fig. 8, high-resolution Mb word line interconnect patterns can be printed using a multilayer resist process. Thanks to a portable conformal mask structure, a photo bleaching material,[6] and a thin imaging layer over thick planarizing resist, the process provides vertical resist profiles with improved exposure latitude and an adequate focus budget.

Figures 8(b) and 8(c) show the results of final pattern transfer into $TaSi_2$/poly-Si and Al layers respectively of a Mb DRAM. This was achieved using automated multistep reactive sputter etch processes optimized to provide precise linewidth control, extremely low defect densities and a high selectivity with respect to underlying thin oxides.

The presence of significant topography in the device structure leads to concerns with uniformity of metal linewidth and metal thickness. The metal linewidth can suffer from "necking" or "notching" effects at steps as a result of undesirable effects during photolithography, such as overexposure, substrate refections and standing waves.

Thickness uniformity over oxide steps, i.e. the step coverage can be marginal with most physical deposition processes such as evaporation and sputtering. Consequent metal thinning causes local enhancements of both electrical and mechanical stresses. Nearly conformal step coverage is achieved with chemical vapor deposition processes due to surface activated nature of the deposition reaction. Practical CVD methods are now available for W.[7,8] Processes exist for both blanket deposition and selective deposition in windows (plugs) (Fig. 9). The W-plug so created planarizes the topography, and provides a diffusion barrier between Al and Si.

Recently, extrusions of Al hillocks, intrusions of Al voids and stress-induced cracks at steps have been the subject of intensive investigations. Due to a large thermal mismatch of Al on Si, Al films develop compressive stresses on heating and tensile stresses on cooling.[9] Relief of compressive stresses through diffusion creep is responsible for hillocks in uncapped Al, whereas relaxation of stresses through a dislocation dominated creep during slow cooling of fine-grained Al,Si/SiN capped films was shown to generate voids at the edges of Al-runners.[10] Mechanical properties of Al films such as the yield strength and plasticity are degraded by excessive Si additions, or N-contamination[11] making the films susceptible to work-hardening and cracking.

GENERIC RELIABILITY

There are generic reliability issues associated with various interconnect materials and processes, issues which are more dependent on process technology than on a specific product type. These include electromigration, corrosion and α-particles.

The electromigration resistance of fine-line Al depends in a complex manner on the film composition as well as on several microstructural inhomogeneity factors such as the grain size, its variance and the degree of preferred (111) orientation in the film.[12] Additions of Cu to Al,Si alloys have been shown to improve both the electromigration resistance[13] and mechanical properties[14] of Al,Si films; however, this can occur at the expense of corrosion resistance. The Cu segregates at Al grain boundaries passivating them against self-diffusion, but it probably also creates electrochemical cells that enhance corrosion. Another interesting effect observed with high quality Al,Si,Cu films is their tendency to develop a bamboo structure in fine-line runners. The bamboo structure, shown in Fig. 10 for a finished Mb DRAM, is known to improve electromigration resistance by drastically reducing the number of grain boundary triple points within the Al runners.

With continued scaling, there is an increasingly greater concern regarding the contact metallurgy. Conventional Al-to-Si contacts require a contact bake at 400°-450°C during which Si dissolves into Al creating pyramidal dissolution pits. Because of a mass action effect, Si pit dimensions increase with decreasing window size. Since small geometry contacts are also associated with shallow junctions, the probability of device failure by enhanced junction leakage increases rather rapidly. Moreover, the current flow through contact windows to shallow junctions is highly non-uniform. Joule heating at the edges of contact windows in conjunction with contact electromigration can also accelerate the dissolution of Si into Al, thereby causing enhanced junction leakage and eventually burned-out contacts.[15,16] Improved contact materials being used to alleviate some or all of these problems include Al-Si alloys, Al/poly-Si, Al/TiN/Silicides and Al/W contacts.

Finally, soft errors in DRAM's have been correlated with trace contaminants of radioactive impurities present in both the packaging material and refractory metals used for interconnects. The problem is particularly relevant with sputtering targets of W, Mo and Ta which must be purified to $\lesssim$ 10 ppb levels of U and Th.

CONCLUSION

In conclusion, we have classified the main driving forces for continued progress in VLSI metallization and interconnects.

(1) The need to control circuit parasitics, particularly the RC time constant has led to poly-silicide and refractory metal interconnects at the gate level. The best RC time constant performance is obtained with multilevel Al interconnects, but at the expense of certain process limitations.

(2) The need to control defect density in VLSI circuits has inspired impressive, new advances in photolithography and dry pattern transfer processes for fine-line interconnects that are best illustrated with reference to the Mb DRAM. Defects associated with poor step coverage have led to developments in CVD processes, especially for refractory metals. Extrusions, intrusions and stress-induced cracks in Al films have led to improved Al alloy materials and processes.

(3) Generic reliability issues such as electromigration in runners as well as in contacts have motivated advances in fine-line Al metallurgy and synthesis of contact structures that incorporate silicides and diffusion barriers such as TiN or W. Soft errors in DRAM's have led to fabrication of ultra-pure sputtering sources for W, Mo and Ta, containing $\ll$ 10 ppb of U and Th.

REFERENCES

1) H. C. Kirsch, D. G. Clemons, S. Davar, J. E. Harman, C. H. Holder, W. F. Hunsicker, F. J. Procyk, J. H. Stefany and D. S. Yaney, ISSCC Digest, p. 256 (1985).

2) A. K. Sinha, J. A. Cooper, Jr. and H. J. Levinstein, IEEE Electron Device Letters, EDL3, 90 (1982).

3) S. P. Murarka, D. B. Fraser, A. K. Sinha and H. J. Levinstein, IEEE Trans. Electron Devices, ED 27, 1409 (1980).

4) K. Okumura and T. Moriya in VLSI Science and Technology/1985, edited by W. M. Bullin and S. Broydo, Electrochemical Society, Pennington, NJ, p. 163.

5) C. Y. Ting, IEDM Technical Digest, IEEE, p.110 (1984).

6) B. F. Griffing and P. R. West, IEEE Electron Device Letters, EDL4, 14 (1983).

7) M. L. Green and R. A. Levy, J. Electrochemical Society, 132, 1243 (1985).

8) E. J. Broadbent and C. J. Ramiller, J. Electrochemical Society, 131, 1427 (1984).

9) A. K. Sinha and T. T. Sheng, Thin Solid Films, 48, 117 (1978).

10) J. T. Yue, W. P. Funsten and R. V. Taylor, Proc IRPS, IEEE, p. 126 (1985).

11) J. Klema, R. Pyle, and E. Domangue, Proc IRPS, IEEE, p.1 (1984).

12) S. Vaidya, D. B. Fraser and A. K. Sinha, Proc IRPS, IEEE, p. 165 (1980).

13) P. S. Ho in VLSI Science and Technology/1985, edited by W. M. Bullis and S. Broydo, The Electrochemical Society, Pennington, NJ, p. 146 (1985).

14) T. Turner and K. Wendel, Proc IRPS, IEEE, p. 142 (1985).

15) P. A. Gargini, C. Tseng and M. H. Woods, Proc IRPS, IEEE, p. 66 (1982).

16) S. Vaidya and A. K. Sinha, Proc IRPS, IEEE, p. 50 (1982).

17) M. H. Woods and B. L. Euzent, IEDM Tech. Digest, IEEE, p. 50 (1984).

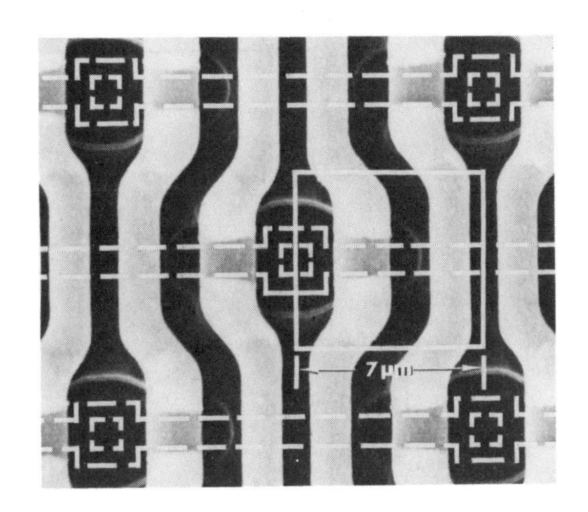

(b)

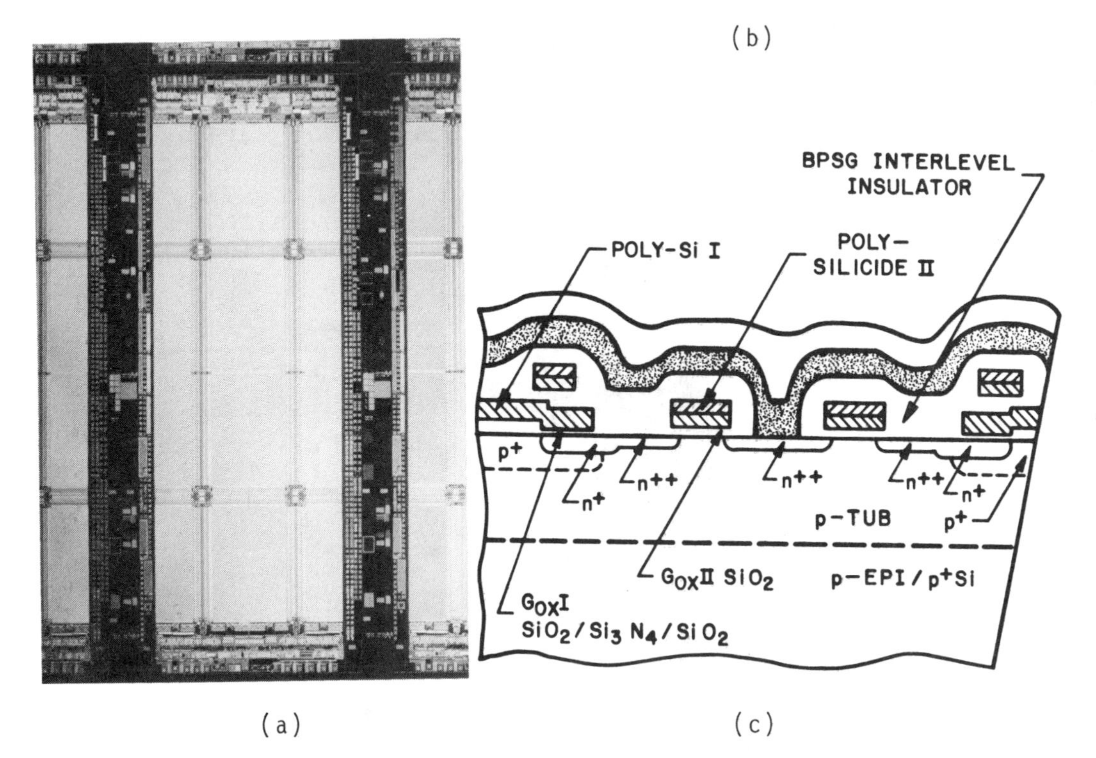

(a) (c)

Fig. 1 (a) Photomicrograph of a 1Mb DRAM, (b) detail and (c) schematic cross-section, through a memory cell.

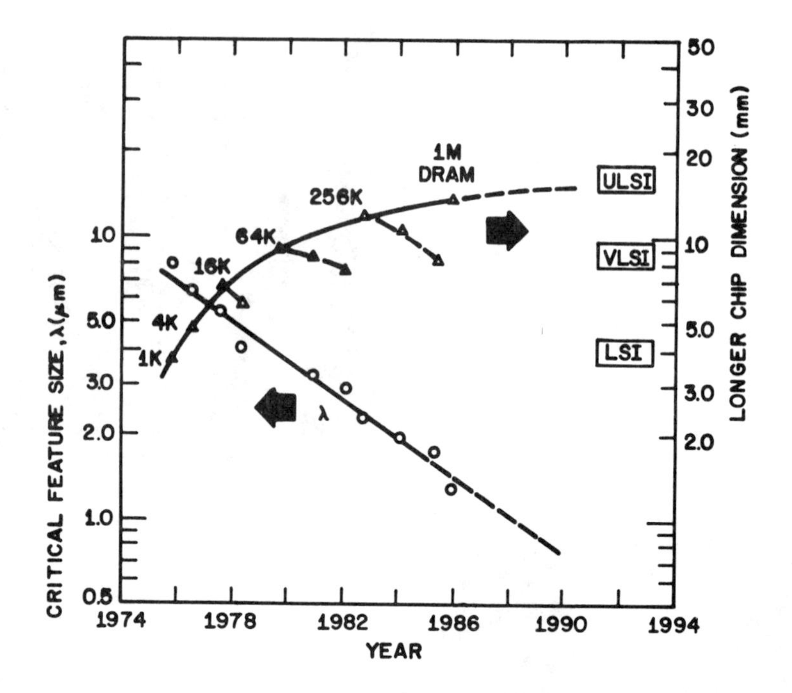

Fig. 2 Decrease in the critical feature size and increase in the longer chip dimension, with time, of DRAM chips.

Fig. 3 Photomicrograph of a 32-bit microprocessor.

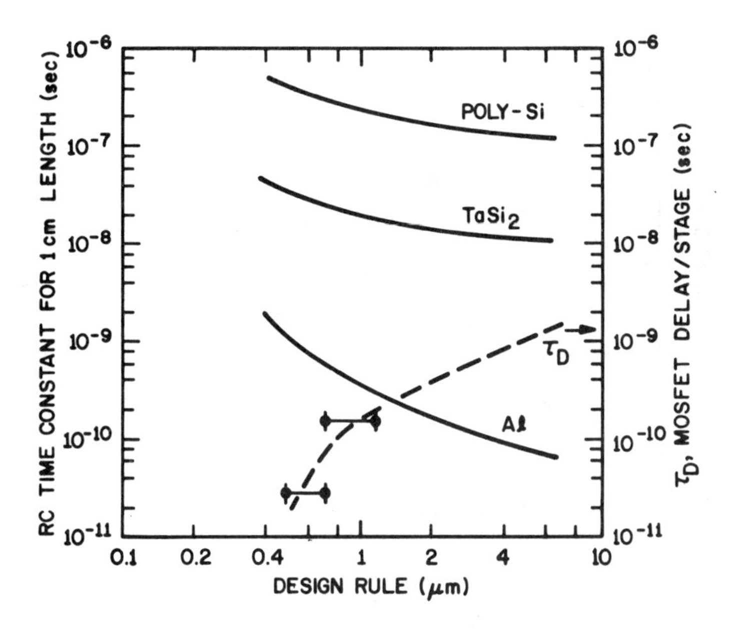

Fig. 4 The RC time constant of various interconnects and MOSFET delay per stage vs. design rule.

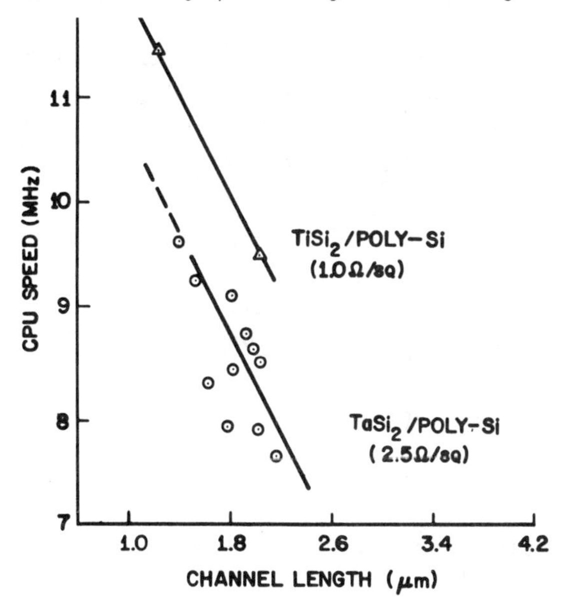

Fig. 5 Effect of MOSFET channel length on the clock rate of a 32-bit microprocessor for two different poly-silicide interconnect materials.

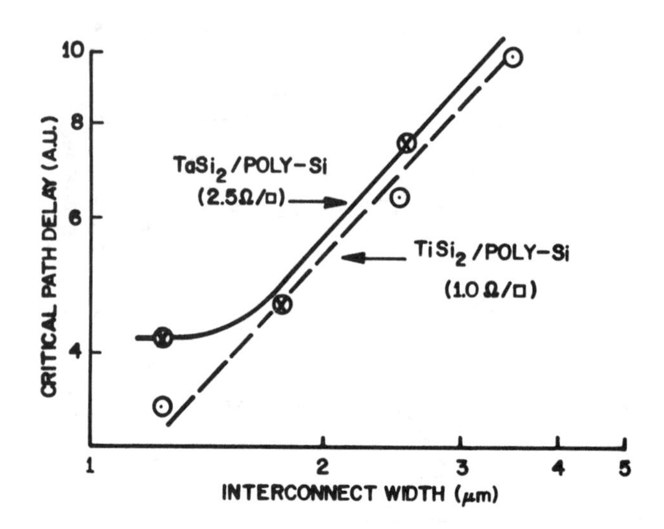

Fig. 6 Relative delay along critical interconnect paths vs. gate level interconnect width for poly-silicides containing $TaSi_2$ and $TiSi_2$.

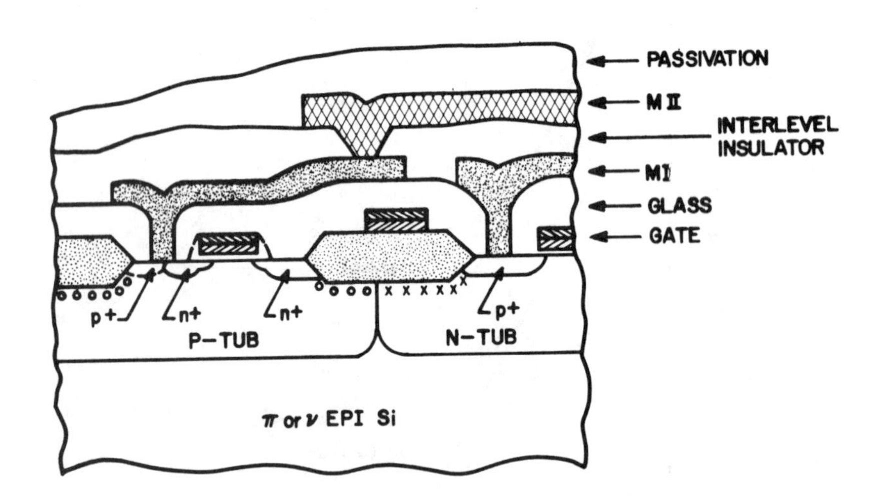

Fig. 7 Schematic cross-section of an advanced CMOS logic circuit utilizing two levels of metal interconnects.

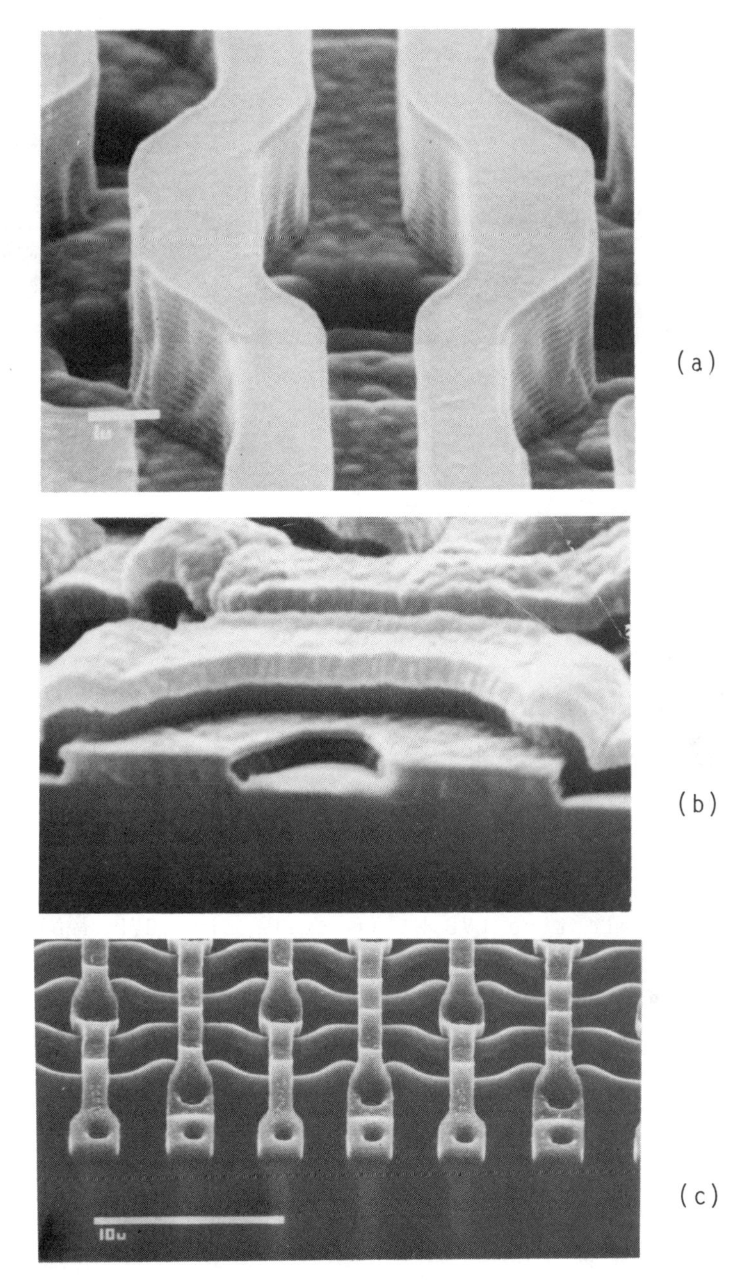

Fig. 8 SEM micrographs of high resolution interconnect patterns in a Mb DRAM. (a) Photoresist over $TaSi_2$, (b) $TaSi_2$/poly-Si, (c) Al over flowed BPSG.

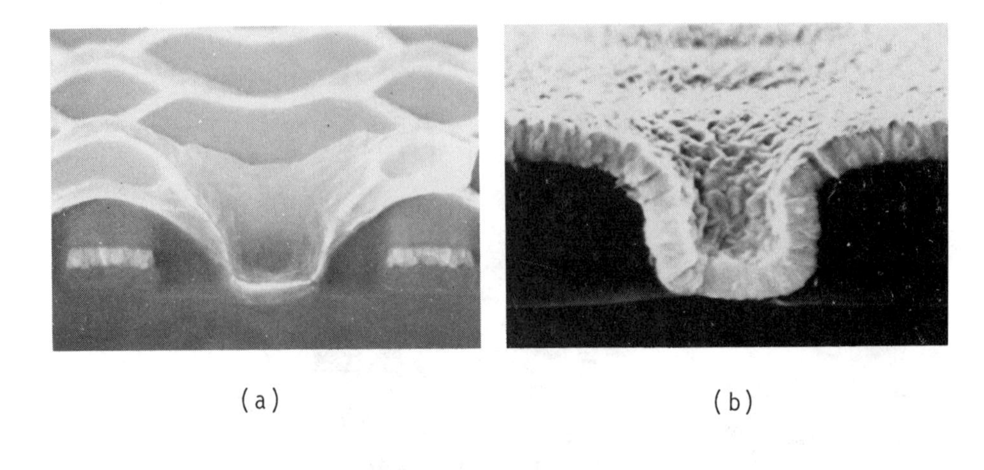

(a) (b)

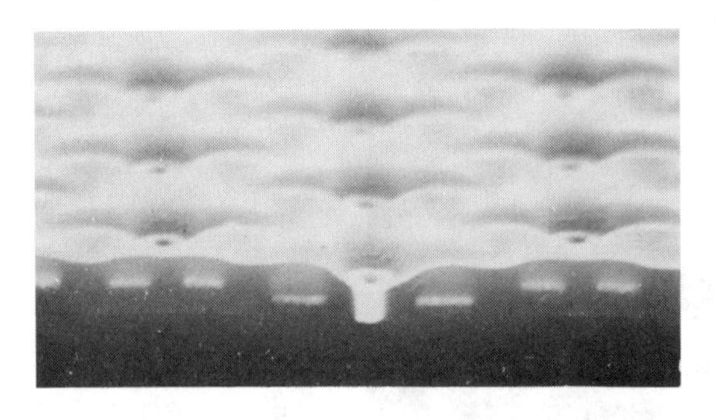

(c)

Fig. 9 Step coverage of various metallizations in 1μ contact windows: (a) S-gun deposited Al, (b) non-selective CVD W, (c) selective CVD W-plug.

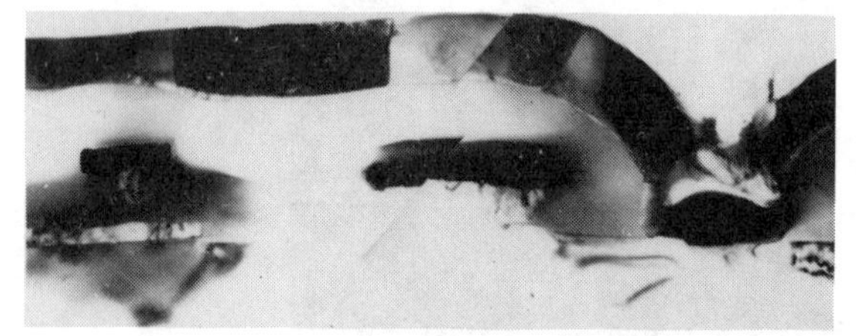

Fig. 10 Transmission electron microscope cross-section through a Mb DRAM showing a chain-shaped grain structure in the Al.

RELIABILITY OF VLSI INTERCONNECTIONS

P. B. Ghate
Texas Instruments, P. O. Box 225012, MS 17
Dallas, Texas 75265

ABSTRACT

As the complexity of very large scale integrated (VLSI) circuits continues to grow with shrinking device geometries, a relatively large portion of the chip area is being allocated for interconnections, and there is an increasing awareness that interconnection reliability will be a major factor limiting VLSI chip reliability. The failure rate of today's commercial ICs is well below 100 FITs (1 FIT = 1 failure in 10^9 device hours), and today's VLSI reliability goals are to surpass the 100 FITs level under normal operating conditions. The predominant failure modes limiting interconnection reliability are (a) electromigration (EM), (b) corrosion, and (c) stress induced deformation. The interrelationship between these failure modes and basic material properties is highlighted. EM failure rates for Ti:W/Al and Ti:W/Al-Cu interconnections are presented to illustrate the interrelationship of current density guidelines and material properties. There is a growing concern that VLSI chip reliability may be adversely impacted due to EM induced failures in these interconnections. A better understanding of the factors governing the failure modes in thin film interconnections is needed for reliability enhancement of VLSI interconnections.

I. INTRODUCTION

Quality and reliability of very large scale integrated (VLSI) circuits are two of the major concerns for end use equipment manufacturers because of the cost resulting from repair, maintenance, and unavailability of equipment. Quality means compliance to specifications at time zero, whereas reliability implies long term performance to specifications in the field. During recent years, the major emphasis has been on quality, and great strides have been made in reducing the incoming PPM defect levels as determined by electrical and visual/mechanical (V/M) screening tests. An example of this trend for three product families from Texas Instruments (TI) - (a) Linear, (b) Transistor-Transistor Logic/Low Power Schottky (TTL/LPS), and (c) 64K DRAMS - as a function of calender years is shown in Figure 1. These data correspond to a product mix of small scale integrated (SSI) and medium scale integrated (MSI) circuits.

0094-243X/86/1380321-18$3.00

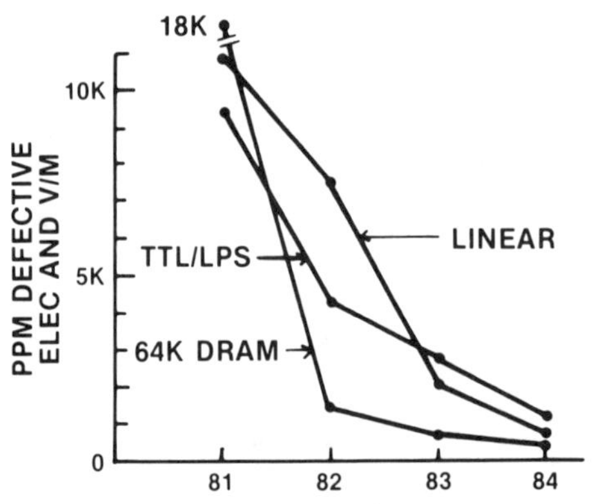

Fig. 1. PPM Defective Electrical Visual/Mechanical vs. Calendar Year

Reliability of Semiconductor devices, the rate at which devices fail as a function of operating time, is represented by the well-known bathtub curve shown in Figure 2. A small percentage of devices with design and process problems escapes initial testing, and these devices exhibit a high failure rate in early life. These infant mortality failures are responsible for the left hand side (decreasing failure rate) of the bathtub curve. All devices fail due to wear-out mechanisms, and the right hand side of the bathtub curve displays this increasing failure rate. The central portion of the curve - almost a flat portion with a slight positive slope - represents the reliability of the devices. The advances in manufacturing, process control, and testing have substantially reduced the infant mortality rates with a considerable improvement in reliability. In the last decade, the failure rates of SSI and MSI circuits have improved from several thousand FITS (1 FIT = 1 failure in 10^9 device hours) to a range of 10 to 100 FITS.

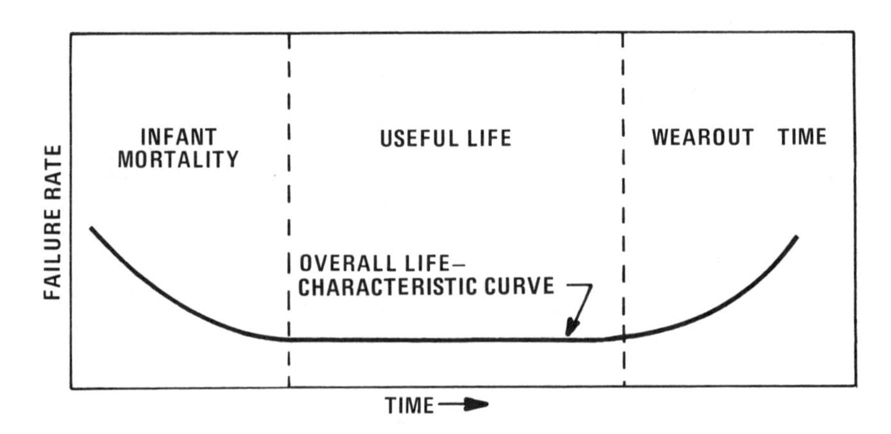

Fig. 2. Bathtub Curve - IC Failure Rate as Function of Time

As an example, failure rate estimates based on in-house reliability testing of TI Standard Plastic encapsulated device families - Advanced Low power Schottky (ALS), Low power Schottky (LS), Schottky Transistor Transistor Logic (STTL), Linear, and 64K DRAMS - are summarized in Table 1. The failure rates are given for a specific temperature, 55°C, and 60% Upper Confidence Limit (U.C.L.). These data illustrate the reliability improvements achieved through design and process optimization.

TABLE 1: TI STANDARD PLASTIC RELIABILITY FAILURE RATE ESTIMATES @ 55°C AND 60% U.C.L. IN FITs*

FAMILY	1982	1984	1985 (1QTR)
ALS	3	3	1
LS	2	6	1
STTL	4	3	1
LINEAR	23	7	6
64K DRAM	257	106	74

*Based on average activation energy of 0.96eV for Bipolar and 0.5eV for MOS

As the VLSI chip area continues to grow with a relatively large portion (approximately two-thirds) of the chip area being utilized by single and multilevel interconnections, there is an increasing awareness that interconnection reliability will be a major factor limiting VLSI chip reliability.[1-4] System designers are demanding that VLSI chips should maintain, if not surpass, the reliability of LSI chips. Today's VLSI reliability goals are to surpass the 100 FITS level under normal operating conditions.

In general, the "normal" operating conditions are defined by the system designers (end users), and these conditions vary from one application to the other. However, the chip temperature under "normal" operating conditions falls within the device operating temperature range.

The field of VLSI interconnections encompasses contacts, barrier metals, gate materials, single and multilevel conductors (interconnections), wire attachment at bond pads, barrier layers between conductors and bumps for flip chip packaging, and any combination of the above. Interconnection reliability is likely to be limited by three classes of problems: circuit designs and layouts, process controls and manufacturing defects, and basic material properties. In this article, the three predominant failure modes - electromigration, corrosion, and internal stresses in multilayered films - are briefly discussed and the interrelationship between these failure modes and basic material properties is highlighted.

II. ELECTROMIGRATION

A) Interconnections

Electromigration (EM) induced failure is one of the primary failure mechanisms limiting the reliability of IC interconnections. Most of the EM studies have been carried out with Al and Al-alloy films, as they happen to be most widely used as IC interconnections.

Mass transport under the influence of an impressed DC electric field is referred to as electromigration. Following Huntington and Grone [5], net atomic flux J_A in a lattice due to current density j can be expressed as

$$J_A = (ND/kT)Z^*eE = (ND/kT)Z^*e\rho j \tag{1}$$

where

N = density of ions
D = $D_o \exp(-Q/kT)$ = self-diffusion coefficient
Z^*e = effective charge on the migrating ion
k = Boltzmann's constant
E = Electric field = ρj
ρ = resistivity of the conductor
T = absolute temperature

In EM studies on single crystals at temperatures fairly close to the melting point, the diffusion coefficient D and the activation energy Q are well defined, and Z*e is a physical parameter describing the momentum exchange between the electrons and diffusing ions.

In the case of vacuum-deposited metal film conductors, the problem is more complex because these film conductors are tested at moderately low temperatures ($0.3T_m < T < 0.5T_m$ where T_m is the melting point of the material), and the mass transport is mainly controlled by grain boundary diffusion. In a film, all grain boundaries are not alike. A detailed theoretical discussion of EM in thin films is very complex and has not been attempted; however, the expression given in Eq. (1) for the atomic flux due to EM in bulk samples has been modified for a film with an ideally textured grain structure, as follows: [6]

$$J_b = N_b \frac{\delta}{d} \cdot \frac{D_b}{kT} \cdot Z^*_b e \rho j \tag{2}$$

where

J_b = flux of metal ions
N_b = local density of ions in the grain boundary
d = average grain size
δ = width of the grain boundary
D_b = grain boundary diffusion coefficient
Z^*_b = effective charge
ρ = resistivity of the film
j = current density

Here it is assumed that mass transport proceeds with the same characteristics in all the grain boundaries. Hence, the grain boundary diffusion coefficient in Eq. (2) is a suitably averaged quantity for the film; other quantities related to the grain boundaries are appropriate averages. It is assumed that the atomic flux in thin films due to EM can be qualitatively expressed by an expression of the type in Eq. (2).

Electromigration alone cannot induce failures in film conductors unless there is a nonvanishing divergence of atomic flux. Two factors, inhomogeneities in the microstructure and temperature gradients, are responsible for flux divergences. The first factor is far more influential than the others. A flux divergence leads to mass depletion and hence to void formation. These voids grow in size and they have been observed to migrate upstream against the electron flow.[7] As these voids coalesce to form large voids, a crack develops which eventually leads to a discontinuity in the conductor.

An example of depletion of matter/void formation in an Al-Si interconnect of an MOS IC is shown in Figure 3(A). An accumulation of matter and whisker growth (seen at random) in a Ti:W/Al interconnect of a bipolar IC is shown in Figure 3(B).

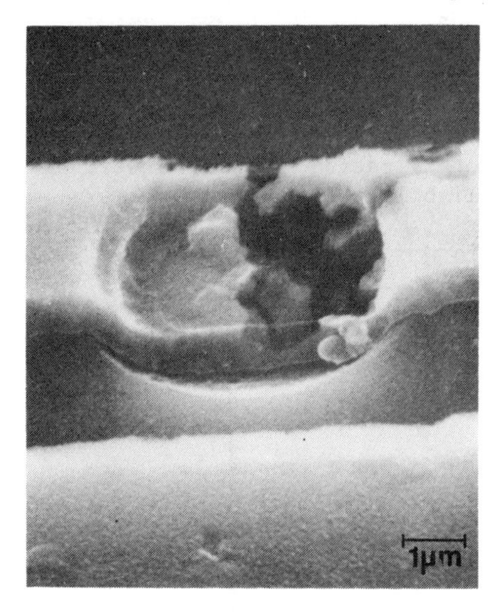

Fig. 3(A). EM-Induced Failures in Al Film Interconnects (MOS Devices)

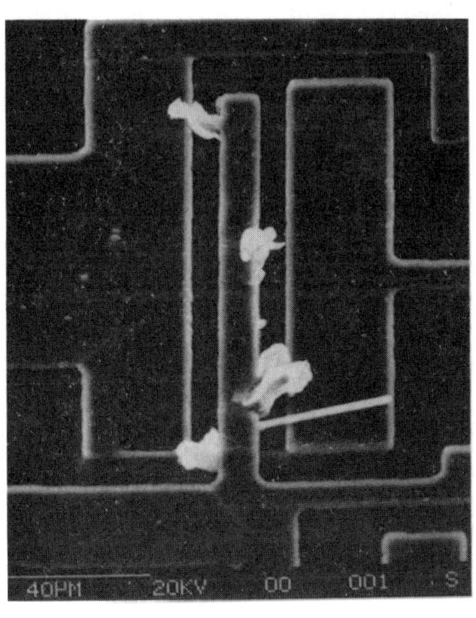

Fig. 3(B). EM-Induced Failures in Ti:W/Al Interconnects (Bipolar Devices)

A general expression for the median time to failure (MTF) is as follows:

$$MTF = Aj^{-n} \exp(Q/kT) \qquad (3)$$

where

A = parameter depending on the geometry, physical characteristics of film, protective coating and substrate

j = the current density in Amperes cm^{-2} ($A\ cm^{-2}$)

n = the exponent : $1 < n < 3$

Q = the activation energy

T = the average temperature of the conductor

k = Boltzmann's Constant

MTF data are observed to obey lognormal distribution with a dispersion parameter σ. Accelerated EM testing is carried out at current densities on the order of $1 - 2 \times 10^6\ A\ cm^{-2}$ and temperatures from 150° to 250°C. Based on these life test data, one may proceed to calculate the failure rates of these film conductors for prescribed current densities and operating temperatures.

EM studies on thin film interconnects suggest that fabrication procedures (film properties, microstructure, composition) and chip operating conditions (temperature, current density, duty cycle) have profound impact on failure rates; for example, replacement of pure Al by Al-Cu films reduces failure rates for a specified IC design layout and operating conditions. The predicted failure rates $\lambda(t)$ for Ti:W/Al and Ti:W/Al-Cu film conductors for 5×10^5 and $2 \times 10^5\ A\ cm^{-2}$ at 85°C are presented in Figure 4.[3] Also, for the same metallization system, lowering operating temperature results in reliability improvement.

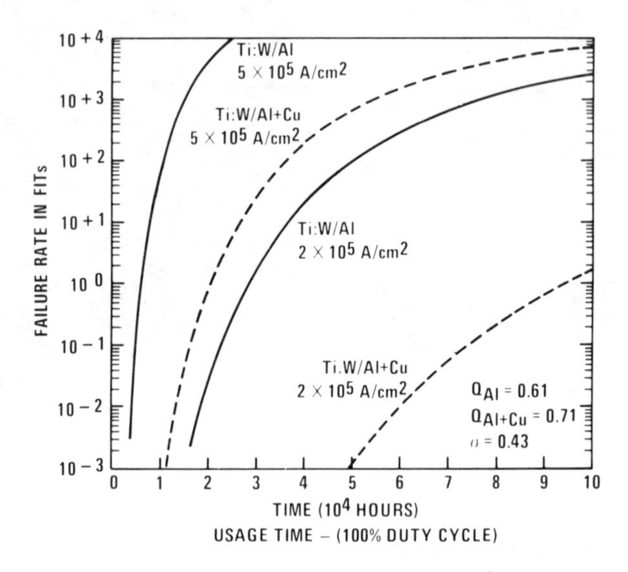

Fig. 4. Predicted Failure Rates of Ti:W/Al and Ti:W/Al-Cu Film Conductors as a Function of Usage Time at 85°C

Recently, it has been reported that MTFs of Al and Al-alloy film conductors decrease with width in the 8 to 2 μm range and suddenly turn around and start increasing below a width of 2μ m.[8] Superior performance of EB-evaporated Al-Cu (0.5 wt%) films has been rationalized on the basis of a "bamboo"- type grain structure.

In another study on Al-Cu-Si film conductors, the superior performance was attributed to very uniform grain size of 0.25μ m and to uniform Cu distribution throughout the films.[9] This study supports the thesis that EM-induced failures resulting from flux divergences due to microstructural inhomogeneities can be avoided by employing uniform grain size distribution across the conductor width, and thereby enhance VLSI interconnect reliability.

Plasma (or reactive ion) etching of Al-Cu films in Cl_2 containing gases to form narrow (1- to 3- μ m wide) interconnects has increased the potential for corrosion-induced failures due to chloride containing residues. Researchers are busy developing proprietary recipes to overcome the corrosion problem for Al-Cu films with superior EM resistance. Concurrently, alternate Al-alloy metallization systems are explored. A recent study suggests that Al-Si-Ti films are compatible with dry etching techniques without the corrosion problem and that their EM resistance is equivalent to that of Al-Cu-Si.[10]

Since the EM induced failure rates depend on the activation energy Q, gold film interconnects with Q=0.90eV can be expected to display a far superior EM resistance in comparison to Al interconnects. Reported life test data on Ti:W/Au interconnects[11] suggest a failure rate of 0.1 FIT for $j=5x10^5$ A cm^{-2} at 85°C operating temperature after 10 years of continuous usage.

During device operation, most of the interconnections are subjected to pulse currents rather than steady state direct currents. Limited data on pulse testing are available.[12-14] In a recent paper, a methodology has been presented whereby simulations of the actual transient current conditions in the interconnect can be converted to a continuous DC EM equivalent for proper determination of the required interconnect width.[15]

As the width and spacings of multilevel interconnections continue to shrink, the EM induced failures are more likely to be due to intra and interlevel shorts rather than the classical open failures. Recent results indicate (a) 80% of the failures on the first level due to shorts vs. 91% of the failures due to opens on the third level in a three level test structure[16,17], and (b) the same activation energy for opens and shorts.[18]

B) Contacts

EM failures are observed also in Si/Al contacts with Al as the contact/interconnection metallization. Studies indicate that Si atoms are transported in the direction of electron flow from contact windows, eroding the contacts; and these Si atoms are later deposited onto the Si contacts where electrons leave the Al interconnections.[19-21] Erosion of contacts leads to increased leakage currents and device failures. These failures depend on junction depth and circuit layouts (i.e., distance of contact window edge from the diffusion window). Contact erosion and growth of precipitates depend on Si surface preparation of contact windows prior to metal deposition and they do not proceed uniformly in contact windows. Experimental data on test structures have been collected to predict contact failure rates vs. interconnection failure rates. These predictions should be used with caution.

Reliability of shallow junction devices is a major concern and this problem is solved by interposing either a silicide/barrier layer or refractory silicide/barrier layer between the contacts and interconnections.[1] Examples include: Si/PtSi/Ti:W/Al, Si/PtSi/TiN/Al, Si/PtSi/Cr/Al, Si/PtSi/Ti/Pt/Au, and Si/W/Al with Ti:W, TiN, Cr, Ti/Pt, and W as barrier layers.

III. CORROSION

The corrosion of IC metallization has been a frequent reliability problem.[22-25] A large number of variables such as metallization, encapsulation, ionic contamination, temperature, humidity, and applied bias affects the corrosion failures. The SC industry has employed hermetic packages to protect the device metallization from corrosion by sealing the devices in an inert ambient with minimum (less that 3000 ppm) moisture inside the package. The current trend is to replace hermetic packages with plastic packages to reduce cost. Small amounts of ion contaminants, mostly chlorides, are present in the encapsulating (molding) materials. As the moisture and contaminants reach device metallization, corrosion reactions set in and they eventually lead to device failures. Possible Al corrosion reactions[21] are presented in Table 2. Examples of bond pad and internal metallization corrosion are presented in Figure 5.

It is a common practice in bipolar IC fabrication to use phosphosilicate glass (PSG) layers as passivation for devices to tie up the mobile Na ionic contamination. Also, in MOS LSI fabrication, a phosphorus-doped CVD SiO_2 layer (4 to 8 wt% phosphorus) is deposited and then reflowed at a high temperature to produce smooth topography for metal coverage on oxide steps. The Al and Al-alloy film conductors are formed on this PSG layer. It is not uncommon to find a CVD SiO_2 layer deposited on top of these interconnects as a protective coating. This SiO_2 layer is doped with phosphorus to reduce internal stresses and thereby eliminate cracks in this layer. Data in the literature lead to the conclusion that corrosion of Al and Al-alloy conductors decreases with a decreasing amount of phosphorus in the PSG layer in the presence of moisture, and that the failure rates are sensitive to the applied bias on these conductors.

TABLE 2: POSSIBLE Al CORROSION REACTIONS

	Reaction	Note
a.	$Al \rightarrow Al^{+3}+3e^-$	E°=1.66 volt
b.	$Al \rightarrow Al^{+1}+1e^-$	Mechanism for hydrogen evolution at anode[2]
c.	$Al^{+1}+2H_2O \rightarrow Al^{+3}+2OH^-+H_2(g)$	
d.	$2Al+6HCl \rightarrow 2AlCl_3+3H_2(g)$	Acidic solutions and reactions involving chloride ions
e.	$Al+3Cl^- \leftrightarrow AlCl_3+3e^-$	
f.	$AlCl_3+3H_2O \rightarrow Al(OH)_3+3HCl$	
g.	$Al+NaOH+H_2O \rightarrow NaAlO_2+\frac{3}{2}H_2(g)$	Basic solutions and reactions involving oxides and hydroxides
h.	$Al+3OH^- \rightarrow Al(OH)_3+3e^-$	
i.	$2Al(OH)_3 \rightarrow Al_2O_3+3H_2O$	
j.	$2AlO_2^-+2H^+ \rightarrow Al_2O_3+H_2O$	

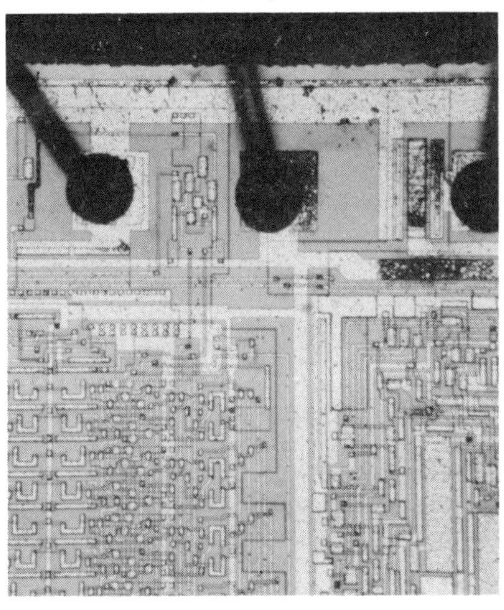

Fig. 5(A). Al Bond Pad Corrosion

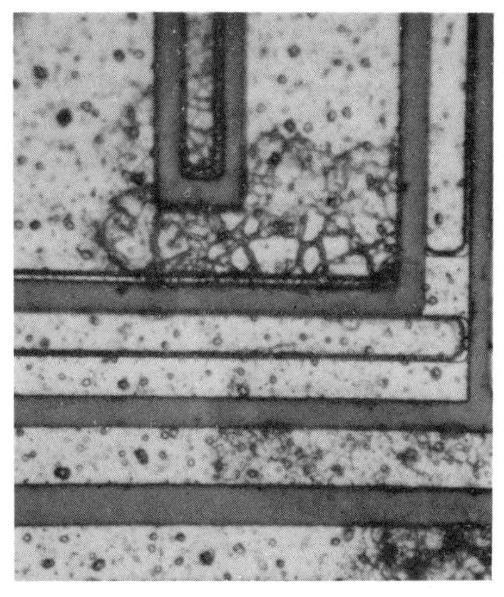

Fig. 5(B). Internal Corrosion

The corrosion susceptibility of plastic encapsulated ICs is assessed by accelerated testing of a small sample of devices at 85% relative humidity (RH) and at 85°C, referred to as 85/85, for periods ranging from 25 to 1000 hrs (in some cases 2000 hrs) with or without bias. The cumulative failures are measured at specified intervals up to 1000 hrs. These data are reported to indicate the reliability of these devices against corrosion, e.g. 1% failures at 2K hrs at 85/85.

Attempts have been made to develop models to use 85/85 data to predict failures for use conditions. The MTF is represented by an equation of the form[27]

$$MTF = A \exp(\beta/H) \cdot \exp(Q/kT) \qquad (4)$$

where A = constant, β = a parameter dependent on humidity, H = relative humidity, Q = activation energy, T = temperature, and acceleration factors (AF) for use conditions are estimated. The parameters in this equation are dependent on molding compounds and device packaging processes. The SC industry is constantly searching for alternate test methods to reduce the 85/85 test time. Highly accelerated stress tests (HAST) with high temperature and humidity are explored and efforts are under way to relate HAST data to 85/85 data.[28-31]

In summary, the corrosion processes are too complex and reliability modeling is far from satisfactory. The biasing conditions (i.e., applied voltages) during device operations add another dimension to this complex corrosion problem. Material selection based on basic principles, and process controls to avoid ionic contamination have been the best methods to minimize corrosion failures and thereby achieve cumulative failures less that 0.5% after 1000 hours at 85/85 (with an estimated AF of 1000, this data corresponds to an average failure rate of 5 FITs under normal operating conditions.) Since VLSI processing involves phosphorus-doped SiO_2 and dissimilar metals from contacts through packaging, stringent process controls to eliminate ionic contamination from the chip surface and moisture from the package appear to be the best remedies against corrosion-induced interconnect failures.

IV. INTERNAL STRESSES

Internal stresses in Al films, conductor/insulator multilayers, protective coatings, and others in the packaged devices can lead to device failures. A few examples are presented below to illustrate the potential reliability problems.

Internal stresses in Al film interconnects depend on film deposition parameters and subsequent processing.[32-34] As the aluminum metallized substrate is heated to a temperature in the 100°-180°C range for curing photoresist used in patterning step, or to a temperature of 450°C for contact sintering and/or to a temperature of 300°-400°C for insulator film deposition, the Al film is subjected to a compressive stress due to the difference in coefficients of thermal expansion of Al and Si substrates. The strain resulting from this compressive stress is relaxed by hillock formation and the height of these hillocks may equal film thickness. Consider the case of Al film interconnections on an oxidized silicon substrate crossing the Si-SiO_2 steps and temperature cycled between 150° and -55°C. This temperature cycling, which is equivalent to mechanical stress cycling, leads to hillock growth, and in some cases to grain boundary sliding at the oxide steps. Also, pronounced slip bands and striations are formed on samples subjected to 1500 temperature cycles.[33] These deformation processes are illustrated in Figure 6.

It is one of the reliability concerns that the hillocks may lead to (a) interlevel shorts due to thinning of the interlevel insulator layer in multilevel interconnections, or (b) pinholes in the protective coatings that may provide potential paths for moisture to reach Al interconnections and cause corrosion failures.[35]

Another reliability problem is that of internal stress generated cracks near the corner of large chips. The stress enhancement may result from any one or a combination of the following factors: (1) the stress distribution in multilayered

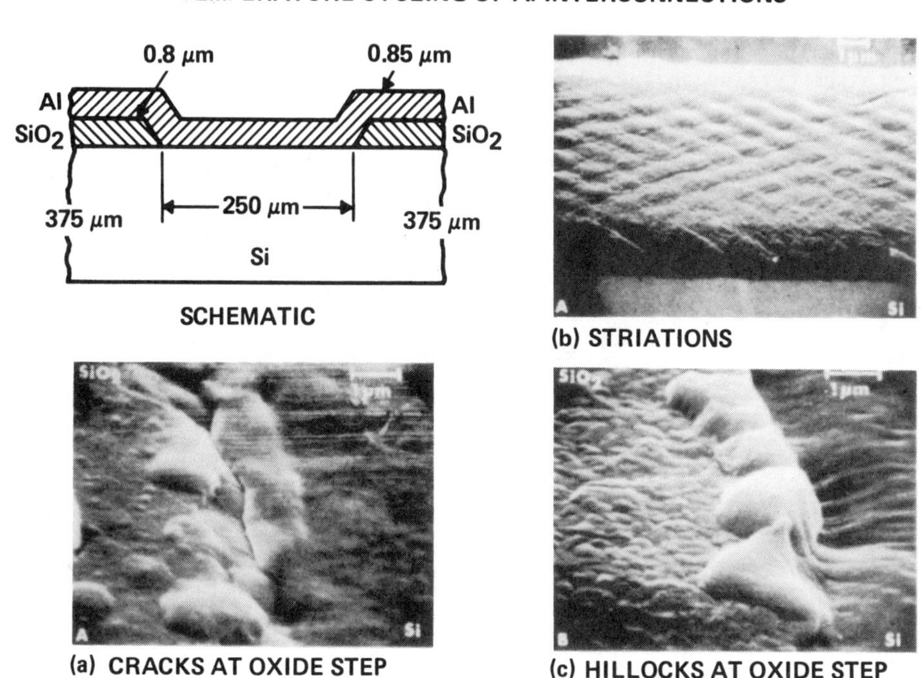

Fig. 6. Deformation Processes in Al Film Interconnections Due to Temperature Cycling

conductor/insulator layers, (2) the thermal mismatch between the chip and packaging material, (3) chip size relative to the package, (4) plastic encapsulation process, and others. Some of the chips with minute cracks may pass initial electrical tests and pose (infant mortality) reliability risks. As the cracks grow, some of the narrow interconnects may develop opens and cause device failures. Examples of a crack in the chip corner and a failure in a narrow polysilicon interconnect in an MOS VLSI circuit are presented in Figures 7A and 7B, respectively. A better understanding of the stress distribution on the chips by finite element stress analysis and better process controls have been helpful in almost eliminating this reliability problem.[36]

Recently, open metal or void failures with Al-Si interconnections on MOS devices have been reported.[37-40] The internal stresses in Al-Si films are tensile in nature and the magnitude depends on Al-Si processing and the stresses in the protective coating. Al films sustain high internal stresses ($2 - 3 \times 10^9$ dynes/cm^2), a factor of 2 to 5 above the bulk breaking strength. When a protective coating with high compressive stress is deposited on the metallized substrates, then the Al films are subjected to a relatively high tensile stress to balance the forces. Also note that the interconnections of varying lengths and widths are pinned by contacts and possibly by Si precipitates. It is conceivable that these interconnections are subjected to varying stress levels depending on their location on the chip. Stress gradients within the film from grain boundaries to the outside surface are possible.

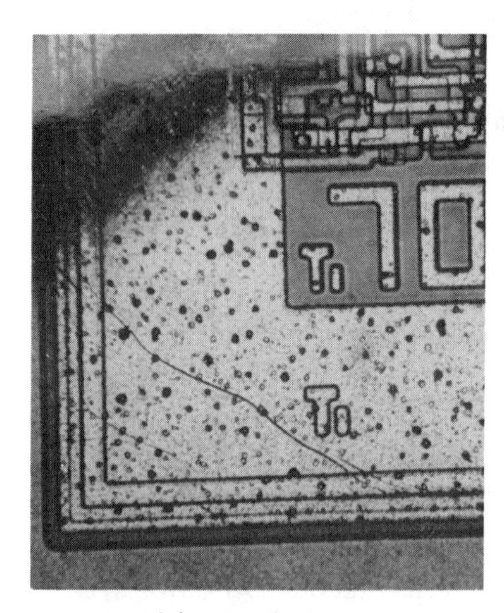

Fig. 7(A). Crack in Chip Corner

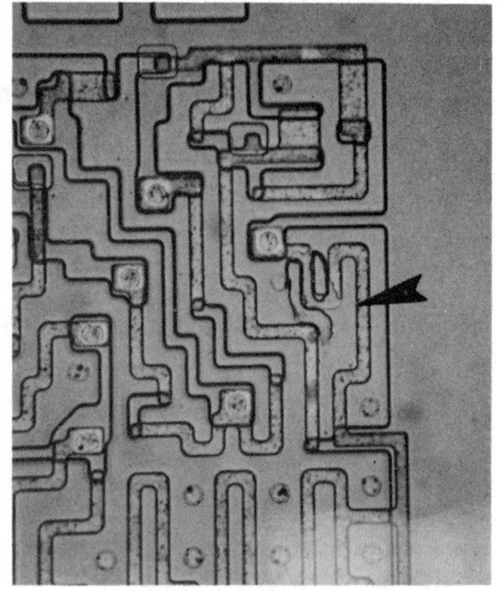

Fig. 7(B). Open in Polysilicon Interconnect

Like all other systems under stresses, these interconnections will try to attain a stress-free state by plastic deformation. The stress is relaxed by the conversion of elastic strain to nonelastic strain. Formally,

$$\dot{\sigma} = -M\dot{\varepsilon} \tag{5}$$

where $\dot{\sigma}$, M, and $\dot{\varepsilon}$ are stress relaxation rate, an appropriate elastic modulus, and the nonelastic strain rate, respectively. The two strain relaxation mechanisms by atom flow through lattice diffusion and grain boundaries are referred to as (lattice) diffusion and grain boundary creep, respectively.[41]

The diffusion creep model was originally developed by Nabarro[42] and Herring[43] for bulk material, and this model was extended to thin films by Gibbs.[44] The stress relaxation by grain boundary migration is described in the framework of Coble Creep.[45] The stress relaxation rates for the two cases may be expressed as follows:

$$\text{(Diffusion Creep)}\quad \dot{\sigma} = \text{constant}\frac{\sigma}{gh} \cdot \frac{1}{kT} \cdot \exp\left(-Q_B/kT\right) \tag{6}$$

$$\text{(Coble Creep)}\quad \dot{\sigma} = \text{constant}\frac{\sigma}{gh^2} \cdot \frac{1}{kT} \cdot \exp\left(-Q_{GB}/kT\right) \tag{7}$$

where g = grain size, h = film thickness, Q_B and Q_{GB} = activation energies for bulk and grain boundary diffusion, respectively. At high temperatures ($> 0.5\ T_m$, where T_m is the melting temperature) diffusion creep is expected to dominate, whereas at low temperatures ($< 0.5\ T_m$) grain boundary diffusion dominates and the grain boundary creep is considered to be operative.

The rupture life or time to failure (t_f) is a complex inverse function of $\dot{\sigma}$; following the literature on creep phenomena in metals,[46] t_f may be expressed as follows:

$$t_f = A_o\ \sigma^{-n} \exp\left(Q_c/kT\right) \tag{8}$$

where A_o, σ, n, and Q_c are material dependent parameter, stress, stress exponent on the order of 4, and the activation energy for creep, respectively.

The important thing to notice is that during the post metal processing, stress relaxation depends on the heating and cooling rates of the first high temperature excursion.[47] Mass transport (or vacancy flow) will proceed both by grain boundary and lattice diffusion to relieve the stress. At higher temperatures the vacancy flow due to stress gradients will be towards the external surfaces (the edges of the interconnections, preferably towards the surface closer to protective coating), and this vacancy flow is considered to be responsible for the void nucleation. The grain boundaries intersecting the interconnection edges at 45° (the direction of maximum shear stress that causes dislocation motion and slip) appear to be the logical sites for void nucleation and growth. Examples of the voids in early stages of growth are illustrated in Figure 8. These void failures are noticeable in temperature cycling (-65° to 150°C) tests. Recent studies[39] on Al film interconnections of varying widths have shown that size and density of voids appear to be maximum for widths in the range of 2 to 3 micrometers.

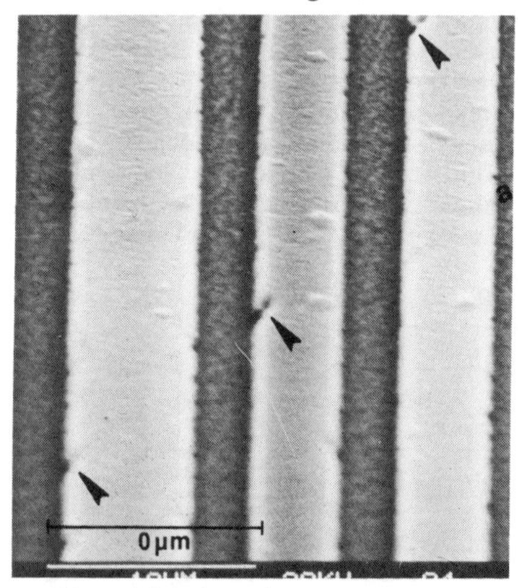

Fig. 8(A). Voids in Al-Si Interconnects

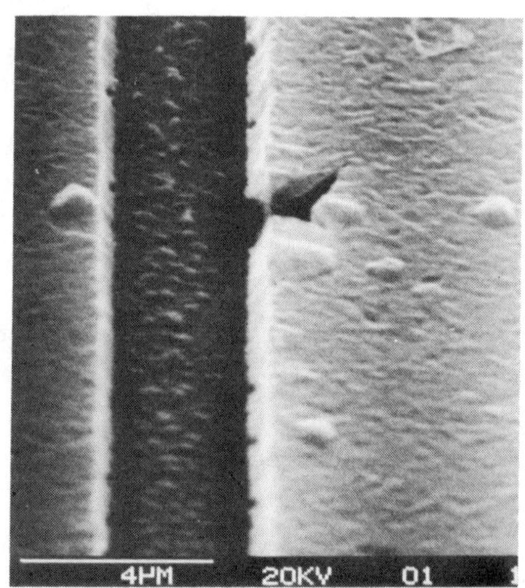

Fig. 8(B). A Void that has Grown to 1/3 the Width of the Interconnect

At the normal device operating temperatures (25°-100°C), the grain boundary diffusion will be a rate controlling step for void growth. For those voids that are already formed, the subsequent growth will be controlled by the supply of vacancies from the interior of the grains, and hence controlled by lattice diffusion. One may be tempted to conclude that lattice diffusion controlled void growth can hardly be considered a potential reliability problem. However, these voids of different sizes cause corresponding increments in the current densities above the design values and lead to early failures in EM testing. One may conclude that these voids increase the dispersion parameter σ of the EM life test data; conversely, a relatively higher value σ is an indicator of the voids developed due to stress relaxation. One of the pragmatic solutions to minimize the reliability risks is to lower the compressive and tensile stresses in the protective oxide and metal films, respectively. Other solutions include the replacement of pure Al films by microstructure tailored alloy films and/or the multilayered films.

V. SUMMARY

In the introduction of this article, the estimated failure rates for various products are presented. They are estimated by using an average activation energy of 0.96eV for Bipolar and 0.5eV for MOS devices. These failure rates are considered to be the sum total of not only the interconnection failure rates, but also those of other failure modes. When a VLSI reliability goal of 100 FITs for 100K power on hours under normal operating conditions is specified, one needs to ask how to tailor the interconnection film properties to achieve a 50 FIT failure rate for the interconnections (author's suggestion only), so that a 50 FIT allowance is comprehended for other failure modes. There is no simple answer to this difficult question. However, a basic understanding of thin film properties and their interactions is a key to identifying the rate controlling processes leading to interconnection device failures.

Reliability of VLSI interconnections is a multivariable problem. Electromigration, corrosion, and voids due to internal stresses are some of the key reliability problems. They are strong functions of device technologies, circuit designs and operating conditions, and basic material properties. In this article, the basic concepts describing the EM induced failures, corrosion reactions, and strain relaxation (creep) induced void formation have been reviewed. Also, it is pointed out that intuitive reasoning and/or simple models based on basic concepts have provided the MTF equations. The parameters in those equations are determined from accelerated life test data. Note the similarity in the MTF equations:

$$\text{EM} \qquad MTF = A\, j^{-n} \exp\,(Q/kT) \tag{9}$$

$$\text{Corrosion} \qquad MTF = A \exp\,(\beta/H) \exp\,(Q/kT)$$

$$\text{Voids (creep)} \qquad MTF = A_o\, \sigma^{-n} \exp\,(Q_c/kT)$$

The activation energies in these equations are not necessarily the same. The interconnection reliability (or the failure rate) for a given set of operating conditions (temperature, current density, humidity, etc.) is a cumulative effect of EM, corrosion, and internal stresses.

Everyone engaged in VLSI efforts intends to design-in reliability by using carefully agreed upon current density design guidelines and optimized processes.[2-4] This approach has been illustrated in Figure 4. However, as the device dimensions continue to shrink and the interconnection widths approach dimensions comparable to the grain size, it appears that a small number of grain boundaries may control the EM induced failures. Also, it is anticipated that the orientation of grain boundaries in narrow interconnects will be strongly influenced by the internal stress levels. EM studies on "bamboo" structure, 1 to 2μ m wide Al interconnects have shown considerable improvement in MTF; however, the limited data show an increase in the dispersion parameter σ, which adversely impacts the time for a small percentage of cumulative failures. Results of another study[48] caution that the dispersion parameter becomes large when the grain size exceeds line width.

An examination of VLSI interconnection statistics shows that VLSI chips use interconnections of varying widths. A bamboo-type microstructure is hardly possible for wider leads. The possibility of fabricating single crystal films on a cost effective basis is far out in the future, and one has to reconcile with the fact that grain boundary diffusion will affect EM induced failures. A number of studies on Al-alloy films, such as Al-Cu, Al-Si-Cu, Al-Si-Ti, and other sandwich layered structures with dopants, have shown the desired improvement in EM tolerance;[49,50] however, a basic understanding of the physical processes leading to this EM improvement is yet to be developed.

The corrosion of VLSI interconnections will continue to be a potential reliability problem because the interconnection system consists of dissimilar materials, and is exposed to trace amounts of ionic contamination and humidity during its fabrication and usage. So long as a monomaterial VLSI circuit is not feasible, (a) material selection based on basic properties of thin film materials, and (b) stringent process controls to eliminate ionic contamination offer the best hope for controlling corrosion induced failures.

The presence of internal stresses in thin films is not new; however, the observation of stress induced void formation on narrow interconnects is very recent. The VLSI interconnection systems use a variety of conductor/insulator multilayers, and the internal stresses in these composite multilayers are highly process dependent. Furthermore, the chip attachment to a package, package size, and the packaging operation will have a significant impact on the stress distribution in the VLSI chip. The finite element stress analysis in packaged units and process optimization will be helpful in controlling the stress induced failure modes.

As the VLSI chips continue to grow in size by integrating an ever increasing number of components, the power dissipation is increasing and the interconnections tend to be operating at relatively higher temperatures with increased susceptibility to EM induced failures. Detailed experimental and theoretical studies on electromigration, corrosion, and internal stresses in thin film interconnects are needed to develop reliability models for failure rate projections. A better understanding of thin film processes can guide the efforts to design-in reliability of VLSI interconnections.

REFERENCES

1. P. B. Ghate, Thin Solid Films 45, 69 (1982).
2. P. S. Ho, IEEE 20th Annual Proc. Rel. Phys., 288 (1982).
3. P. B. Ghate, IEEE 20th Annual Proc. Rel. Phys., 292 (1982).
4. J. R. Black, IEEE 20th Annual Proc. Rel. Phys., 300 (1982).
5. H. B. Huntington and A. R. Grone, J. Phys. Chem. Solids 20, 76 (1961).
6. F. M. d'Heurle and P. S. Ho, "Electromigration in Thin Films," Physics of Thin Films: Interdiffusion and Reactions, ed. J. M. Poate, K. N. Tu and J. W. Mayer (John Wiley, N.Y., 1978), p. 243.
7. R. W. Thomas and D. W. Calabrese, IEEE 21st Annual Proc. Rel. Phys., 1 (1983).
8. S. Vaidya, D. B. Fraser and A. K. Sinha, IEEE 18th Annual Proc. Rel. Phys., 165 (1980).
9. P. B. Ghate, IEEE 19th Annual Proc. Rel. Phys., 243 (1981).
10. F. Fisher and F. Neppl, IEEE 22nd Annual Proc. Rel. Phys., 190 (1984).
11. J. C. Blair, C. R. Fuller, P. B. Ghate and C. T. Haywood, J. Appl. Phys. 43, 307 (1972).
12. A. T. English, K. L. Tai and P. A. Turner, Appl. Phys. Lett. 21, 397 (1972).
13. C. J. Wu and M. J. McNutt, IEEE 21st Annual Proc. Rel. Phys., 24 (1983).
14. J. M. Towner and E. P. van de Ven, IEEE 21st Annual Proc. Rel. Phys., 36 (1983).
15. J. W. McPherson and P. B. Ghate, Proc. Symp. on Electromigration of Metals and First Int. Symp. on Multilevel Metallization and Packaging, The Electrochemical Soc. Inc. 85-6, 64 (1985).
16. J. R. Lloyd and P. M. Smith, J. Vac. Sci. Technol. 1, 455 (1983).
17. J. R. Lloyd, IEEE 21st Annual Proc. Rel. Phys., 208 (1983).
18. J. M. Towner, IEEE 23rd Annual Proc. Rel. Phys., 81 (1985).
19. J. R. Black, IEEE 16th Annual Proc. Rel. Phys., 233 (1978).
20. P. B. Ghate, IEEE 19th Annual Proc. Rel. Phys., 243 (1981).
21. S. Vaidya and A. K. Sinha, IEEE 20th Annual Proc. Rel. Phys., 50 (1982).
22. S. C. Kolesar, IEEE 12th Annual Proc. Rel. Phys., 155 (1976).
23. W. H. Paulson and R. W. Kirk, IEEE 12th Annual Proc. Rel. Phys., 172 (1976).
24. S. P. Sim and R. W. Lawson, IEEE 17th Annual Proc. Rel. Phys., 103 (1979).
25. V. Bhide and J. M. Eldridge, IEEE 21st Annual Proc. Rel. Phys., 44 (1983).
26. T. Wada, H. Higuchi and T. Ajiki, IEEE 23rd Annual Proc. Rel. Phys., 159 (1985).

27. K. M. Striny and A. W. Schelling, IEEE Trans. Components Hybrids and Manu. Technol. CHMT-4, 476 (1981).
28. K. Ogawa, J. Suzuki and K. Sano, Proc. Int. Symp. Testing and Failure Analysis, 75 (1981).
29. J. E. Gunn, S. K. Malik and P. M. Mazumdar, IEEE 19th Annual Proc. Rel. Phys., 48 (1981).
30. J. E. Gunn, R. E. Camenga and S. Malik, IEEE 21st Annual Proc. Rel. Phys., 66 (1983).
31. R. P. Merrett, J. P. Bryant and R. Studd, IEEE 21st Annual Proc. Rel. Phys., 73 (1983).
32. F. d'Heure, L. Berenbaum and R. Rosenberg, Trans. AIME 242, 502 (1968).
33. P. B. Ghate, IEEE Trans. PHP-7, 134 (1971).
34. P. B. Ghate and L. H. Hall, J. Electrochem. Soc. 119, 491 (1972).
35. P. B. Ghate and C. R. Fuller, "Multilevel Interconnections for VLSI," Semiconductor Silicon 1981, Proc. 81-5, ed. H. R. Huff, R. J. Kriegler and Y. Takeishi (The Electrochemical Society, Inc., Pennington, N.J., 1981), p. 680.
36. S. Groothuis, W. Schroen and M. Murtuza, IEEE 23rd Annual Proc. Rel. Phys., 184 (1985).
37. J. Curry, G. Fitzgibbon, Y. Guan, R. Mullo, G. Nelson and A. Thomas, IEEE 22nd Annual Proc. Rel. Phys., 6 (1984).
38. S. J. O'Donnell, J. W. Bartling and G. Hill, IEEE 22nd Annual Proc. Rel. Phys., 9 (1984).
39. L. D. Yau, C. C. Hong and D. L. Crook, IEEE 23rd Annual Proc. Rel. Phys., 115 (1985).
40. T. Turner and K. Wendell, IEEE 23rd Annual Proc. Rel. Phys., 142 (1985).
41. M. Murakami, CRC Critical Reviews in Solid State and Materials Sciences 11, 317 (1984).
42. F. R. N. Nabarro, Report of a Conf. on the Strength of Solids (Physical Society, London, 1948), p. 75.
43. C. Herring, J. Appl. Phys. 21, 437 (1950).
44. G. B. Gibbs, Philos. Mag. 13, 589 (1966).
45. R. L. Coble, J. Appl. Phys. 34, 1679 (1963).
46. See Frank Garafalo, "Fundamentals of Creep and Creep-Rupture in Metals," MacMillan Series in Materials Science, 210 (1966).
47. M. S. Jackson and C. Y. Li, Acta Metall. 30, 1993 (1982).
48. J. M. Schoen, J. Appl. Phys. 51, 513 (1980).
49. J. K. Howard, J. F. White and P. S. Ho, J. Appl. Phys. 49, 4083 (1978).
50. D. S. Gardner, T. L. Michalka, K. C. Saraswat, T. W. Barbee, Jr., J. P. McVilte and J. D. Meindl, IEEE J. S. S. Circuits SC-20, 94 (1985).

AIP Conference Proceedings

		L.C. Number	ISBN
No. 1	Feedback and Dynamic Control of Plasmas – 1970	70-141596	0-88318-100-2
No. 2	Particles and Fields – 1971 (Rochester)	71-184662	0-88318-101-0
No. 3	Thermal Expansion – 1971 (Corning)	72-76970	0-88318-102-9
No. 4	Superconductivity in *d*- and *f*-Band Metals (Rochester, 1971)	74-18879	0-88318-103-7
No. 5	Magnetism and Magnetic Materials – 1971 (2 parts) (Chicago)	59-2468	0-88318-104-5
No. 6	Particle Physics (Irvine, 1971)	72-81239	0-88318-105-3
No. 7	Exploring the History of Nuclear Physics – 1972	72-81883	0-88318-106-1
No. 8	Experimental Meson Spectroscopy –1972	72-88226	0-88318-107-X
No. 9	Cyclotrons – 1972 (Vancouver)	72-92798	0-88318-108-8
No. 10	Magnetism and Magnetic Materials – 1972	72-623469	0-88318-109-6
No. 11	Transport Phenomena – 1973 (Brown University Conference)	73-80682	0-88318-110-X
No. 12	Experiments on High Energy Particle Collisions – 1973 (Vanderbilt Conference)	73-81705	0-88318-111–8
No. 13	π-π Scattering – 1973 (Tallahassee Conference)	73-81704	0-88318-112-6
No. 14	Particles and Fields – 1973 (APS/DPF Berkeley)	73-91923	0-88318-113-4
No. 15	High Energy Collisions – 1973 (Stony Brook)	73-92324	0-88318-114-2
No. 16	Causality and Physical Theories (Wayne State University, 1973)	73-93420	0-88318-115-0
No. 17	Thermal Expansion – 1973 (Lake of the Ozarks)	73-94415	0-88318-116-9
No. 18	Magnetism and Magnetic Materials – 1973 (2 parts) (Boston)	59-2468	0-88318-117-7
No. 19	Physics and the Energy Problem – 1974 (APS Chicago)	73-94416	0-88318-118-5
No. 20	Tetrahedrally Bonded Amorphous Semiconductors (Yorktown Heights, 1974)	74-80145	0-88318-119-3
No. 21	Experimental Meson Spectroscopy – 1974 (Boston)	74-82628	0-88318-120-7
No. 22	Neutrinos – 1974 (Philadelphia)	74-82413	0-88318-121-5
No. 23	Particles and Fields – 1974 (APS/DPF Williamsburg)	74-27575	0-88318-122-3
No. 24	Magnetism and Magnetic Materials – 1974 (20th Annual Conference, San Francisco)	75-2647	0-88318-123-1

No. 25	Efficient Use of Energy (The APS Studies on the Technical Aspects of the More Efficient Use of Energy)	75-18227	0-88318-124-X
No. 26	High-Energy Physics and Nuclear Structure – 1975 (Santa Fe and Los Alamos)	75-26411	0-88318-125-8
No. 27	Topics in Statistical Mechanics and Biophysics: A Memorial to Julius L. Jackson (Wayne State University, 1975)	75-36309	0-88318-126-6
No. 28	Physics and Our World: A Symposium in Honor of Victor F. Weisskopf (M.I.T., 1974)	76-7207	0-88318-127-4
No. 29	Magnetism and Magnetic Materials – 1975 (21st Annual Conference, Philadelphia)	76-10931	0-88318-128-2
No. 30	Particle Searches and Discoveries – 1976 (Vanderbilt Conference)	76-19949	0-88318-129-0
No. 31	Structure and Excitations of Amorphous Solids (Williamsburg, VA, 1976)	76-22279	0-88318-130-4
No. 32	Materials Technology – 1976 (APS New York Meeting)	76-27967	0-88318-131-2
No. 33	Meson-Nuclear Physics – 1976 (Carnegie-Mellon Conference)	76-26811	0-88318-132-0
No. 34	Magnetism and Magnetic Materials – 1976 (Joint MMM-Intermag Conference, Pittsburgh)	76-47106	0-88318-133-9
No. 35	High Energy Physics with Polarized Beams and Targets (Argonne, 1976)	76-50181	0-88318-134-7
No. 36	Momentum Wave Functions – 1976 (Indiana University)	77-82145	0-88318-135-5
No. 37	Weak Interaction Physics – 1977 (Indiana University)	77-83344	0-88318-136-3
No. 38	Workshop on New Directions in Mossbauer Spectroscopy (Argonne, 1977)	77-90635	0-88318-137-1
No. 39	Physics Careers, Employment and Education (Penn State, 1977)	77-94053	0-88318-138-X
No. 40	Electrical Transport and Optical Properties of Inhomogeneous Media (Ohio State University, 1977)	78-54319	0-88318-139-8
No. 41	Nucleon-Nucleon Interactions – 1977 (Vancouver)	78-54249	0-88318-140-1
No. 42	Higher Energy Polarized Proton Beams (Ann Arbor, 1977)	78-55682	0-88318-141-X
No. 43	Particles and Fields – 1977 (APS/DPF, Argonne)	78-55683	0-88318-142-8
No. 44	Future Trends in Superconductive Electronics (Charlottesville, 1978)	77-9240	0-88318-143-6
No. 45	New Results in High Energy Physics – 1978 (Vanderbilt Conference)	78-67196	0-88318-144-4
No. 46	Topics in Nonlinear Dynamics (La Jolla Institute)	78-57870	0-88318-145-2

No. 47	Clustering Aspects of Nuclear Structure and Nuclear Reactions (Winnepeg, 1978)	78-64942	0-88318-146-0
No. 48	Current Trends in the Theory of Fields (Tallahassee, 1978)	78-72948	0-88318-147-9
No. 49	Cosmic Rays and Particle Physics – 1978 (Bartol Conference)	79-50489	0-88318-148-7
No. 50	Laser-Solid Interactions and Laser Processing – 1978 (Boston)	79-51564	0-88318-149-5
No. 51	High Energy Physics with Polarized Beams and Polarized Targets (Argonne, 1978)	79-64565	0-88318-150-9
No. 52	Long-Distance Neutrino Detection – 1978 (C.L. Cowan Memorial Symposium)	79-52078	0-88318-151-7
No. 53	Modulated Structures – 1979 (Kailua Kona, Hawaii)	79-53846	0-88318-152-5
No. 54	Meson-Nuclear Physics – 1979 (Houston)	79-53978	0-88318-153-3
No. 55	Quantum Chromodynamics (La Jolla, 1978)	79-54969	0-88318-154-1
No. 56	Particle Acceleration Mechanisms in Astrophysics (La Jolla, 1979)	79-55844	0-88318-155-X
No. 57	Nonlinear Dynamics and the Beam-Beam Interaction (Brookhaven, 1979)	79-57341	0-88318-156-8
No. 58	Inhomogeneous Superconductors – 1979 (Berkeley Springs, W.V.)	79-57620	0-88318-157-6
No. 59	Particles and Fields – 1979 (APS/DPF Montreal)	80-66631	0-88318-158-4
No. 60	History of the ZGS (Argonne, 1979)	80-67694	0-88318-159-2
No. 61	Aspects of the Kinetics and Dynamics of Surface Reactions (La Jolla Institute, 1979)	80-68004	0-88318-160-6
No. 62	High Energy e^+e^- Interactions (Vanderbilt, 1980)	80-53377	0-88318-161-4
No. 63	Supernovae Spectra (La Jolla, 1980)	80-70019	0-88318-162-2
No. 64	Laboratory EXAFS Facilities – 1980 (Univ. of Washington)	80-70579	0-88318-163-0
No. 65	Optics in Four Dimensions – 1980 (ICO, Ensenada)	80-70771	0-88318-164-9
No. 66	Physics in the Automotive Industry – 1980 (APS/AAPT Topical Conference)	80-70987	0-88318-165-7
No. 67	Experimental Meson Spectroscopy – 1980 (Sixth International Conference, Brookhaven)	80-71123	0-88318-166-5
No. 68	High Energy Physics – 1980 (XX International Conference, Madison)	81-65032	0-88318-167-3
No. 69	Polarization Phenomena in Nuclear Physics – 1980 (Fifth International Symposium, Santa Fe)	81-65107	0-88318-168-1
No. 70	Chemistry and Physics of Coal Utilization – 1980 (APS, Morgantown)	81-65106	0-88318-169-X

No. 71	Group Theory and its Applications in Physics – 1980 (Latin American School of Physics, Mexico City)	81-66132	0-88318-170-3
No. 72	Weak Interactions as a Probe of Unification (Virginia Polytechnic Institute – 1980)	81-67184	0-88318-171-1
No. 73	Tetrahedrally Bonded Amorphous Semiconductors (Carefree, Arizona, 1981)	81-67419	0-88318-172-X
No. 74	Perturbative Quantum Chromodynamics (Tallahassee, 1981)	81-70372	0-88318-173-8
No. 75	Low Energy X-Ray Diagnostics – 1981 (Monterey)	81-69841	0-88318-174-6
No. 76	Nonlinear Properties of Internal Waves (La Jolla Institute, 1981)	81-71062	0-88318-175-4
No. 77	Gamma Ray Transients and Related Astrophysical Phenomena (La Jolla Institute, 1981)	81-71543	0-88318-176-2
No. 78	Shock Waves in Condensed Mater – 1981 (Menlo Park)	82-70014	0-88318-177-0
No. 79	Pion Production and Absorption in Nuclei – 1981 (Indiana University Cyclotron Facility)	82-70678	0-88318-178-9
No. 80	Polarized Proton Ion Sources (Ann Arbor, 1981)	82-71025	0-88318-179-7
No. 81	Particles and Fields –1981: Testing the Standard Model (APS/DPF, Santa Cruz)	82-71156	0-88318-180-0
No. 82	Interpretation of Climate and Photochemical Models, Ozone and Temperature Measurements (La Jolla Institute, 1981)	82-71345	0-88318-181-9
No. 83	The Galactic Center (Cal. Inst. of Tech., 1982)	82-71635	0-88318-182-7
No. 84	Physics in the Steel Industry (APS/AISI, Lehigh University, 1981)	82-72033	0-88318-183-5
No. 85	Proton-Antiproton Collider Physics –1981 (Madison, Wisconsin)	82-72141	0-88318-184-3
No. 86	Momentum Wave Functions – 1982 (Adelaide, Australia)	82-72375	0-88318-185-1
No. 87	Physics of High Energy Particle Accelerators (Fermilab Summer School, 1981)	82-72421	0-88318-186-X
No. 88	Mathematical Methods in Hydrodynamics and Integrability in Dynamical Systems (La Jolla Institute, 1981)	82-72462	0-88318-187-8
No. 89	Neutron Scattering – 1981 (Argonne National Laboratory)	82-73094	0-88318-188-6
No. 90	Laser Techniques for Extreme Ultraviolt Spectroscopy (Boulder, 1982)	82-73205	0-88318-189-4
No. 91	Laser Acceleration of Particles (Los Alamos, 1982)	82-73361	0-88318-190-8
No. 92	The State of Particle Accelerators and High Energy Physics (Fermilab, 1981)	82-73861	0-88318-191-6

No. 93	Novel Results in Particle Physics (Vanderbilt, 1982)	82-73954	0-88318-192-4
No. 94	X-Ray and Atomic Inner-Shell Physics – 1982 (International Conference, U. of Oregon)	82-74075	0-88318-193-2
No. 95	High Energy Spin Physics – 1982 (Brookhaven National Laboratory)	83-70154	0-88318-194-0
No. 96	Science Underground (Los Alamos, 1982)	83-70377	0-88318-195-9
No. 97	The Interaction Between Medium Energy Nucleons in Nuclei – 1982 (Indiana University)	83-70649	0-88318-196-7
No. 98	Particles and Fields – 1982 (APS/DPF University of Maryland)	83-70807	0-88318-197-5
No. 99	Neutrino Mass and Gauge Structure of Weak Interactions (Telemark, 1982)	83-71072	0-88318-198-3
No. 100	Excimer Lasers – 1983 (OSA, Lake Tahoe, Nevada)	83-71437	0-88318-199-1
No. 101	Positron-Electron Pairs in Astrophysics (Goddard Space Flight Center, 1983)	83-71926	0-88318-200-9
No. 102	Intense Medium Energy Sources of Strangeness (UC-Sant Cruz, 1983)	83-72261	0-88318-201-7
No. 103	Quantum Fluids and Solids – 1983 (Sanibel Island, Florida)	83-72440	0-88318-202-5
No. 104	Physics, Technology and the Nuclear Arms Race (APS Baltimore –1983)	83-72533	0-88318-203-3
No. 105	Physics of High Energy Particle Accelerators (SLAC Summer School, 1982)	83-72986	0-88318-304-8
No. 106	Predictability of Fluid Motions (La Jolla Institute, 1983)	83-73641	0-88318-305-6
No. 107	Physics and Chemistry of Porous Media (Schlumberger-Doll Research, 1983)	83-73640	0-88318-306-4
No. 108	The Time Projection Chamber (TRIUMF, Vancouver, 1983)	83-83445	0-88318-307-2
No. 109	Random Walks and Their Applications in the Physical and Biological Sciences (NBS/La Jolla Institute, 1982)	84-70208	0-88318-308-0
No. 110	Hadron Substructure in Nuclear Physics (Indiana University, 1983)	84-70165	0-88318-309-9
No. 111	Production and Neutralization of Negative Ions and Beams (3rd Int'l Symposium, Brookhaven, 1983)	84-70379	0-88318-310-2
No. 112	Particles and Fields – 1983 (APS/DPF, Blacksburg, VA)	84-70378	0-88318-311-0
No. 113	Experimental Meson Spectroscopy – 1983 (Seventh International Conference, Brookhaven)	84-70910	0-88318-312-9

No. 114	Low Energy Tests of Conservation Laws in Particle Physics (Blacksburg, VA, 1983)	84-71157	0-88318-313-7
No. 115	High Energy Transients in Astrophysics (Santa Cruz, CA, 1983)	84-71205	0-88318-314-5
No. 116	Problems in Unification and Supergravity (La Jolla Institute, 1983)	84-71246	0-88318-315-3
No. 117	Polarized Proton Ion Sources (TRIUMF, Vancouver, 1983)	84-71235	0-88318-316-1
No. 118	Free Electron Generation of Extreme Ultraviolet Coherent Radiation (Brookhaven/OSA, 1983)	84-71539	0-88318-317-X
No. 119	Laser Techniques in the Extreme Ultraviolet (OSA, Boulder, Colorado, 1984)	84-72128	0-88318-318-8
No. 120	Optical Effects in Amorphous Semiconductors (Snowbird, Utah, 1984)	84-72419	0-88318-319-6
No. 121	High Energy e^+e^- Interactions (Vanderbilt, 1984)	84-72632	0-88318-320-X
No. 122	The Physics of VLSI (Xerox, Palo Alto, 1984)	84-72729	0-88318-321-8
No. 123	Intersections Between Particle and Nuclear Physics (Steamboat Springs, 1984)	84-72790	0-88318-322-6
No. 124	Neutron-Nucleus Collisions – A Probe of Nuclear Structure (Burr Oak State Park - 1984)	84-73216	0-88318-323-4
No. 125	Capture Gamma-Ray Spectroscopy and Related Topics – 1984 (Internat. Symposium, Knoxville)	84-73303	0-88318-324-2
No. 126	Solar Neutrinos and Neutrino Astronomy (Homestake, 1984)	84-63143	0-88318-325-0
No. 127	Physics of High Energy Particle Accelerators (BNL/SUNY Summer School, 1983)	85-70057	0-88318-326-9
No. 128	Nuclear Physics with Stored, Cooled Beams (McCormick's Creek State Park, Indiana, 1984)	85-71167	0-88318-327-7
No. 129	Radiofrequency Plasma Heating (Sixth Topical Conference, Callaway Gardens, GA, 1985)	85-48027	0-88318-328-5
No. 130	Laser Acceleration of Particles (Malibu, California, 1985)	85-48028	0-88318-329-3
No. 131	Workshop on Polarized ^{3}He Beams and Targets (Princeton, New Jersey, 1984)	85-48026	0-88318-330-7
No. 132	Hadron Spectroscopy–1985 (International Conference, Univ. of Maryland)	85-72537	0-88318-331-5
No. 133	Hadronic Probes and Nuclear Interactions (Arizona State University, 1985)	85-72638	0-88318-332-3
No. 134	The State of High Energy Physics (BNL/SUNY Summer School, 1983)	85-73170	0-88318-333-1

No. 135	Energy Sources: Conservation and Renewables (APS, Washington, DC, 1985)	85-73019	0-88318-334-X
No. 136	Atomic Theory Workshop on Relativistic and QED Effects in Heavy Atoms	85-73790	0-88318-335-8
No. 137	Polymer-Flow Interaction (La Jolla Institute, 1985)	85-73915	0-88318-336-6